The Alatzomouri Rock Shelter

An Early Minoan III Deposit in Eastern Crete

A selection of White-on-Dark Ware vessels from the Alatzomouri Rock Shelter. Watercolor by D. Faulmann.

PREHISTORY MONOGRAPHS 58

The Alatzomouri Rock Shelter
An Early Minoan III Deposit in Eastern Crete

edited by

Vili Apostolakou, Thomas M. Brogan, and Philip P. Betancourt

with contributions from

Vili Apostolakou, Philip P. Betancourt, Thomas M. Brogan, Joanne E. Cutler,
Heidi M.C. Dierckx, Susan C. Ferrence, Floyd W. McCoy,
Evi Margaritis, Dimitra Mylona, and Eleni Nodarou

Published by
INSTAP Academic Press
Philadelphia, Pennsylvania
2017

Design and Production
INSTAP Academic Press, Philadelphia, PA

Library of Congress Cataloging-in-Publication Data

Names: Apostolakou, Vilē, editor of compilation. | Brogan, Thomas M., editor of compilation. | Betancourt, Philip P., 1936- editor of compilation.
Title: The Alatzomouri rock shelter : an early Minoan III deposit in eastern Crete / edited by Vili Apostolakou, Thomas M. Brogan, and Philip P. Betancourt ; with contributions from Vili Apostolakou, Philip P. Betancourt, Thomas M. Brogan, Joanne E. Cutler, Heidi M.C. Dierckx, Susan C. Ferrence, Floyd W. McCoy, Evi Margaritis, Dimitra Mylona, and Eleni Nodarou.
Description: Philadelphia, Pennsylvania : INSTAP Academic Press, 2017. | Includes bibliographical references and index.
Identifiers: LCCN 2017010701 (print) | LCCN 2017016960 (ebook) | ISBN 9781623034153 (ebook) | ISBN 9781931534932 (hardcover : alkaline paper)
Subjects: LCSH: Crete (Greece)—Antiquities. | Pacheía Ámmos (Greece)—Antiquities. | Caves—Greece—Pacheía Ámmos. | Minoans—Greece—Pacheía Ámmos—Material culture. | Pottery, Minoan—Greece—Pacheía Ámmos. | Excavations (Archaeology)—Greece—Pacheía Ámmos.
Classification: LCC DF221.C8 (ebook) | LCC DF221.C8 A43 2017 (print) | DDC 939/.18—dc23

Front cover: watercolor by Douglas Faulmann of a selection of White-on-Dark Ware vessels from the Alatzomouri Rock Shelter.

Back cover: photo by Thomas Brogan from Alatzomouri Hill looking northeast across the Gulf of Mirabello to the island of Pseira; photos by Chronis Papanikolopoulos of three White-on-Dark Ware vessels from the Alatzomouri Rock Shelter.

Copyright © 2017
INSTAP Academic Press
Philadelphia, Pennsylvania
All rights reserved
Printed in the United States of America

Table of Contents

List of Tables in the Text... vii

List of Figures... ix

List of Plates... xiii

Preface.. xvii

List of Abbreviations.. xix

THE EVIDENCE

    1. Introduction, *Vili Apostolakou, Thomas M. Brogan, and Philip P. Betancourt*..................... 1

    2. Geology and Geoarchaeology, *Floyd W. McCoy*... 3

    3. Excavation and Stratigraphy, *Thomas M. Brogan*.. 9

    4. Pottery, *Philip P. Betancourt and Thomas M. Brogan*... 37

    5. Chipped and Ground Stone Tools, *Heidi M.C. Dierckx*... 69

    6. Textile Tools, *Thomas M. Brogan and Joanne E. Cutler*.. 81

    7. Marine Animals, *Dimitra Mylona*... 85

    8. Archaeobotanical Remains, *Evi Margaritis*.. 89

    9. Miscellaneous Objects, *Thomas M. Brogan and Philip P. Betancourt*.......................... 91

10. Pottery Statistics, *Philip P. Betancourt, Thomas M. Brogan, and Susan C. Ferrence*............ 93

11. Ceramic Petrography, *Eleni Nodarou*... 99

DISCUSSION

12. Evidence for Chronology, *Philip P. Betancourt, Thomas M. Brogan, and Vili Apostolakou*... 107

13. Interpretation of the Rock Shelter, *Vili Apostolakou, Thomas M. Brogan, Susan C. Ferrence, and Philip P. Betancourt*... 111

References.. 115

Concordance A. Hagios Nikolaos Museum Numbers and Catalog Numbers........................... 125

Concordance B. Field and Catalog Numbers... 129

Index... 133

Figures

Plates

List of Tables in the Text

Table 5.1.	Number of stone implements by context.	70
Table 6.1.	Early Minoan III discoid and cuboid loomweights: weight and thickness and the resulting thread count per cm when used with threads requiring different tensions in a tabby weave.	82
Table 7.1.	Taxonomic representation of marine invertebrates from the rock shelter.	86
Table 7.2.	Modification of limpets from the rock shelter.	86
Table 7.3.	Edibility of the shellfish recorded on site.	87
Table 8.1.	Complete seeds/fragments.	90
Table 10.1.	Statistics of the pottery assemblage based on fabric, decoration, date, and vessel type.	94
Table 11.1.	Concordance of petrographic samples with vessel shapes and petrographic fabric groups.	101

List of Figures

Figure 1.   Map of Crete.

Figure 2.   Map of East Crete. Contour interval is 300 m.

Figure 3.   Map of the Ierapetra Isthmus. Contour interval 100 m.

Figure 4.   Topographic map of the southeastern slope of Alatzomouri Hill and surrounding area showing position and approximate outline of the Alatzomouri Rock Shelter (solid black) and archaeological field grids (2007–2008).

Figure 5.   Classification scheme for carbonate rocks used in this study.

Figure 6.   Stratigraphic section A–A' in Figure 4 along the southeastern slope of Alatzomouri Hill.

Figure 7.   Generalized lithologic and stratigraphic section of the Pacheia Ammos Formation at Alatzomouri Hill.

Figure 8.   Plan of the trenches and road located west of Pacheia Ammos.

Figure 9.   Schematic south–north section of the rock shelter stratigraphy showing loci and finds by layer or level.

Figure 10. Harris Matrix of Trench A600.

Figure 11. Small finds drawing of Stratum 1, Layer 1: A600.1.

Figure 12. Small finds drawing of Stratum 2 upper, Layer 2: A601.1 and A602.1.

Figure 13. Small finds drawing of upper portion of Stratum 2 upper: Layers 2–4.

Figure 14. Small finds drawing of upper portion of Stratum 2 interior and exterior: objects from Layers 2–9 and stones from Layers 4–9.

Figure 15. Small finds drawing of the lower portion of Stratum 2 interior and exterior: Layers 10–12.

Figure 16. Small finds drawing of Stratum 3a: Layers 13–14.

Figure 17. Small finds drawing of Stratum 3b: Layers 15–16.

Figure 18. Small finds drawing of Stratum 3c: Layers 17–19.

Figure 19. Small finds drawing of Stratum 3d: Layers 20–21.

Figure 20. Stratigraphic section running E–W through the rock shelter.

Figure 21. Plain pottery and pottery with slight decoration found outside the cave.

Figure 22. Funal Neolithic to EM I cup (**1**); EM I pyxis lid (**2**); EM IIB Vasiliki Ware: shallow bowls (**3**, **4**) and jugs (**5**–**7**); EM IIB South Coast jug (**8**); and EM III East Creten White-on-Dark Ware shallow bowls and basins (**9**–**17**).

Figure 23. East Cretan White-on-Dark Ware: shallow bowls and basins (**18**–**22**) and conical bowls (**23**, **24**).

Figure 24. East Cretan White-on-Dark Ware: conical bowls (**25**–**41**).

Figure 25. East Cretan White-on-Dark Ware: conical bowls (**42**, **43**), conical bowls with frying pan handles (**44**, **45**), and spouted conical jars (**46**–**50**).

Figure 26. East Cretan White-on-Dark Ware: spouted conical jars (**51**–**60**), jars (**61**, **62**), and straight-sided or rounded cup (**63**).

Figure 27. East Cretan White-on-Dark Ware: cylindrical cup (**64**), rounded cup with tiny lugs or handles (**65**–**67**), rounded cups (**68**–**70**), spouted cup (**71**), and bridge-spouted jars (**72**–**77**).

Figure 28. East Cretan White-on-Dark Ware: bridge-spouted jars (**78**–**82**).

Figure 29. East Cretan White-on-Dark Ware: bridge-spouted jar (**83**), jugs or jars (**84**–**86**), and jugs (**87**–**90**).

Figure 30. East Cretan White-on-Dark Ware: jug (**91**), jug or jar (**92**), collared jar (**93**), vessel with elliptical base (**94**), closed vessel (**95**), and teapot (**96**).

Figure 31. Plain pottery and pottery with slight decoration: basins and bowls (**97**–**113**).

Figure 32. Plain pottery and pottery with slight decoration: basins with scoring inside (**114**–**120**).

Figure 33. Plain pottery and pottery with slight decoration: spinning bowl (**121**) and cups (**122**–**130**).

Figure 34. Plain pottery and pottery with slight decoration: cups (**131**–**134**) and jugs (**135**–**137**).

Figure 35. Plain pottery and pottery with slight decoration: jugs (**138**–**141**).

Figure 36. Plain pottery and pottery with slight decoration: jugs (**142**–**150**).

Figure 37. Plain pottery and pottery with slight decoration: wide-mouthed jugs (**151**, **152**) and pithoi (**153**–**161**).

Figure 38. Plain pottery and pottery with slight decoration: pithoi (**162**–**171**).

Figure 39. Plain pottery and pottery with slight decoration: pithoi (**172**–**174**) and pithoi or jars (**175**–**187**).

Figure 40.	Plain pottery and pottery with slight decoration: pithoi or jars with drains (**188–191**), spouted jars (**192–197**), and collared jars (**198–200**).
Figure 41.	Plain pottery and pottery with slight decoration: collared jars (**201–203**), small vessel with knobs (**204**), bridge-spouted jars (**205–207**), and hole-mouthed jars (**208, 209**).
Figure 42.	Plain pottery: hole-mouthed jars (**210–213**).
Figure 43.	Plain pottery: hole-mouthed jars (**214–224**).
Figure 44.	Plain pottery: tripod cooking pots (**225–231**).
Figure 45.	Plain pottery: tripod cooking pots (**232–241**).
Figure 46.	Plain pottery: tripod cooking pots and probable tripod cooking vessels (**242–254**).
Figure 47.	Plain pottery: tripod cooking pots (**255–266**).
Figure 48.	Plain pottery: tripod cooking pots (**267–279**).
Figure 49.	Plain pottery: tripod cooking pots (**280–284**) and cooking dishes (**285–299**).
Figure 50.	Plain pottery: cooking dishes (**300–329**).
Figure 51.	Plain pottery and pottery with slight decoration: large vats (**330–332**), flat lids (**333–338**), unknown shapes (**339, 340**), and closed pottery vessels (**341–345**).
Figure 52.	Plain pottery: closed vessels (**346–364**).
Figure 53.	Plain pottery: closed vessels (**365–381**).
Figure 54.	Plain pottery: closed vessels (**382–391**).
Figure 55.	Chipped stone: prismatic blades (**392–397**), retouched flakes (**398, 400**), retouched blade (**399**), and débitage (**401–408**).
Figure 56.	Pounders or hammer stones (Type 1: **409–417**).
Figure 57.	Pounders or hammer stones (Type 1: **418–421**) and pounder-abraders (Type 2: **422–428**).
Figure 58.	Pounder-abraders (Type 2: **429–438**).
Figure 59.	Choppers/hammers (Type 3: **439–444**), faceted tool (Type 4: **445**), and abraders/grinders (Type 5: **446, 447**).
Figure 60.	Abraders/grinders (Type 5: **448–453**) and whetstones (Type 6: **454, 455**).
Figure 61.	Whetstones (Type 6: **456, 457**), pumice abrader/polishers (Type 7: **458, 459**), polishers (Type 8: **460–463**), pestles (Type 9: **464, 465**), scraper and piercer (Type 10: **466, 467**), and weights (Type 11: **468, 469**).
Figure 62.	Weight (Type 11: **470**) and querns (Type 12: **471–473, 475, 477–479, 481**).
Figure 63.	Mortar (Type 13: **482**) and working slab (Type 14: **483**).
Figure 64.	Early Minoan III loomweights (n=9): type, weight, and thickness.
Figure 65.	Textile tools from the rock shelter (**501–503, 505–510**).
Figure 66.	Clay potter's batts (**511–513**).
Figure 67.	Clay potter's batt (**514**) and a clay triangle (**515**). Potter's batt from Myrtos after Warren 1972, 245, fig. 98:14.

Figure 68. View, from above, of a teapot from Koumasa (HM 4107) showing the symmetry based on the vessel's structural parts.

Figure 69. Motifs of White-on-Dark Ware from EM IIB Myrtos: (a) Warren 1972, 203, fig. 87:P 675; (b) Warren 1972, pl. 39D; (c) Warren 1972, pl. 62A.

Figure 70. Motifs of White-on-Dark Ware from the EM III Alatzomouri Rock Shelter: (a) **23**; (b) **24**; (c) **46**; (d) **61**; (e) **62**; (f) **71**; (g) **56**; (h) **72**; (i) **96**.

Figure 71. Motifs of White-on-Dark Ware from MM IA.

List of Plates

Plate 1. The village of Pacheia Ammos and the region west of it.

Plate 2A. Area of excavation near Pacheia Ammos.

Plate 2B. Alatzomouri Hill from the south.

Plate 3A. Alatzomouri Hill from the east.

Plate 3B. Pacheia Ammos from the west.

Plate 4A. Pacheia Ammos from the south.

Plate 4B. Hand colored map of "Pachianamo" by Boschini, published in 1641, courtesy of American School of Classical Studies at Athens, Gennadion Library.

Plate 5A. East side of Trenches A500–A800 during surface cleaning before excavation, from the north.

Plate 5B. The hole-mouthed jar (**GRAVE 1A**) and jug (**GRAVE 1B**) recovered by the bulldozer.

Plate 6A. Stratum 1: cleaning plants growing in the trenches on the northwest side of Trench A700, from the north.

Plate 6B. Stratum 1: after digging A600.1–2, from the south..

Plate 7A. Hole-mouthed jar (**208**) in A600.1.

Plate 7B. Stratum 1: after digging A600/700.1.

Plate 8A. Stratum 1: after digging A700.1, from the north.

Plate 8B.	Stratum 2 upper: after digging A601.1 and A602.1, from the north.
Plate 9A.	Stratum 2 upper: deposit of broken vessels in A601.2–5 (**46**, **208**, **209**, **347**), from the north.
Plate 9B.	Stratum 2 upper: after digging A601.7 and A604.1, from the north.
Plate 10A.	Stratum 2 upper: deposit of broken vessels in A604.2–5 (**8**, **96**, **208**, **209**, **229**, **231**, **347**), from the east.
Plate 10B.	Stratum 2 upper: EM IIB jug from the south coast of Crete (**8**) in situ.
Plate 11A.	Stratum 2 upper: a tripod cooking pot (**231**) and a teapot (**96**) in situ.
Plate 11B.	Stratum 2 upper: a tripod cooking pot (**229**) in situ.
Plate 11C.	Stratum 2 lower interior: after digging A604.11–13, from the west.
Plate 12A.	Stratum 2 lower interior: after digging A604.13, from the west.
Plate 12B.	Stratum 2 lower interior: spouted conical jar (**46**) in situ.
Plate 13A.	Stratum 2 lower interior: closed vessel (**347**) in situ.
Plate 13B.	Stratum 2 lower interior: tripod cooking pot (**229**) in situ.
Plate 14A.	Stratum 2 lower interior: tripod cooking pot (**230**), teapot (**96**), and tripod cooking pot (**231**) in situ left to right.
Plate 14B.	Stratum 2 lower interior: after digging A604.17, from the east.
Plate 15A.	Stratum 2 lower interior: hole-mouthed jar (**208**).
Plate 15B.	Stratum 2 lower interior: hole-mouthed jars (**209**, **214**).
Plate 16A.	Stratum 2 lower interior: after excavating A604.19–20, from the west.
Plate 16B.	Stratum 2 lower interior: after excavating A604.19–20, from the north.
Plate 17A.	Stratum 2 lower interior: after excavating A604.27–29, from the west.
Plate 17B.	Stratum 2 lower interior: after excavating A604.27–29, from the north.
Plate 18A.	Stratum 2 lower interior: tripod cooking pot (**230**) at left.
Plate 18B.	Stratum 2 lower interior: vat (**330**).
Plate 19A.	Stratum 2 lower interior: hole-mouthed jar (**210**).
Plate 19B.	Stratum 2 lower interior: jug (**87**), conical bowl (**23**), spouted conical jar (**47**), rounded cup (**125**), and loomweight (**509**), from the west.
Plate 20A.	Stratum 2 lower interior: conical bowl with spout (**24**).
Plate 20B.	Stratum 2 lower exterior: after digging A604.7, from the north.
Plate 21A.	Stratum 2 lower exterior: after digging A604.6–7, from the south.
Plate 21B.	Stratum 2 lower exterior: after digging A604.11–12, from the north.
Plate 22A.	Stratum 2 lower exterior (**388**, **316**, **171**), from the south.
Plate 22B.	Stratum 2 lower exterior: loomweight (**505**), from the northwest.
Plate 23A.	Stratum 2 lower exterior: loomweight (**506**), from the northwest.

LIST OF PLATES

Plate 23B.	Stratum 2 lower exterior: after digging A604.9, from the north.
Plate 24A.	Stratum 2 lower exterior: after digging A604.11, from the west.
Plate 24B.	Stratum 2 lower exterior: after digging A604.19, from the west.
Plate 25A.	Stratum 2 lower exterior: after digging A604.19, from the north.
Plate 25B.	Stratum 2 lower exterior: spindle whorl (**510**), from the east.
Plate 26A.	Stratum 2 lower exterior: after digging A604.29, from the north.
Plate 26B.	Stratum 2 lower exterior: after digging A604.29, from the south.
Plate 27A.	Stratum 3a fill: after digging A604.46–47, from the north.
Plate 27B.	Stratum 3a fill: after digging A604.46–47, from the south.
Plate 28A.	Stratum 3a fill showing a sherd of a bowl (**26**), from the northeast.
Plate 28B.	Stratum 3a fill showing a basin (**114**), from the west.
Plate 29A.	Stratum 3a fill with the sherd of a basin (**97**), from the east.
Plate 29B.	Stratum 3a fill: detail of southeast corner of the rock shelter, which continues under the modern road, from the northwest.
Plate 30A.	Stratum 3b fill: after digging A604.56, from the north.
Plate 30B.	Stratum 3b fill: after digging A604.55–58, from the north.
Plate 31A.	Stratum 3b fill: after digging A604.55–58, from the south.
Plate 31B.	Detail of Stratum 3b fill, from the west.
Plate 32A.	Stratum 3c fill: after digging A604.59–64, from the north.
Plate 32B.	Stratum 3c fill: after digging A604.59–64, from the south.
Plate 33A.	Stratum 3c fill: after digging A604.59–64, from the west. .
Plate 33B.	Stratum 3c fill: pottery concentration in the middle of the fill after digging A604.59–64, from the west.
Plate 34A.	Stratum 3c fill: pottery in A604.61 in the southeast corner of the chamber.
Plate 34B.	Stratum 3d fill: after digging A604.65–66, from the north.
Plate 35A.	Stratum 3d fill: after digging A604.65–66, from the south.
Plate 35B.	Stratum 3d fill: after digging A604.65–66, from the west.
Plate 36A.	Stratum 3d fill: southeast corner of the chamber after digging A604.65–66, from the west.
Plate 36B.	Stratum 3d fill: after digging A604.67, from the north.
Plate 37A.	Stratum 3d fill: after digging A604.67, from the south.
Plate 37B.	Stratum 3d fill: after digging A604.67, from the west.
Plate 38A.	Stratum 3d fill: after digging A604.65–66 (**151**), from the north.
Plate 38B.	Stratum 3d fill: after digging A604.65–66 (**137, 380**) from the north.

Plate 39.	South Coast Class of pottery: jug (**8**). East Cretan White-on-Dark Ware: conical bowls (**23–25**) and spouted conical jars (**46–50**).
Plate 40.	East Cretan White-on-Dark Ware: spouted conical jar (**52**), spouted two-handled cup (**71**), bridge-spouted jar (**80**), and teapot (**96**). Plain pottery and pottery with slight decoration: basin (**97**) and rounded cup (**124**).
Plate 41.	Pottery with slight decoration: rounded cups (**125, 126**), jugs (**136, 137**), and spouted jar (**192**). Plain pottery: wide-mouthed jugs (**150, 151**).
Plate 42.	Plain pottery: hole-mouthed jars (**208, 209**).
Plate 43.	Plain pottery: tripod cooking pots (**225, 227–232**).
Plate 44A.	"Nicking" and breaking of the limpets' lip (A604.1).
Plate 44B.	A limpet with broken tip (A604.65).
Plate 44C.	Perforated fragment of a *Pinna nobilis* shell (A604.67).
Plate 44D.	Olive pit (left) from locus A604.22 and grape seed (right) from locus A604.53.
Plate 45A.	Fabric Group 1, the jar fabric, sample PAR 09/29 (**207**), x25.
Plate 45B.	Fabric Group 1, the jar fabric, sample PAR 09/08 (**73**), x50.
Plate 46A.	Fabric Group 1, the jar fabric, sample PAR 09/14 (**183**), x25.
Plate 46B.	Fabric Group 2, Subgroup A: the cooking fabric, sample PAR 09/28 (**246**), low fired, x25.
Plate 47A.	Fabric Group 2, Subgroup B: the cooking fabric, sample PAR 09/26 (**262**), high fired with black mica (biotite), x25.
Plate 47B.	Fabric Group 2, Subgroup C: the cooking fabric, sample PAR 09/25 (**276**), with weathered granodiorite, x25.
Plate 48A.	Sample PAR 09/06 (**72**), x25.
Plate 48B.	Sample PAR 09/01 (**6**), x25.
Plate 49A.	Sample PAR 09/02 (**11**), x50.
Plate 49B.	Clay sample CS 13/26 from experimental briquette.

Preface

In January of 2007, a bulldozer operator was creating a parking area next to the Greek National Highway for a church near the village of Pacheia Ammos, in northeast Crete. He stopped immediately when his machine uncovered two Minoan clay jars and a human skull, and he immediately reported the discovery to the Greek Archaeological Service. Near this Minoan burial was a small artificial rock shelter dug into the soft bedrock. Vili Apostolakou, director of the 24th Ephorate of Prehistoric and Classical Antiquities, directed the rescue excavation of the small cave. Such artificial cavities in the soft rock, called rock shelters, are commonly used for burial by members of the Minoan culture of Bronze Age Crete.

The cave contained pottery, loomweights, and other items from a Minoan household that existed near the end of the Early Bronze Age, during the period called Early Minoan (EM) III. It contained no bones (the skull was outside the rock shelter), which raises important questions about the nature of the ancient context and the possible reasons for the absence of the defining item for a burial. The pottery from the assemblage is particularly important because it forms the largest and finest corpus of clay vases known from EM III.

The objects were excavated during 2007 with financial assistance provided jointly by the Institute for Aegean Prehistory (INSTAP) and the Greek Archaeological Service, and they were subsequently studied at the INSTAP Study Center for East Crete. Kostas Chalikias was the trench supervisor, assisted by Alekos Nikakis, Nearchos Nikakis, and Matina Papadaki. Conservation was done by Stephania Chlouveraki with the help of assistant conservators Matina Tzari and Cindy Lee Scott and techniti Matina Papadaki. The registrar was Mary Betancourt with the help of Florence C.S. Hsu and Marie Nicole Pareja. Drawings were made by Kostas Chalikias, Doug Faulmann, Tanya McCullough, Susan C. Ferrence, Heather Hicks, Lily Bonga, and Philip P. Betancourt. Computer processing of illustrations was accomplished in the Tyler School of Art Digital Laboratory (Temple University) by Florence C.S. Hsu, Marie Nicole Pareja, and Amie Gluckman. Other studies were made by the authors of the relevant sections in this volume. Matina Papadaki conducted the processing of the soil samples, the sorting of the residues, and the organizing of the material for the next stages of analysis.

Field mapping and laboratory work for the geological study was accomplished with funding from INSTAP using facilities at the INSTAP Study Center for East Crete. Portions of this work were performed at the Wiener Laboratory of the American School of Classical Studies at Athens (ASCSA) and at the University of Hawaii.

List of Abbreviations

ARS	Alatzomouri Rock Shelter
ASCSA	American School of Classical Studies at Athens
ca.	circa
CS	chipped stone
cm	centimeter(s)
CTR	Center for Textile Research, Copenhagen
d.	diameter
DAI	Deutsches Archäologisches Institut
dGPS	differential global positioning system
dim.	dimension
EDM	electronic distance measuring device
EM	Early Minoan
FN	Final Neolithic
g	gram(s)
GS	ground stone
GSS	Gournia survey site
h.	height
ha	hectare(s)
HM	Archaeological Museum, Herakleion
HNM	Archaeological Museum, Hagios Nikolaos
IFZ	Ierapetra Fault Zone
IGME	Institute for Geology and Mining Exploration, Greece
kg	kilogram(s)
km	kilometer(s)
LM	Late Minoan
m	meter(s)
m asl	meters above sea level
max.	maximum
mm	millimeter(s)
MM	Middle Minoan
MM	Modified Mercelli scale of relative earthquake damage (Ch. 2 only)
μm	micrometer(s)
MNI	minimum number of individuals
MNV	minimum number of vessels
Mw	moment magnitude of an earthquake
mybp	million years before present
PAR	Pacheia Ammos Rock Shelter (accession number written on the artifacts)
pres.	preserved

PT	possible tool	w.	width
rest.	restored	WOD	White-on-Dark Ware
SEM	scanning electron microscopy	wt.	weight
th.	thickness	yr	year

1

Introduction

by
Vili Apostolakou, Thomas M. Brogan, and Philip P. Betancourt

The Alatzomouri Rock Shelter (ARS) preserves an important deposit of pottery and other objects from the Early Minoan (EM) III period, a crucial formative phase in Minoan cultural development that precedes the construction of the Middle Minoan (MM) palaces. The objects are of interest both for the light they shed on Cretan pottery and ceremonial beliefs and for their artistic merit. The information from this deposit helps to solve essential chronological problems for the development of Cretan pottery at a crucial point in time. The deposit's fine pottery illustrates the initial steps in the development of Minoan painted ceramics in the light-on-dark technique, a tradition that would soon spread to other parts of the island to become the main technique of vase painting in Middle Minoan Crete—the EM III–MM I painted pottery is the ancestor of the brilliant ceramic art objects of the Kamares Ware tradition. The fine painted vases from this deposit, the largest known assemblage from a single context from this period, allow one to examine the light-on-dark style in its early stages in more detail than has previously been possible.

The archaeological site, in northeast Crete near the southeasternmost coast of the Gulf of Mirabello (Figs. 1–3; Pl. 1), was discovered by accident in 2007. The burial was displaced by the bulldozer. It was close to the rock shelter, but its exact configuration could not be reconstructed. The context consisted of a small rock shelter carved into the soft limestone and closed with a wall at its front. Before placing objects in the space, the floor had been covered with a deposit of rubble that included many sherds as well as some small stones. The inside was then packed with objects up to the ceiling, including pottery, stone tools, loomweights, turntables for making pottery, and a few other objects. One assumes that organic objects that have perished were present as well. The items had been deliberately placed inside the small man-made rock shelter. Some of them were sorted, and they may have been originally in bags or baskets (e.g., groups of stone tools were together, and some of the clay loomweights were together in one place). Almost every piece had been deliberately broken, and some broken pieces from the same vessel were found in disparate places. Many items had missing pieces that were not put into this deposit. The mouth of the chamber was closed with stones at its entrance after it was filled with the objects, and erosion from soil washing down from the upper parts of the hill to the north eventually covered it completely. Two preliminary reports have been published (Apostolakou, Betancourt, and Brogan 2007–2008, 2010).

2

Geology and Geoarchaeology

by
Floyd W. McCoy

Exposed sedimentary rocks in eastern Crete provide a remarkable sedimentary record of the Miocene–Pliocene world and its oceans 23.0–1.6 million years ago (23.0–1.6 mybp). In particular, the transition between these two time epochs 5.3 million years ago (5.3 mybp) is documented in rock exposures on Alatzomouri Hill. Here thick sequences of white calcareous breccias, sandstones, marls, and limestones, with isolated exposures of gypsum, describe an exceptional geological event: the complete isolation of the Mediterranean Sea from the global ocean and its subsequent desiccation, followed by refilling from the Atlantic Ocean. Tectonism then reconfigured that world of five million years ago, uplifted the Mio-Pliocene seafloor and its deposits, and then added those sediments, now lithified into rocks, to create modern Crete.

Alternating soft and hard rock units in this sedimentary succession were exploited by ancient and modern cultures in Crete for a variety of purposes, one being caves created either by natural erosional processes or underground spaces excavated by man for use as rock shelters (Apostolakou, Betancourt, and Brogan 2007–2008; Ferrence 2008). The Alatzomouri Rock Shelter is located on Alatzomouri Hill west and above the coastal town of Pacheia Ammos in a complex geological setting typical of Crete. This small cave was cut into soft Miocene marls in order to contain an assortment of EM III archaeological materials. The placement of this rock shelter took advantage of local geologic factors—and it may also have led to both its preservation and destruction.

Field and Laboratory Techniques

Topography (Fig. 4) was mapped using differential global positioning system (dGPS) techniques. Sedimentological criteria for establishing descriptions and nomenclature of sediments and sedimentary rocks utilized standard field and laboratory techniques; the classification scheme for carbonate rocks and sediments used in this study is outlined in Figure 5. For rocks, hand specimens were studied using a hand lens; thin sections were made from selected rocks and analyzed following standard methods using petrographic microscopes. For sediments and poorly indurated sedimentary rocks, smear slides were made and studied using both petrographic (for high magnifications) and dissecting (for lower magnifications) microscopes; sediment texture (gravel:sand:silt+clay) was determined by wet-sieving using standard U.S. sieve sizes. Total carbonate content was determined by the acid weight-loss technique.

Geologic Setting of the Mio-Pliocene World

It was during the Miocene epoch (Cenozoic era), 23.0–5.3 mybp, that foundations were laid for the modern Mediterranean in terms of tectonic, geomorphic, oceanic, and climatic conditions (Potter and Szatmari 2009). And it was at the close of this epoch that geological activity commenced to later form a natural setting at Pacheia Ammos to be utilized by man some five or more million years afterward for a rock shelter. The Eurasian convergent orogenic belt, which is approximately 13,000 km long and stretches from Spain to Vietnam, was one of two enormous and dominant tectonic lineaments on this planet (the second was the Andean-Cordillerian belt measuring 16,000 km long). Crete, along with the ancestral Mediterranean Sea known as the Tethys Ocean or Seaway, was at the western end of this huge Eurasian tectonic alignment. The Tethys Seaway was a remnant of an earlier landscape of supercontinents and oceans produced during former orogenies, and it was now being remolded apparently as a consequence of deep earth processes activated by two super-plumes in the earth's mantle—one beneath Africa and one beneath the central-southwestern Pacific Ocean. Both super-plumes remain active today (Larson 1991; Montelli et al. 2006; Potter and Szatmari 2009).

At the beginning of the Miocene (23.0 mybp), Tethys was a major connector between the Atlantic and Indian Oceans, providing significant control on global oceanic properties and circulation. By the Late Miocene (ca. 10 mybp), progressive closure between Africa and Europe had closed the eastern gateway to the Indian Ocean and formed the Levant. By the end of the Miocene (5.3 mybp), convergence closed the western connection to the Atlantic (then positioned across southern Spain). Tethys was now an isolated sea.

The consequences were profound for the Mediterranean region as well as for global climate and oceanic circulation: the Tethys Ocean, the precursor to the Mediterranean Sea, dried up. What had been sea was now a series of salt lakes progressively evaporating to leave thick accumulations of salt deposits including gypsum and halite. For approximately 630,000 years the Mediterranean was a desert (Ryan and Cita 1978; Hsu 1983; Roveri et al. 2008). Continuing tectonic motion then created a new connection to the Atlantic Ocean through today's Strait of Gibraltar. The Mediterranean rapidly refilled—it is at this time that a small basin peripheral to the eastern Mediterranean Sea, extending northward where Crete is today, was inundated to form the sedimentary rocks exposed today on Alazomouri Hill at the rock shelter. The regional desiccation episode is known as the Messinian Salinity Crisis (Messinian stage of the Miocene epoch), a unique geological event that had worldwide repercussions after the withdrawal of so much salt from the global ocean budget (Ryan and Cita 1978; Hsu 1983; Roveri et al. 2008; Potter and Szatmari 2009).

The refilling of the sea was recorded by changing seafloor sediment types and fossil assemblages: shallow water deposits (calcarenites) extended offshore into deeper water where carbonate sediments (fossiliferous marls and limestones) accumulated. As the sea filled and deepened, loose sediment in the form of sand mixed with large blocks and boulders cascaded off steeper slopes into deeper water as underwater avalanches (marl breccias). Elsewhere sediment on steep slopes slid downslope to form deformation structures. Both avalanching and sliding were likely triggered by earthquakes in response to loading of the earth's crust by the weight of this new saltwater and to continuing tectonic activity. Subsequent tectonic uplift brought these sediments above sea level to form Alatzomouri Hill—a location to be sought for excavating a rock shelter in the late Early Bronze Age.

Tectonism continues today, announced by seismic activity. Alatzomouri Hill represents an upthrown fault block on the western side of the Ierapetra graben (forming today's valley and isthmus between Pacheia Ammos and Ierapetra). On the east side of the graben and valley is the Ierapetra Fault Zone (IFZ), a 25 km long zone of north-northeast to south-southwest trending crustal segments marked by a steep escarpment up to 800 m high (Armijo, Lyon-Caen, and Papanastassiou 1992; ten Veen and Kleinspehn 2003; Caputo et al. 2010). Offshore, the IFZ extends another 25 km to the north-northeast and ca. 200 km to the southwest. This constitutes one of the major tectonic features along the Hellenic tectonic arc. As such, it represents an active zone of seismicity with average slip rates for vertical motion in the range of 0.1 to 2–3 cm/year during the past 10,000 years (Armijo, Lyon-Caen, and Papanastassiou 1992). Caputo and colleagues (2010) estimate a mean recurrence rate of earthquakes along the IFZ as between 812 to 271 years, with the latter the more probable repeat time, for the past 13,000 years.

During the past 230 years, Papadopoulos (2011, 399–405) lists six strong earthquakes in the Ierapetra region with relative magnitudes of Mw = 6.4–5.5, with an average of 6.1 (Mw is the moment magnitude of an earthquake, a relative intensity measurement of ground motion, and replaces the now-disused Richter scale). In the Modified Mercelli scale (a relative measurement of damage to structures; abbreviated as MM), three earthquakes were classified as "Violent" to "Destructive" (MM = IX–VII) and two as "Rather Strong" (MM = V), thus they were damaging to local buildings and potentially destructive in the Pacheia Ammos area (see also Ambraseys 2009; for a map depicting historic earthquake epicenters in the vicinity of the IFZ, see McCoy 2013, 23, fig. 3.7). Although evidence for roof collapse was not found during excavation of the remaining portion of the rock shelter preserved today, it is clear that earthquakes in the Ierapetra region have been strong enough to produce ground motion that could cause roof collapse at the rock shelter, assuming contemporary seismicity is a proxy for ancient seismicity. As noted by Caputo and colleagues (2010, 124): ". . . the fault segments of the IFZ are probably the most seismogenic structures affecting the whole island [Crete]."

It remains unclear if uplift or subsidence might be dominant in the prehistoric-to-modern tectonic signal at Alatzomouri Hill. It is estimated that the regional uplift is about 0.3–0.5 mm/yr in this area of Crete today (Caputo et al. 2010, 113). Certainly uplift is indicated by incised Pleistocene alluvial fans at nearby Gournia. Subsidence, however, seems indicated from archaeological criteria in the nearby village of Pacheia Ammos (Seager 1916), which is within the graben forming the Ierapetra Isthmus. Both uplift and subsidence would be expected in a horst and graben setting such as between Alatzomouri Hill (horst) and Pacheia Ammos village (graben). Eustatic sea level changes have not been considered and remain unstudied locally. Rates of sea level change in the Aegean, independent of tectonic influences over the past 6,000 years, is on the order of 1.0 mm ± 0.2 mm per year (Lambeck and Purcell 2005).

Lithostratigraphy

The stratigraphic section exposed on the east-facing slope of Alatzomouri Hill encompassing the Alatzomouri Rock Shelter and environs is a lithostratigraphic sequence of marls, marl breccias, calcarenites, and limestones (Figs. 6, 7). Fortuin (1977, 119–121, 139–141) identifies this as the Pacheia Ammos Formation ("Pakhiammos") of earliest Pliocene Age, the lowermost portion of the Finikia Group. Calcarenites of the latest Miocene Ammoudares Formation (Hellenikon Group) may be present in the lowermost portion of the section in the seacliff at Pacheia Ammos. This entire section overlies upper Cretaceous limestones along an unconformable contact (IGME 1959; Fortuin 1977, 119–121; Fassoulas 2000, fig. 6.1), the hiatus indicated by the unconformable contact presumably representing the Messinian Salinity Crisis.

Prior mapping has identified this stratigraphic sequence variously as: (1) undifferentiated Pliocene marine marls and limestones (IGME 1959); (2) middle–upper Miocene lacustrine-shallow marine limestone breccias, sands, clays, and marls (Creutzburg 1977); (3) early–late Miocene shallow marine marls and conglomerates of the Kalamavka Formation/SP Series of the Tefeli Group

(Postma, Fortuin, and van Wamel 1994); (4) late Miocene–early Pliocene marls, marl breccias, and limestones (van Hinsbergen and Meulenkamp 2006; Zachariasse, van Hinsbergen, and Fortuin 2008).

All interpretations, however, agree on lithologic types, substantiated in this study: marls, marl breccias, limestones, and calcarenites. Rock layers vary in induration: some are well cemented and hard, thus resisting erosion and thereby forming hardgrounds and ledges in exposure; others are poorly indurated and softer, thus easily eroded and forming undercut areas along steep slopes. Some rock units are massive without internal structure; others are prominent by numerous thin laminations. The marl breccias were poorly understood and interpreted until careful mapping and biostratigraphic work by van Hinsbergen and Meulenkamp (2006) and Zachariasse, van Hinsbergen, and Fortuin (2008). These breccias are particularly well exposed in the cut along the national highway at the sharp bend in the road near the Alatzomouri Rock Shelter.

One layer is particularly prominent and significant in this regional stratigraphic sequence, here termed the Pefka layer (Figs. 6, 7). It is one of numerous hard limestone/marl layers within this stratigraphic sequence, but in exposure today it characteristically forms a hardground on the surface (Fig. 5) that is bare without vegetation and creates prominent ledges exposed in escarpments. Erosion seems to break the rock into large slabs that litter the surface immediately above the Alatzomouri Rock Shelter (Fig. 7). Exposures of the Pefka layer today are increasingly broken and eroded in response to construction, agriculture, and animal husbandry. Where it is thin or absent, rapid erosion of the underlying softer marl layer occurs. Locally the Pefka layer, and some of the other similarly well-indurated layers, grade along strike into marl breccias. The Pefka layer forms the roof at the Alatzomouri Rock Shelter.

Microfossils are ubiquitous (foraminifera, calcareous nannoplankton, and spicules, among others) and form the basis for geological dating. Megafossils are less common, although one of the marls contains complete and broken shells of two or more types of bivalve mollusk shells up to one cm or more in size, in addition to large sponge spicules and tubelike worm burrows. Initial age/stage assignments relied on mega- and microfossil assemblages. Earlier geologic mapping focused on larger megafossil finds, and where these were rare or absent, an age was established from suggested correlations to nearby exposures of similarly appearing rock types. Later work has utilized key index species from microfossil criteria, providing good definition of Mediterranean stage boundaries for the Mio-Pliocene.

It is these microfossil criteria that van Hinsbergen and Meulenkamp (2006) and Zachariasse, van Hinsbergen, and Fortuin (2008) utilized to constrain the dating of the stratigraphic succession at ARS to the earliest Pliocene. Fortuin (1977, 139–140) noted, however, that palaeontological criteria indicated that the Pacheia Ammos Formation, certainly the lowermost section, "straddles the Miocene-Pliocene boundary." The presence of the mollusk *Cyprideis pannonica* found in one of the marl units (Fig. 7) indicates deposition under extreme salinity conditions, suggesting a latest Miocene Age—this would suggest that the stratigraphic sequence at the Alatzomouri Rock Shelter records the initial refilling of the Mediterranean Sea, ending the Messinian Salinity Crisis during the Pliocene.

The marl breccias thus may be the result of soft sediment avalanches off steeper slopes as the basin was inundated by the sea, the avalanches perhaps triggered by seismicity associated with both tectonism and crustal response to the added weight of seawater as the Mediterranean filled. Soft sediment deformation structures occur in the lower portion of the stratigraphic section (Fig. 7) that also are indicative of submarine slope failure. Deposition was on an eroded basin floor of older rocks, limestones of Upper Cretaceous Age (IGME 1959; Fortuin 1977, 119–121; Fassoulas 2000, fig. 6.1). No Miocene gypsum or other evaporative salts are present in exposures in the Pacheia Ammos area indicating a depositional site removed from the zone of salt deposition. Accordingly, the stratigraphic section exposed on the eastern and northern slopes of Alatzomouri Hill likely represents sedimentation in a Mio-Pliocene basin marginal to the newly filling Mediterranean Sea with—once filling was completed—shallow water depths on the order of a few hundred meters.

Sediments

Sediments from archaeological contexts within the Alatzomouri Rock Shelter—the two layers of fill found during excavation (see Ch. 3)—are marly chalks (see Fig. 5 for classification scheme) with total carbonate contents averaging 75% (range is 50%–86%). Colors are dominantly pale brown to light grays (10YR 7/2–8/4), light yellow (2.5Y 8/3), and very pale brown (10YR 7/4) reflecting the higher concentrations of calcium carbonate. Noncarbonate components are fine-grained clays and quartz mixed with opaque material, likely iron oxide minerals. Textural criteria identify these as silty clays mixed with sandy silty clays.

Similarity between these geoarchaeological sediments and sediments external to Alatzomouri Rock Shelter (silty clays with average carbonate content = 68%, range = 67%–71%) suggest derivation of geoarchaeological sediments from local sources as fill within the rock shelter. Scattered well-rounded granules, pebbles, and cobbles composed of limestones and marls occur in the upper portion of the marl layer that was excavated to form the rock shelter, but they do not exist at adequately high concentrations to suggest a source for the numerous "sea pebbles" in the geoarchaeological sediments filling the shelter (Layers 5–9). Accordingly it may be assumed that these rounded "sea pebbles" were transported into the rock shelter from nearby sources such as streams and beaches.

Five pieces of pumice were found in the shelter fill (see Ch. 5). They are all within pebble sizes, well-rounded, and disk-to-equant shapes, with a stained outer zone (light gray, 10YR 7/2) surrounding a white (2.5Y 8/1) interior. Derivation of the pumice via either ocean rafting or import for cultural uses from one of the six major volcanic fields along the Hellenic island arc chain is suggested (major element analyses of the glass components in the pumice were not done to characterize a specific volcanic source). Given the cultural dates determined for the use of the rock shelter, this pumice was not a product of the Late Bronze Age eruption of Santorini.

Geomorphology

Alatzomouri Hill is an uplifted fault block (a horst), and it represents an inverted topography—marine basin to subaerial hill. As noted previously, the eastern boundary of Alatzomouri Hill is an eroded fault trace of the horst (IGME 1959; Creutzburg 1977), one of many such horsts associated with the Ierapetra Fault Zone (IFZ) that crosses Crete and contributes to the fault-block tectonic fabric of Crete (Armijo, Lyon-Caen, and Papanastassiou 1992; ten Veen and Kleinspehn 2003; Caputo et al. 2010). The modern relief of Alatzomouri Hill and the surrounding area west of Pacheia Ammos has been partially determined by the presence or absence of indurated hard marl/limestone layers, such as the Pefka layer. While exposures of this layer are discontinuous and variable in thickness, at the top of the hill it is up to 2 m thick, thus forming a capping layer that minimizes erosion and contributes to the preservation of this area through post-Pliocene time as a hill. Erosion of the eastern slope of the hill, with the ARS, is controlled by local dips of the bedrock (Fig. 6). Drainage networks are poorly established due to the high porosity of the local bedrock that minimizes runoff and local exposures of the low porosity Pefka layer.

Geoarchaeology

Given the geologic control on landscape evolution at Alatzomouri Hill and the surrounding area, it is likely that the Bronze Age landscape was similar to that seen today, but with a slightly altered hill slope and rock exposure prior to highway construction. Bronze Age inhabitants took advantage of a location where natural tectonic and erosional processes had uncovered the softer marl layer that was easily dug to create the rock shelter—there seems to be no geological process that might erode a cave in just this area. The two harder marls or limestones, overlying and underlying the softer interbed, formed a good floor and ceiling for the shelter (Fig. 6). Geoarchaeological sediments found in the ARS excavation as partial fill suggest a source from local marls. The shelter may have been partially buried by collapse of the roof in response to seismic activity. No natural sedimentary processes are obvious that would have completely filled the cave after its abandonment from cultural use, suggesting that if this were the case, additional fill may be anthropogenic in origin.

Shelters and pits dug into the hill during modern times have continued to utilize this alternating hard-soft-hard lithologic sequence as shelters excavated by farmers and animals. The most extensive of these is the underground bunker network dug along the ridge during World War II by the Italian army (Ferrence 2008). It is also seen elsewhere on Crete where large animals will seek shade and in the process abrade undercut shelters in similar topographic and lithologic settings.

It remains unclear from limited archaeological evidence whether the shelter went out of use from cultural or natural causes. If the latter, and destruction by an earthquake is speculated by Apostolakou, Betancourt, and Brogan (2007–2008, 35), it is easy to envision the collapse of the roof by seismic activity especially if the shelter were enlarged beyond structural stability of the overlying marl. Evidence for roof collapse was absent during excavations, and if once present it may have been removed by the bulldozer that uncovered the rock shelter. An intact roof over the western portion of the rock shelter is pictured in Apostolakou, Betancourt, and Brogan (2007–2008, fig. 3); thus, if collapse occurred, it would have involved only the eastern portion of the shelter.

3

Excavation and Stratigraphy

by
Thomas M. Brogan

Topography of the Alatzomouri Hill and Its Bronze Age Settlement History

The Alatzomouri Hill is a small promontory on the north coast of Crete roughly 65 km east of Herakleion and 13 km north of Ierapetra (Figs. 1, 2; Pls. 1, 2). It provides clear views of the Gulf of Mirabello to the north, the small valley and hill of Pera Alatzomouri to the west, the wide beach of Pacheia Ammos and its eponymous village to the east, and the low valley between the Thriphti and Dictaean mountain ranges (the Isthmus of Ierapetra) to the south (Pl. 3).

McCoy's geological study traces the formation of the hill to the tectonic uplifts of the Mio-Pliocene seafloor 5 million years ago, which followed the Messinian Salinity crisis (see Ch. 2). He notes that its present condition likely differs little from the Early Bronze Age landscape discussed in this volume. The hill today is a bare expanse of uncultivated land, with the exception of the small modern cemetery (and its associated church) and the hairpin stretch of the National Road on the east slope above Pacheia Ammos (Pl. 2A). Deep trenches on Alatzomouri Hill were made by the Italians during World War II (Pl. 2B; Ferrence 2008). The modern village of Pacheia Ammos, located to the east at the foot of the hill, began as the port facility and customs house for commerce in the North Isthmus of Ierapetra during the late 19th century. Tourism has been responsible for its growth since the 1960's (population 545 in 2011; Pl. 3B). Two images capture earlier views of both the village and the beach: one photo shows Pacheia Ammos in 1904, while a hand colored map by Boschini provides the earliest illustration of the hillside and beach identified as "Pachianamo" in the 17th century (Pl. 4; Boschini 1651, pl. 35). Both the beach and hill are noticeably deserted in Venetian times.

Never the locus of a major settlement in earlier periods, the hillside of Alatzomouri and the beach of Pacheia Ammos were frequently exploited by the occupants of the area. The recent publication of the Gournia Survey Project provides a detailed report of activity in this area from the Early

Bronze Age and also examines how these finds fit into the wider patterns of settlement in the North Isthmus from the Final Neolithic through the Roman periods (Pl. 1; Watrous 2012, 97–102). The analysis of the cluster of sites in the immediate area of Pacheia Ammos is of particular interest (Watrous 2012, 105–114).

The survey identified no Final Neolithic material from this area (Pl. 1), though one sherd of this date was recovered from the recent excavations. In the EM I–IIA period a small village was established at Halepa on the eastern end of the beach near the church of Hagia Photeini (Watrous 2012, 112–113, site 23, map 9; Betancourt 2013). Two small EM II farms were also identified approximately 40 m south of Halepa (Watrous 2012, 112, site 22, map 9). This expansion into areas with fertile land and water sources fits the wider EM II pattern in the Mirabello area, which saw a number of settlements almost reach the size of villages, including Prinatikos Pyrgos, Halepa, Vasiliki, Kavousi, and Mochlos (Hayden 2004, 46–48, 71; Watrous and Schultz 2012a, 26–31).

By Middle Minoan I, the EM I–II sites near Halepa were abandoned and a new site was established on the east slope of Alatzomouri "in a low saddle hidden from the sea by the hill summit to the north" (Watrous 2012, 111, site 17, tables 2–5, maps 2, 13, 14, 17, 20, 21, 23, 24). The scatter of sherds at Site 17 covers a minimum area of 150 x 150 m, and Watrous identifies this location as a village of 1.88 ha. Two isolated scatters of pottery 70 m north of the modern cemetery and Church of Hagios Demetrios (Watrous 2012, 111) were identified as possible tombs associated with Site 17. No contemporary sites were identified in the area immediately to the east and south of Site 17; rather, its appearance probably reflects the growth and expansion of the cluster of sites located around Gournia to the west (Watrous 2012, map 13 marking the off-site pottery densities around these sites). By MM IB–II Site 17 had grown in size to 2.25 ha, and its inhabitants had begun burying their dead in jars in the cemetery excavated by Seager on the beach in front of the village of Pacheia Ammos (Seager 1916; Watrous 2012, 111, site 18). Moreover, Watrous has suggested that the appearance of nine smaller sites (hamlets, farms, and fieldsites) south and east of Site 17 may reflect the formation of a settlement cluster around that site.

In MM III to Late Minoan (LM) I the settlement dynamic changed again with the abandonment of Site 17 and half of the smaller sites in the local cluster. The surviving habitations were probably hamlets, with those near Pacheia Ammos still burying their dead by the beach (Watrous 2012, 110–112, sites 16, 18, 20). Watrous connects the abandonment of these sites with the emergence of a regional center at Gournia and the expansion of that town's agricultural catchment in MM III–LM IA (Watrous and Schultz 2012c, 55–57). Haggis has noted a similar decline in the overall number of settlements in the neighboring area of Kavousi in MM III–LM I, as well as the growth of settlements near the coast or inland that were positioned to mobilize resources on behalf of regional elites (Haggis 2005, 120–143). After the LM I destructions the Alatzomouri Hill and Pacheia Ammos had no known habitation sites for more than 1,000 years, although LM III tombs (Ferrence 2008) suggest nearby residents in that period; it was not until the early Roman period that they again served as a hamlet and small harbor (Vogeikoff-Brogan 2012, 83–85).

Excavation of the Rock Shelter

The rock shelter is located on a flat strip of soil on the west side of the National Road, 100 m south of the modern cemetery (Pls. 2A, 5A). In January 2007 a bulldozer began removing topsoil next to the road to create a parking area for the nearby cemetery when it uncovered a jar (**GRAVE 1A**), a jug (**GRAVE 1B**), and a fragment of a human skull, which appeared to form part of a Bronze Age burial assemblage (Pl. 5B). As the ancient context of these objects was not certain, the 24th Ephorate of the Greek Ministry of Culture undertook a small rescue excavation. Work began on 20 February 2007 and was completed on 26 March of the same year.

Before the excavation began, G. Damaskinakis surveyed the area with an EDM Total Station (Fig. 8). He provided a datum point for elevations on the nearby road surface at 37.06 m asl and laid out trenches (Fig. 8). The trenches were numbered A100–A800; the level path cut by the bulldozer lay in the eastern half of A500, A600, and A800 (Pl.

5A). In the northern part of Trench A700, plants were growing in what appeared to be tracks from earlier bulldozing in the area (Pl. 6A). The project examined the path of the bulldozer in all four trenches and opened a fifth test trench, A900, a short distance to the southwest; trenches A100–A400 were not explored. A handful of sherds were recovered in Trenches A500, A700, A800, and A900, and the rock shelter lay in Trench A600. Digging here removed the entire assemblage from within the rock shelter and a portion of the deposit from what appears to have been its entrance on the east side. Excavation was not completed outside the entrance because it lay underneath the modern highway.

A form of the locus system, which assigns a locus number to each feature (e.g., a deposit of soil or a wall) in the trench, was used to document the excavation. In Trench A600, Loci 1, 2, and 4 represent deposits of soil (e.g., Trench A600, Locus 1 = A601; Pl. 6B), while Locus 3 was a layer of stones (A603) and Locus 5 a wall (A605). The deposits of soil (e.g., A601) were removed in a series of horizontal passes or pails (e.g., Trench A600, Locus 1, Pail 1 = A601.1 or Trench A600, Locus 1, Pail 7 = A601.7). In each pass ancient material (e.g., pottery, bone, shell, and carbon) was collected by hand, and soil characteristics were recorded (the *Munsell Color Chart* was used to identify soil color [Kollmorgen 1992]). The stratigraphy of Trench A600 can be seen in both schematic section and a Harris Matrix (Figs. 9, 10). Standard soil samples were collected from across each pass, and additional intensive soil samples were taken from specific locations, where appropriate, within passes (e.g., near well-preserved vessels). In Trench A600 the project recorded two standard and 12 intensive samples from Locus A601, three intensive samples from Locus A602, and 18 standard and 31 intensive samples from A604. The soil samples provide an important record for the context because wet weather made it impossible to use a dry sieve during the first half of the excavation. The organic record (e.g., seeds and animal bones) from the context was, however, very poor.

Two different numbering systems were used to record objects. Ground and chipped stone tools and clay tools, such as loomweights, were given object numbers assigned sequentially for the trench (e.g., no. 36 in pass A604.11), and their precise position was recorded on 1:20 plans of in situ small finds. Complete and broken ceramic vessels were recorded as concentrations of pottery within a pass (e.g., A600.2, A600.3, A600.4, A600.5, and A600.6 represent five separate pots found within pass A600.1). The stratigraphy in Trench A600 is illustrated as a Harris Matrix to avoid any confusion between the passes and objects (Figs. 9, 10). The location and elevation of each vessel was also noted on the 1:20 plans of in situ small finds. After cleaning and conservation, all objects were given separate accession numbers with the initials PAR (an abbreviation for Pacheia Ammos Rock shelter).

All finds were taken to the INSTAP Study Center for East Crete for processing, conservation, cataloging, illustration, and study. When this work was finished, the material was transferred to the storerooms of the 24th Ephorate of Prehistoric and Classical Antiquities in Ierapetra. For reasons of road safety, the rock shelter was back filled in 2008.

The chamber and entrance of the rock shelter were located in Trench A600, and much smaller amounts of material were recovered to the north in Trenches A600/700 and A700. The bulldozer had removed most of the superstructure of the rock shelter, but various details, including a small section of ceiling preserved at the western edge of the bedrock and sloping down to the east, suggest that the rock shelter would have resembled a small cave. The entrance on the northeast side was partially blocked by an irregular wall of rubble that probably served two purposes. The lower course retained a fill of soil which provided an earth floor for the artifacts deposited inside the shelter; additional courses may also have served to close the entrance completely.

The excavation revealed three stratigraphic layers in the rock shelter: a disturbed surface layer created by the bulldozer (Stratum 1), the main deposit of objects placed within and outside the rock shelter in EM III (Stratum 2), and a lower deposit of soil brought in to create a level floor (Stratum 3). The main deposit (Stratum 2) is divided into two parts, upper and lower, with the latter separated into interior and exterior areas in relation to the wall blocking its entrance.

Because the excavation took place in February and March in an exposed setting, it was not possible to leave well-preserved objects in the ground for long periods of time in order to produce photographs and drawings of the complete assemblage.

The situation was also complicated by the fact that the rock shelter did not have a level floor but employed an irregular surface of soil and bedrock that sloped upward on the south and west sides. Objects appear to have been placed wherever they would fit, in some cases piled on top of each other. Within the strata, they were removed in a series of artificial and broadly horizontal layers throughout the deposit—21 layers (or levels) in all beginning from Layer 1 at +36.57 m asl down to Layer 21 at +35.72 m asl (Fig. 9). We use the term layer or level in various chapters (layer here; level in Ch. 4), but the terms means the same thing. The position of hand-collected artifacts was carefully noted in each layer; well-preserved artifacts were drawn at 1:20 scale, and the positions of distinctive sherds were marked with x in 12 drawings. Two impressive layers of artifacts were drawn at 1:10, and details from additional drawings were added to them (compressing multiple excavation layers) in an attempt to illustrate the complete assemblage. The photographs provide a general record of each layer and also demonstrate the chaotic state of the artifacts within them.

Each stratum is discussed as a separate unit, and its finds are presented in two lists (by layer within the stratum and by object type, together with soil samples and environmental data). Many vases were mended with sherds from more than one stratum (e.g., the lower fill and the main deposits of the rock shelter), an important factor for understanding the depositional processes in EM III. This information is presented in the lists of objects by stratum and layer, as well as in the individual catalog entries presented in the pottery chapter.

Stratum 1 (Layer 1: A600.1–2, A600/700.1, and A700.1)

No antiquities or other ancient features were visible on the surface to guide the excavation, so the project probed the area of the level, bulldozed track with test trenches. The surface layer, which lay at an elevation of +36.57–36.42 m asl, was removed in four adjacent areas of different size (Fig. 8; Pl. 5A): A600.1 at the south end (2.0 x 4.0 m), A600.2 (1.0 x 1.0 m), A600/700 (2.15 x 0.60 m), and A700 at the north end (2.10 x 2.0 m). The soil in A600.1 and A600.2 (Pl. 6B) was soft and pale yellow (2.5Y 8/3) and probably represented the eroded marl slope that had been exposed by the bulldozer on the west side. The first objects lay on the west side of A600.1, including the upper portion of a large hole-mouthed jar (**208**), which was broken in antiquity (Figs. 11–14; Pl. 7A). Digging in subsequent layers exposed this jar completely, and it was finally removed in Layer 8 of Stratum 2. Additional sherds from Layer 1 joined with vessels later removed in lower strata of the rock shelter (e.g., hole-mouthed jar **214** was found in Layer 1, and Layers 2–9 and 12 of the interior portion of Stratum 2). A stone weight (**468**) and two hand tools (a pounder **409** and a pounder-polisher **422**) were recovered from A600.2, while a small triangular clay tool (**515**) was found in the small strip of soil to the north, A600/700.1. The soil in Trench A700.1 was slightly darker (2.5Y 7/4, pale yellow) than that of A600.1–2, which likely indicates that it lay outside the rock shelter (Pls. 7B, 8A, 8B). Work in this trench exposed a flat, rectangular patch of stones (1.30 x 0.80 m) in the southeastern corner, and a possible stone tool (**484**) was recovered.

After removing Stratum 1, it was still not possible to discern the circular plan of the rock shelter; however, the angular profile of its poorly preserved superstructure was beginning to appear on the western side of Trench A600.1 (Pl. 6B where it is visible on the northern part of the cutting made by the bulldozer). This evidence provided the first clue to the nature of the context and served as an important guide for later work. A patch of small stones that appeared along the east side of A600/700.1 and A700.1 probably represents tumble from the blocking wall (Pls. 7B, 8A), which appeared in the lower part of Stratum 2. Although the exact date of the collapse cannot be determined (i.e., whether ancient or modern), we are certain that it took place before the bulldozing in 2007.

No soil samples were collected in Stratum 1.

Small Finds by Layer and Pass within the Stratum

Layer 1: A600.1–2, A600/700.1, A700.1
  A600.1
    Plain pottery and pottery with slight decoration
      Two hole-mouthed jars (**208** from A600.1, A605.1; **214** from A600.1, A601.3, A604.2, 11, 17, 28)
  A600.2
    Ground stone

Hand tools: one pounder (**409**) and one pounder-polisher (**422**)
One weight (**468**)
A600/700.1
Ceramic object
One triangular tool (**515**)
A700.1
Ground stone
One possible hand tool (**484**)

Summary of Stratum 1: Small Finds by Type

Plain pottery and pottery with slight decoration
Two hole-mouthed jars (**208** from A600.1, A605.1; **214** from A600.1, A601.3, A604.2, 11, 17, 28)
Ceramic object
One triangular tool (**515** from A600/700.1)
Ground stone
Four hand tools: one pounder (**409** from A600.2), one pounder-polisher (**422** from A600.2), and one possible hand tool (**484** from A700.1)
One weight (**468** from A600.2)

Stratum 2 Upper (Layers 2–4) and Lower (Layers 5–12)

Stratum 2 includes the main assemblage of the rock shelter, and it is divided into two parts. The upper portion contains the top three layers of the deposit (Layers 2–4), which extend over both the interior and exterior of the rock shelter. The lower portion of Stratum 2 includes Layers 5–12, which were removed in separate parts because clear indications were visible (even before the lower blocking wall appeared) as to whether the recovered material lay inside or outside the rock shelter. In effect, the distinction between the upper and lower part of Stratum 2 is an artificial but important control for understanding the physical context of the assemblage.

STRATUM 2 UPPER (LAYERS 2–4): A601.1, A601.7 (ACROSS TRENCH AND COVERING INTERIOR AND EXTERIOR OF ROCK SHELTER); A601.2–5, A602.1, A604.2–5 (INSIDE ROCK SHELTER)

The upper portion of Stratum 2 was removed in a sequence of three layers (Layers 2–4). Layer 2 (Fig. 12) included a large rectangular space (4.7 × 3.0 m) excavated as Locus A601.1, which lay at an elevation of +36.45 to +36.36 m asl and extended across the trench covering the interior and exterior of the rock shelter below. Layer 2 also contained a smaller triangular extension on the west side (Locus A602.1), inside the rock shelter. The excavation of Locus A602 revealed a hollow cavity with a height of 0.70 m in the vertical scarp of limestone cut by the bulldozer (Pls. 8B, 9A), and this feature thus allowed us to identify the context as a rock shelter. The curving southern limit of its chamber began to appear after the digging of A601.1.

The soil at the southern end of Locus A601.1 was soft and yellow (2.5Y 8/4), while the soil on the east side was reddish yellow (7.5YR 6/8). In the northeastern corner of Locus A601.1, two flat areas of small stones were revealed (Fig. 12; Pl. 8B), which probably represent additional tumble from the blocking wall that lay lower down in Stratum 2. More fragments of four broken vessels were uncovered on the west side of Locus A601.1 east of Locus A602 (Fig. 12; Pl. 9A,). These objects were later removed in Layer 7 as A601.2–6 and include two intact hole-mouthed jars (**208**, **209**; Figs. 11–14; Pls. 7A, 9A, 9B, 10A), a spouted conical jar (**46**; Figs. 11–14; Pl. 9A), and a closed vessel sherd (**347**; Figs. 12–14; Pls. 9A, 9B, 10A). Among the pottery fragments were a few pieces of two well-preserved vessels that were later removed in lower layers (hole-mouthed jar **214** found in Stratum 2 and bridge spouted jar **80** recovered in both Stratums 2 and 3). Layer 2 also contained a pounder (**410**) and one more possible stone tool (**486**); a small number of shells were found in the northeastern corner of Locus A601.1.

Layer 3 was removed as Locus A601.7. It extended the area of Locus A601.1 by 0.30 m to the north, thus exploring a space measuring 5.0 × 3.0 m, which extended across the trench and covered the interior and exterior of the rock shelter below. Most of the digging took place on the eastern side of the locus in order to remove the patches of stones revealed in Layer 2. The soil texture and color was the same as in the previous layer. No new objects were recorded on the plan, but one complete hole-mouthed jar (**211**) was mended from the broken pottery. Twenty fragmentary objects were cataloged from this layer, and there were joins to a nearly complete hole-mouthed jar (**213**), which was recovered primarily from the lower part of Stratum 2 and the fill of Stratum 3.

The third pass (Locus A604.1) across the same space removed Layer 4, which has an elevation of +36.36 to 36.25 m asl. The complete southern outline of the rock shelter was now visible, and the upper course of stones from the wall blocking its entrance appeared in the southeastern corner of the locus (Fig. 13; Pl. 9B). The soil was a dark reddish brown (2.5YR 8/4), and its removal exposed a new cluster of broken vessels and one possible stone tool (**487**) at the southern end (Fig. 13; Pl. 10A). The shapes included an EM IIB jug from the south coast of Crete that probably represents an heirloom in the deposit (**8**; Figs. 13, 14; Pls. 10A, 10B), two tripod cooking pots (**229, 231**; Figs. 12–14; Pls. 10A, 11), and a teapot (**96**; Figs. 13, 14; Pls. 10A, 11A). These vessels were removed in Layer 8 as Loci A604.2–5. The locations of three fragmentary vessels were also recorded (Fig. 14), including two tripod cooking pots (**253, 263**), and one conical bowl (**38**).

The project cataloged a large selection of fragmentary objects collected from the pottery lots of the upper portion of Stratum 2 in order to show the range of shapes and decoration in the ceramic assemblage and conduct chemical analysis of their contents. Their exact numbers are not of particular significance and are certainly not meant to represent a minimum number of vessels because in some cases the fragments probably are non-joining pieces of the same vessel.

In addition to the EM IIB jug imported from the south coast of Crete (**8**), the deposit contained several vessels produced in the EM III tradition of White-on-Dark Ware. Among the open vessels were four shallow conical bowl sherds (**12, 13, 17, 28**), four sherds of conical bowls with or without spouts (**38**, Fig. 14; **36, 40, 41**), one rounded cup (**68**), and one cylindrical cup (**64**). The closed shapes include a sherd of a bridge-spouted jar (**72** and a more complete example **80**), three spouted conical jars (**46, 54, 55**; Figs. 11–14), two jug sherds (**89, 90**), one jar sherd (**61**), and one teapot (**96**; Figs. 13, 14; Pls. 10A, 11A).

The other major group of pottery is the plain or slightly decorated type. The open shapes are represented by sherds of four conical bowls/basins (**97, 98, 102, 113**), two basins with scoring on the interior (**118, 120**), one rounded cup (**127**), and one cup with straight side (**122**). The closed shapes include one jug (**139**), five pithoi (**154, 156, 157, 159, 160**), two pithoi/jars (**175, 187**), one collared jar (**93**), six hole-mouthed jars, several of them almost complete (**208, 209, 211, 213, 214, 221**; Figs. 11–14; Pls. 7A, 9A, 10A), seven tripod cooking pots and sherds (**229, 231, 247, 253, 257, 260, 263**; Figs. 12–14; Pls. 10A, 11), 11 closed vessel sherds (**343, 347, 348, 359–363, 365, 379, 386**), and pieces of three cooking dishes (**289, 306, 323**), and one flat lid (**334**).

The organic remains were modest. A handful of shellfish were collected primarily along the eastern side of the excavated area.

Many objects in the rock shelter were mended from fragments collected in more than one layer, and this taphonomy is important for understanding the construction of the rock shelter and the deposition of the assemblage inside it. From the upper part of Stratum 2, five vessels have revealing biographies. Two vases, a jug (**8**; Figs. 13, 14; Pls. 10A, 10B) and a hole-mouthed jar (**214**), were recovered exclusively within the upper and lower parts of Stratum 2. It also appears that the fragments of both vessels were collected from inside the rock shelter, suggesting that they were deposited there as complete pots. Pieces of a bridge-spouted jar (**80**), a hole-mouthed jar, and a basin (**97**; Figs. 15, 16) were found in a more scattered pattern within Strata 2 and 3. This distribution suggests that they were already broken when the blocking wall and lower fill were introduced in order to create a shelf on which to place objects inside the rock shelter.

Small Finds by Layer and Pass within the Stratum

Layer 2: A601.1–6 and A602
    A601.1 (no finds in the area inside the rock shelter)
        White-on-Dark Ware pottery
            One sherd of a conical bowl with or without spout (**36** from A601.1, A604.20)
        Ground stone
            Two hand tools: one pounder (**410**) and one possible tool (**486**)
        Soil samples: one standard, three intensive
        Marine invertebrates
            Six seashells: one *Patella caerulea*, one *Bollinus brandaris*, one *Columbella rustica*, three *Pisania* sp.
    A601.2
        White-on-Dark Ware pottery
            One cylindrical cup (**64**)
        Plain pottery and pottery with slight decoration
            One hole-mouthed jar (**208** from A601.2, 5)

Soil samples: four intensive
A601.3
    Plain pottery and pottery with slight decoration
        Two hole-mouthed jars (**209** from A601.3; **214** from A600.1, A601.3, A604.2, 11, 17, 28)
    Soil samples: three intensive
    Marine invertebrates
        One seashell: one *Columbella rustica*
A601.4
    White-on-Dark Ware pottery
        One spouted conical jar (**46**)
A601.5
    South Coast Fabric of EM IIB
        One jug (**8** from A601.5, A604.5, 36)
    White-on-Dark Ware pottery
        One shallow conical bowl (**17** from A601.5, A604.20)
        Two bridge-spouted jars (**72** from A601.5, A604.13, 27, 62; **80** from A601.5, A604.13, 27, 28, 62)
        One spouted conical jar (**55**)
    Plain pottery and pottery with slight decoration
        Two hole-mouthed jars and sherds (**221**; **208** from A601.2, 5)
        Six sherds of closed vessels (**343, 347, 361, 365, 379, 386**)
A601.6 (no finds)
A602 (no finds)
Layer 3: A601.7 (and A601.2–5, which continue in it)
A601.7
    White-on-Dark Ware pottery
        Two sherds of shallow bowl (**12, 13**)
        Three sherds of conical bowls with or without spouts (**38, 40, 41**)
    Plain pottery and pottery with slight decoration
        Two sherds of conical bowls or basins (**102, 113**)
        Two sherds of basins with scoring inside (**118, 120**)
        Two pithos sherds (**159, 160**)
        One pithos or jar sherd (**175**)
        One sherd of collared jar (**93**)
        Two hole-mouthed jars (**211** from A601.7; **213** from A601.7 and A604.13, 38, 56, 58, 59, 62, 64, 65, 67, 69)
        Four tripod cooking pots and sherds (**247, 253, 257, 263**)
        Two sherds of closed vessels (**348, 363**)
        One sherd of cooking dish (**323**)
    Soil samples: one intensive
Layer 4: A604.1–5 (and A601.2–7, which continue in it)
A604.1
    White-on-Dark Ware pottery
        One sherd of a conical bowl, with or without spout (**28**)
        One sherd of spouted conical jar (**54**)
        One rounded cup (**68**)
        Two jug sherds (**89, 90**)
    Plain pottery and pottery with slight decoration
        One sherd of conical bowl or basin (**98**)
        One sherd of rounded cup (**127**)
        One jug sherd (**139**)
        Three pithoi sherds (**154, 156, 157**)
        One pithos or jar sherd (**187** from A604.1, 61, 62)
        One sherd of tripod cooking pot (**260**)
        Two sherds of closed vessels (**359, 362**)
        Two sherds of cooking dish (**289, 306**)
    Ground stone
        One possible hand tool (**487**)
    Soil samples: one standard
A604.2
    White-on-Dark Ware pottery
        One teapot (**96** from A604.2, 5)
    Plain pottery and pottery with slight decoration
        One hole-mouthed jar (**214** from A600.1, A601.3, A604.2, 11, 17, 28)
        One tripod cooking pot (**229**)
A604.3
    Plain pottery and pottery with slight decoration
        One sherd of closed vessel (**360** from A604.3, 13)
A604.4
    Plain pottery and pottery with slight decoration
        One sherd of tripod cooking pot (**231**)
    Marine invertebrates: one cuttlebone fragment of *Sepia officinalis*
    Soil samples: one intensive
A604.5
    South Coast Fabric of EM IIB
        One jug (**8** from A601.5, A604.5, 36)
    White-on-Dark Ware pottery
        One jar fragment (**61**)
        One teapot (**96** from A604.2, 5)
    Plain pottery and pottery with slight decoration
        One basin (**97** from A604.5, 7, 12, 23, 26, 29, 46, 53)
        One flat lid sherd (**334**)
    Marine invertebrates
        Five seashells: one *Patella* sp., one *Fasciolaria lignaria*, two *Bollinus brandaris*, one *Pecten jacobaeus*

Summary of Stratum 2 Upper: Small Finds by Type, Including Both Complete and Fragmentary Cataloged Objects

South Coast Fabric of EM IIB
    One jug (**8** from A601.5, A604.5, A604.36)
White-on-Dark Ware pottery
    Four shallow conical bowls (**12** and **13** from A601.7; **17** from A601.5, A604.20; **28** from A604.1)
    Four conical bowls with or without spout (**36** from A601.1, A604.20; **38, 40, 41** from A601.7)
    One cylindrical cup (**64** from A601.2)
    One rounded cup (**68** from A604.1)

Two bridge-spouted jars (**72** from A601.5, 604.13, 27, 62; **80** from A601.5, A604.13, 27, 28, 62)

Three spouted conical jars (**46** from A601.4; **54** from A604.1; **55** from A601.5;)

Two jugs (**89**, **90** from A604.1)

One jar (**61** from A604.5)

One teapot (**96** from A604.2, 5)

Plain pottery and pottery with slight decoration

Four conical bowls and basins (**97** from A604.5, 7, 12, 23, 26, 29, 46, 53; **98** from A604.1; **102** and **113** from A601.7)

Two basins with scoring inside (**118**, **120** from A601.7)

One rounded cup (**127** from A604.1)

One cup with straight side (**122** from A604.1)

One jug (**139** from A604.1)

Five pithoi (**154**, **156**, **157** from A604.1; **159**, **160** from A601.7)

Two pithoi or jars (**175** from A601.7; **187** from A604.1, 61, 62)

One collared jar (**93** from A601.7)

Six hole-mouthed jars (**208** from A601.2, 5; **209** from A601.3; **211** from A601.7; **213** from A601.7, A604.13, 38, 56, 58, 59, 62, 64, 65, 67, 69; and **214** from A600.1, A601.3, A604.2, 11, 17, 28; **221** from A601.5)

Seven tripod cooking pots (**229** from A604.2; **231** from A604.4; **247**, **253**, **257**, **263** from A601.7; **260** from A604.1)

11 closed vessels (**348**, **363** from A601.7; **359**, **362** from A604.1; **360** from A604.3, 13; **343**, **347**, **361**; **365**, **379**, **386** from A601.5)

Three cooking dishes (**289**, **306** from A604.1; **323** from A601.7)

One flat lid (**334** from A604.5)

Ground stone

One pounder (**410** from A601.1)

Two possible hand tools (**486** from A601.1; **487** from A604.1)

Soil samples

Standard: A601.1 (1), A604.1 (1)

Intensive: A601.1 (3), A601.2 (4), A601.3 (3), A601.7 (1), A604.4 (1)

Marine invertebrates

12 seashells: 2 *Patella caerulea*, 1 *Fasciolaria lignaria*, 3 *Bollinus brandaris*, 2 *Columbella rustica*, 3 *Pisania* sp., 1 *Pecten jacobaeus* (A604.5)

One cuttlebone fragment of *Sepia officinalis* (A604.4)

STRATUM 2 LOWER INTERIOR (LAYERS 5–12): A604.8, A604.10, A604.13, A604.17–18, A604.20, A604.22–24, A604.27–28, A604.30–45

The lower part of Stratum 2 includes both the rectangular interior chamber of the rock shelter and a smaller exterior space in the northeast corner of Trench A600 (Pl. 11C). The excavation recorded eight layers of material (Layers 5–12) in each area. Inside the rock shelter, digging exposed a thick deposit of pottery and stone tools (Pls. 12A–20A). Some well-preserved objects were initially revealed in Layers 2–4, and then they were removed in Layers 5–12. Because of the location of the excavation near the national road and the wet weather conditions, it was not possible to excavate and illustrate the complete deposit at the same time; it would have required leaving the objects in the ground for a long period. Instead, the objects were recorded in two levels. The first includes material from Layers 2–9. Several objects in this level (e.g., **8**, **46**, **96**, **208**, **209**, **229**, **239**; Figs. 11–14; Pls. 7A, 9A, 10A, 10B, 11A, 12B, 13B, 14A, 15A, 15B) had already been partially exposed in the upper portion of Stratum 2, and they were removed with the material exposed in Layers 5–9 of the lower part of Stratum 2 (Fig. 14). A second level of material was revealed in Layers 10–12 (Fig. 15).

Layers 5–9 consisted of a thin deposit of pale yellow soil (2.5Y 7/6), which lay at an elevation of +36.25–36.14 m asl. Layer 5 was removed with Locus A604.8, which exposed more of the southern end of the trench; a tripod cooking pot (**230**; Fig. 14; Pls. 14A, 14B, 16A, 16B, 17A, 17B, 18A) was uncovered upside down on its rim (Fig. 14; Pl. 18A) beside a teapot (**96**; Figs. 13, 14; Pls. 10A, 11A, 14A). Layers 6 and 7 involved work in the southeast corner of the rock shelter with Locus A604.10 and the southern part of A604.11. A pounder (**413**) and another possible stone tool (**489**) were recorded in Layer 6 (Fig. 14), while more fragments of a hole-mouthed jar (**214**; Pl. 15B) and a possible stone tool (**491**; Fig. 14) were recovered in Layer 7.

Locus A604.13 consisted of a substantial pass across the interior of the rock shelter (Layer 8; Fig. 14; Pls. 11C, 12A), and it revealed five additional stones of the wall that ran along its northeast side. The work also clarified the position of several objects exposed in previous layers, which were then removed (**46** in Pl. 12B, **347** in Pl. 13A, **96** in Pl. 14A, **229** in Pl. 13B, and **231** in Pl. 14A). New finds (Fig. 14) included a pumice polisher near the southern edge of the chamber (**458**), a limestone polisher (**460**) next to the South Coast jug (**8**), and an obsidian blade (**394**) beside a spouted conical jar (**46**; Pl. 12B). Sherds from this layer also joined with a substantially preserved scored basin (**114**;

Fig. 16), found broken across the interior and exterior of Stratum 2 Lower, and a spouted jar (**192**), a hole-mouthed jar (**213**), and a tripod cooking pot (**236**), which were recovered in both Stratum 2 Lower and the fill of Stratum 3.

Layer 9 consisted of a shallow pass, which cleared objects along the west side of the rock shelter with Locus A604.17 (Fig. 14) and the south side with Locus A604.18. Two hole-mouthed jars (**208, 209**; Pl. 15A, 15B), first noted in Layer 2, were finally removed. No new objects were found in this layer, though fragments of a scored basin (**114**), recovered in both loci, joined sherds from various layers of Strata 2 and 3. Parts of a jug (**137**), a hole-mouthed jar (**210**; Fig. 15; Pl. 19A), and two tripod cooking vessels (**228, 233**) share a similar distribution. Two more vessels that contained fragments from A604.17 (spouted conical jar **49** and hole-mouthed jar **214**; Pl. 15B) were found only in layers of Stratum 2 inside the rock shelter.

Locus A604.17 also produced a single seashell. While this find might appear unremarkable at first glance, it is one of only two shells recorded from the main deposit inside the rock shelter, and it could be intrusive (i.e., from the northern limit of the locus, which may have extended into the exterior of Stratum 2 Lower where many shells were recorded). The other shell was found at the bottom of Stratum 2 in Locus A604.28.

Layers 10–12 formed the lower portion (+36.14–36.05 m asl) of this vessel deposit inside the rock shelter. Their removal revealed the irregular line of the low wall (A605), which had been built in short segments across the northeast side of the chamber (Fig. 15; Pls. 16A–17B). The first section of the wall employs small, irregular stones laid in two courses. It begins at the southeast corner of the chamber and runs to the northwest for 1.55 m. The second segment is made of both larger and smaller stones, again in two courses. This section runs for 1.70 m to the northwest corner of the chamber and appears to have created a small portion of covered space inside the rock shelter immediately to the north. The scatter of similar stones in Layers 2–9 to the northeast of this wall suggests that the original construction of A605 may have been more regular and at least one course higher. The wall rested on a thick layer of cultural material in Stratum 3, and it appears to have served two purposes. First, it was used to retain an earth terrace against the sloping marl floor of the chamber after a mixed fill of soil, pottery, and stones had been set inside. The wall probably also blocked access to the chamber after all the objects had been placed within it.

The soil in Layers 10–12 varied from pale yellow (2.5Y 8/3) on the west side to pale brown (10YR 7/4) on the east side. The removal of the uppermost layer (10), as Locus A604.20, revealed a new deposit of vases and broken pottery. Three stone tools were recovered inside wall A605 (Fig. 15): a pounder (**416**), a possible tool (**492**), and a quern (**471**). The next pass across the chamber (Locus A604.22, Layer 11) exposed more vases and smaller clusters of pottery fragments, which were recorded as A604.23–24 and A604.30–35. They include three spouted conical jars (**48, 49, 53**); additional fragments joined vases mended from other layers of Strata 2 and 3 (jug **136**, spouted jar **192**, and three tripod cooking pots, **228, 233, 236**). A grinder (**493**; Fig. 15) and a quern (**472**; Fig. 15) were recovered next to wall A605. Fragments of carbonized olive stones were also recorded in this layer; they represent the only botanical remains found inside this stratum.

The next level (Layer 12) contained the bulk of the objects from the lower interior of Stratum 2. Locus A604.27 focused on the west side of the chamber and uncovered a hole-mouthed jar (**212**) next to a basin (**97**, recorded in A604.29). These vases rested directly on the sloping bedrock surface at the western limit of the rock shelter. Immediately to the south lay another collection of objects, including a tripod cooking pot (**230**; Pl. 18A), a jar sherd (**86**), more fragments of basin **97**, a kantharos rim sherd (**71**), and a large vat (**330**; Pl. 18B). Two additional vessels were mended from the fragmentary pottery (jug **135**, hole-mouthed jar **224**). Locus A604.28 removed Layer 12 from the east side of the chamber. This locus fully exposed vessels from Layer 11 and revealed new vases and pottery clusters, which were recorded as A604.36–38 and A604.41–45. Two hole-mouthed jars (**210, 213**; Fig. 15; Pl. 19A) lay at the southern edge of the chamber. The lower part of a jug (**87**), a conical bowl (**23**), a spouted conical jar (**47**), a rounded cup (**125**), and a loomweight (**509**) were exposed in the center of the chamber (Fig. 15; Pl. 19B). A conical bowl (**24**; Fig. 15; Pl. 20A) and the base of a closed vessel (**384**) lay on the east side of the rock shelter. After these objects had been removed, the remaining pottery, a weight (**469**), a limestone pounder (**434**), and a possible tool (**494**) were collected as Locus A604.40.

To examine the distribution of artifacts within the rock shelter, we need to combine the data from the upper and lower levels of Stratum 2. It appears that some care was taken to structure or order the position of the vessels by type. Storage jars (particularly hole-mouthed jars **208–210**, **212**; Pl. 15A, 15B), basin **97**, and vat **330** were placed in the deepest part of the chamber along the west side where the rising ceiling provided ample room. In front or to the east of these vessels lay several pouring vessels (teapot **96**, spouted kantharos **71**, two spouted conical jars **46** and **47**, and part of jug **87**), and to the east (in front) of these vessels lay a group of cups and bowls (two conical bowls **23** and **24** and rounded cup **125**). Behind these drinking vessels to the south lay a cluster of cooking equipment (tripod cooking pots **229–231**; Pls. 13B, 14A), with the South Coast jug to the east (**8**). Most of the stone tools from this stratum lay immediately inside the enclosure wall A605.

The pottery from the lowest layer (Figs. 17, 18; Pls. 16, 17) joined with several broken vessels recorded in other layers of Strata 2 and 3 (Pl. 14B). These objects include two partly preserved spouted conical jars (**49**, **50**), a scored basin (**114**), two jugs (**136**, **137**), a spouted jar (**192**), three hole-mouthed jars (**210**, **213**, **214**), and five tripod cooking pots (**225**, **228**, **232**, **233**, **236**).

A large selection of fragmentary pottery from the chamber was cataloged with the more complete objects to show the range of shapes and decoration in the interior of Stratum 2. Again, the exact numbers are not meant to represent a minimum number of vessels in the deposit but rather to illustrate the variety. A pair of objects date to the EM IIB period: the jug (**8**) imported from the south coast and a sherd of a Vasiliki Ware jug (**6**). A large amount of EM III White-on-Dark Ware was cataloged as both sherds and vases, including examples of four shallow conical bowls or basins (**14**, **17–19**), four conical bowls with or without spouts (**23–25**, **27**), one conical bowl with a frying pan handle (**44**), eight spouted conical jars (**47–53**, **58**), one jar (**62**), one rounded cup with tiny lug handles (**67**), one rounded cup (**70**), one spouted kantharos (**71**), two bridge-spouted jars (**72**, **80**), one jar or jug (**86**), one jug (**87**), and one closed vessel (**84**). More sherds belong to the class of the plain and slightly decorated pottery. The range of shapes includes one conical bowl (**112**), five basins with scoring inside (**114–117**, **119**), three rounded cups (**125**, **129**, **130**), eight jugs (**135–137**, **141**, **143**, **144**, **148**, **152**), four pithoi (**158**, **163**, **166**, **170**), two pithoi or jars (**184**, **185**), one spouted jar (**193**), one collared jar (**202**), six hole-mouthed jars (**210**, **212–214**, **222**, **224**), 20 tripod cooking pots (**225–228**, **230**, **232–234**, **236**, **242**, **252**, **254**, **262**, **264**, **267–269**, **274**, **275**, **283**), 13 closed vessels (jar, cooking pot, or jug; **342**, **346**, **357**, **364**, **371**, **372**, **374**, **376**, **378**, **380**, **384**, **385**, **391**), 12 cooking dishes (**291**, **295**, **299**, **300**, **302**, **303**, **305**, **317**, **318**, **320**, **327**, **329**), two vats (**330**, **332**), and one flat lid (**333**). The other objects in the stratum included a piece of a potter's disk (**511**) and one loomweight (**509**). Among the ground stone tools were three hammer stones (**413**, **416**, **458**), two abraders (**434**, **446**), one polisher (**460**), one grinder (**453**), four broken querns (**471–474**), and one stone weight (**469**). One partial obsidian blade was also recovered in the assemblage (**394**).

Small Finds by Layer and Pass within the Stratum

Layer 5: A604.8, A604.14 (and A601.2–5 and A604.3–5, which continue in it)
    A604.8
        White-on-Dark Ware pottery
            One cup with a vertical handle, not cataloged
        Plain pottery and pottery with slight decoration
            One fragmentary closed vessel (**385**)
        Ceramic object
            One potter's disk (**513**)
    A604.14
        Plain pottery and pottery with slight decoration
            One tripod cooking pot (**230**)
Layer 6: A604.10 (and A601.2–7, A604.3–5, A604.8, and A604.14, which continue in it)
    A604.10
        White-on-Dark Ware pottery
            One sherd of a shallow basin (**19**)
            One sherd of a conical bowl with or without spouts (**27**)
        Plain pottery and pottery with slight decoration
            One sherd of a cooking dish (**327**)
        Ground stone
            Two hand tools: one pounder (**413**) and one possible tool (**489**)
Layer 7: A604.11 (A601.2–7, A604.3–5, A604.10, A604.14, which continue in it)
    A604.11
        Plain pottery and pottery with slight decoration

One hole-mouthed jar (**214** from A600.1, A601.3, A604.2, 11, 17, 28)
One tripod cooking pot (**254**)
Ground stone
    One possible tool (**491**)

Layer 8: A604.13 (A601.2–7, A604.3–5, A604.8, A604.10–11, and A604.14, which continue in it)
  A604.13
    White-on-Dark Ware pottery
      Two bridge-spouted jars and fragments (**72** from A601.5, A604.13, 27, 62; **80** from A601.5, A604.13, 27, 28, 62)
      One jug sherd (**91** from A604.13, 20, 53, 57, 65)
    Plain pottery and pottery with slight decoration
      Two partly preserved basins with scoring inside (**114** from A604.7, 13, 17, 18, 21, 25, 26, 28, 39, 40, 46, 47, 49, 50, 53, 55, 59, 67; **117**)
      One sherd of a pithos with drain (**189** from A604.13, 15, 17, 55, 57, 62)
      Two spouted jars: one sherd and one complete (**192** from A604.13, 22, 27, 29, 40, 46, 47, 53, 55, 59, 67; **193**)
      Two hole-mouthed jars: two sherds (**213** from A601.7, A604.13, 38, 56, 58, 59, 62, 64, 65, 67, 69; **222**)
      Five fragments of tripod cooking pots (**226** from A604.6, 7, 13, 17–19; **236** from A604.13, 22, 28, 29, 40, 42, 47, 55–57, 59, 61, 62; **267, 269, 283**)
      One sherd of closed vessel (**364**)
      One sherd of cooking dish (**300**)
    Ground stone
      Two hand tools: one pumice abrader/polisher (**458**) and one polisher (**460**)
    Chipped stone
      One blade (**394**)
    Soil samples: two standard, one intensive

Layer 9: A604.17–18 (A601.2–7, A604.3–5, A604.8, A604.10–11, and A604.13–14, which continue in it)
  A602.3
    Soil samples: one intensive
  A604.17
    White-on-Dark Ware pottery
      One spouted jar (**49** from A604.17, 22, 40, 42, 43)
    Plain pottery and pottery with slight decoration
      One basin with scoring inside (**114** from A604.7, 13, 17, 18, 21, 25, 26, 28, 39, 40, 46, 47, 49, 50, 53, 55, 59, 67)
      One jug (**137** from A604.6, 17, 40, 48, 53, 56, 57, 62, 64, 65, 69)
      One pithos sherd (**158**)
      One sherd of a pithos with drain (**189** from A604.13, 15, 17, 55, 57, 62)
      Two hole-mouthed jar sherds and vases (**210** from A604.17, 22, 27, 28, 32, 37, 40, 44, 46, 48, 53, 59, 64, 65, 67; **214** from A600.1, A601.3, A604.2, 11, 17, 28)
      Three tripod cooking vessels (**226** from A604.6, 7, 13, 17–19; **233** from A604.17, 22, 40, 46, 53, 59, 64, 65; **252** from A604.7, 15–17, 19–21, 28, 40, 46, 55–57, 59, 61, 62, 64, 65, 67)
      Three sherds of closed vessel (**342, 357, 378**)
      Two sherds of cooking dishes (**318, 320**)
    Soil samples: one standard, three intensive
    Marine invertebrates
      One *Monodonta* sp. seashell
  A604.18
    Plain pottery and pottery with slight decoration
      One basin with scoring inside (**114** from A604.7, 13, 17, 18, 21, 25, 26, 28, 39, 40, 46, 47, 49, 50, 53, 55, 59, 67)
      Two tripod cooking pots (**226** from A604.6, 7, 13, 17, 18, 19; **228** from A604.18, 20, 22, 28, 42, 46, 55, 64, 65)

Layer 10: A604.20
  A604.20
    White-on-Dark Ware pottery
      One shallow conical bowl (**17** from A601.5, A604.20)
      Two sherds of conical bowls with or without spouts (**25**; **36** from A601.1, A604.20)
      One sherd of a conical bowl with frying pan handle (**44**)
      Two sherds of spouted conical jars (**51**; **52** from A604.20, 45, 46)
      One jug sherd (**91** from A604.13, 20, 53, 57, 65)
      One sherd of a closed vessel (**84**)
    Plain pottery and pottery with slight decoration
      One sherd of a conical bowl (**112**)
      One jug sherd (**141**)
      One pithos sherd (**170**)
      One sherd of a collared jar (**202**)
      One sherd of a hole-mouthed jar (**212** from A604.20, 22, 40, 46, 64)
      Three sherds of a tripod cooking pot and leg (**228** from A604.18, 20, 22, 28, 42, 46, 55, 64, 65; **252** from A604.7, 15–17, 19–21, 28, 40, 46, 55–57, 59, 61, 62, 64, 65, 67; **268**)
      Three sherds of a closed vessel (**346, 376**; **391** from A604.20, 46, 47, 53, 55, 56)
      One sherd of a cooking dish (**291**)
      One sherd of a flat lid (**333**)
    Ground stone
      Two hand tools: one pounder (**416**) and one possible tool (**492**)
      One quern fragment (**471**)

Layer 11: A604.22–24, 30–35
  A604.22
    White-on-Dark Ware pottery
      Four fragments of spouted conical jars (**47** from A604.22, 33; **49** from A604.17, 22,

40, 42, 43; **50** from A604.22, 40, 42, 46, 47, 53, 64–66; **53**)
Plain pottery and pottery with slight decoration
Three jug fragments (**136** from A604.12, 22, 28, 40, 49, 53, 56, 62, 64; **144**; **148** from A604.22, 46, 47, 53, 56, 59, 62, 64)
One pithos or jar sherd (**184**)
One spouted jar (**192** from A604.13, 22, 27, 29, 40, 46, 47, 53, 55, 59, 67)
Two hole-mouthed jars, one of them complete (**210** from A604.17, 22, 27, 28, 32, 37, 40, 44, 46, 48, 53, 59, 64, 65, 67; **212** from A604.20, 22, 40, 46, 64)
Five tripod cooking pots and fragments (**227** from A604.22, 59; **228** from A604.18, 20, 22, 28, 42, 46, 55, 64, 65; **233** from A604.17, 22, 40, 46, 53, 59, 64, 65; **236** from A604.13, 22, 28, 29, 40, 42, 47, 55–57, 59, 61, 62; **264**)
One closed vessel sherd (**374**)
Ground stone
One hand tool: grinder (**493**)
One quern fragment (**472**)
Soil samples: one standard
Botanical remains: 4/2 *Olea europaea* stones
A604.23
Plain pottery and pottery with slight decoration
One basin (**97** from A604.5, 7, 12, 23, 26, 29, 46, 53)
A604.24
White-on-Dark Ware pottery
One jug or jar sherd (**86** from A604.24, 59, 62, 64, 69)
A604.30
Plain pottery and pottery with slight decoration
One closed vessel sherd (**384**)
A604.31
White-on-Dark Ware pottery
One conical bowl with spout (**24**)
A604.32
White-on-Dark Ware pottery
One spouted conical jar (**48** from A604.32, 47)
Plain pottery and pottery with slight decoration
One rounded cup (**125**)
One hole-mouthed jar (**210** from A604.17, 22, 27, 28, 32, 37, 40, 44, 46, 48, 53, 59, 64, 65, 67)
A604.33
One spouted conical jar (**47** from A604.22, A604.33)
A604.34
One conical bowl (**23**)
A604.35
One jug fragment (**87**)
Layer 12: A604.27–28, 36–38, 40, 41–45
A604.27

White-on-Dark Ware pottery
Two fragmentary bridge-spouted jars (**72** from A601.5, A604.13, 27, 62; **80** from A601.5, A604.13, 27, 28, 62)
Plain pottery and pottery with slight decoration
One sherd of rounded cup (**129**)
Three jugs and sherds (**135** from A604.21, 27; **143**, **152**)
One pithos or jar sherd (**184**)
One spouted jar (**192** from A604.13, 22, 27, 29, 40, 46, 47, 53, 55, 59, 67)
Two hole-mouthed jars (**210** from A604.17, 22, 27, 28, 32, 37, 40, 44, 46, 48, 53, 59, 64, 65, 67; **224**)
One sherd of tripod cooking pot (**275**)
One cooking dish (**305**)
Two sherds of closed vessel (**371**, **372**)
Soil samples: one standard
A604.28
White-on-Dark Ware pottery
One bridge-spouted jar (**80** from A601.5, A604.13, 27, 28, 62)
Plain pottery and pottery with slight decoration
Two basins with scoring inside: one complete, one sherd (**114** from A604.7, 13, 17, 18, 21, 25, 26, 28, 39, 40, 46, 47, 49, 50, 53, 55, 59, 67; **116**)
One sherd of rounded cup (**130**)
One jug (**136** from A604.12, 22, 28, 40, 49, 53, 56, 62, 64)
Two hole-mouthed jars (**210** from A604.17, 22, 27, 28, 32, 37, 40, 44, 46, 48, 53, 59, 64, 65, 67; **214** from A600.1, A601.3, A604.2, 11, 17, 28)
Four tripod cooking pots and sherds (**228** from A604.18, 20, 22, 28, 42, 46, 55, 64, 65; **234** from A604.28, 40, 47, 53, 56, 59, 64, 65, 67; **236** from A604.13, 22, 28, 29, 40, 42, 47, 55–57, 59, 61, 62; **252** from A604.7, 15–17, 19–21, 28, 40, 46, 55–57, 59, 61, 62, 64, 65, 67)
One sherd of a cooking dish (**302**)
Ceramic object
One loomweight (**509**)
Ground stone
One abrader fragment (**446**)
Two quern fragments (**473**, **474**)
Soil samples: one standard
Marine invertebrates
One *Patella* sp. seashell
A604.36
South Coast Fabric of EM IIB
One jug (**8** from A601.5, A604.5, 36)
Plain pottery and pottery with slight decoration
One sherd of a basin with scoring inside (**119**)
A604.37

One hole-mouthed jar (**210** from A604.17, 22, 27, 28, 32, 37, 40, 44, 46, 48, 53, 59, 64, 65, 67)
A604.38
   Plain pottery and pottery with slight decoration
      One hole-mouthed jar (**213** from A601.7, A604.13, 38, 56, 58, 59, 62, 64, 65, 67, 69)
      One closed vessel (**380** from A604.38, 62, 69)
A604.40
   Vasiliki Ware pottery
      One jug (**6**)
   White-on-Dark Ware pottery
      Two shallow bowl sherds (**14, 18**)
      One jar sherd (**62**)
      One rounded cup sherd (**70**)
      One rounded cup with tiny handles (**67**)
      Three fragments of spouted conical jars (**49** from A604.17, 22, 40, 42, 43; **50** from A604.22, 40, 42, 46, 47, 53, 64–66; **58**)
   Plain pottery and pottery with slight decoration
      One basin with scoring inside (**114** from A604.7, 13, 17, 18, 21, 25, 26, 28, 39, 40, 46, 47, 49, 50, 53, 55, 59, 67)
      One sherd of a basin with scoring inside (**115**)
      One fragment of rounded cup (**126** from A604.40, 46, 47, 53, 65)
      Two jugs (**136** from A604.12, 22, 28, 40, 49, 53, 56, 62, 64; **137** from A604.6, 17, 40, 48, 53, 56, 57, 62, 64, 65, 69)
      Two pithos fragments (**163, 166**)
      One spouted jar (**192** from A604.13, 22, 27, 29, 40, 46, 47, 53, 55, 59, 67)
      Two hole-mouthed jars (**210** from A604.17, 22, 27, 28, 32, 37, 40, 44, 46, 48, 53, 59, 64, 65, 67; **212** from A604.20, 22, 40, 46, 64)
      Eight tripod cooking pots and sherds (**225** from A604.40, 46, 53, 59, 64, 65; **232** from A604.21, 29, 40, 46, 53, 59, 61, 64, 65, 67; **233** from A604.17, 22, 40, 46, 53, 59, 64, 65; **234** from A604.28, 40, 47, 53, 56, 59, 64, 65, 67; **236** from A604.13, 22, 28, 29, 40, 42, 47, 55–57, 59, 61, 62; **242**; **252** from A604.7, 15–17, 19–21, 28, 40, 46, 55–57, 59, 61, 62, 64, 65, 67; **262, 274**)
      Five cooking dish sherds (**295, 299, 303, 317, 329**)
      One vat spout sherd (**332**)
   Ground stone
      Two hand tools: one pounder-abrader fragment (**434**), one possible tool (**494**)
      One weight (**469**)
A604.41
   White-on-Dark Ware pottery
      One kantharos (**71** from A604.41, A604.44)

A604.42
   White-on-Dark Ware pottery
      Three spouted conical jars (**48** from A604.42, 47; **49** from A604.17, 22, 40, 42, 43; **50** from A604.22, 40, 42, 46, 47, 53, 64–66)
   Plain pottery and pottery with slight decoration
      Two tripod cooking pots (**228** from A604.18, 20, 22, 28, 42, 46, 55, 64, 65; **236** from A604.13, 22, 28, 29, 40, 42, 47, 55–57, 59, 61, 62)
   Ceramic object
      One potter's disk sherd (**511**)
A604.43
   White-on-Dark Ware pottery
      One spouted conical jar (**49** from A604.17, 22, 40, 42, 43)
   Plain pottery and pottery with slight decoration
      One vat (**330**)
A604.44
   White-on-Dark Ware pottery
      One fragment of a spouted kantharos (**71** from A604.41, 44)
   Plain pottery and pottery with slight decoration
      One hole-mouthed jar (**210** from A604.17, 22, 27, 28, 32, 37, 40, 44, 46, 48, 53, 59, 64, 65, 67)
A604.45
   White-on-Dark Ware pottery
      One fragment of a spouted conical jar (**52** from A604.20, 45, 46)

Summary of Stratum 2 Lower Interior: Small Finds by Type Including Both Complete and Fragmentary Cataloged Objects

South Coast Fabric of EM IIB
   One jug with handle attached by thrusting through the shoulder (**8** from A601.5, A604.5, 36)
Vasiliki Ware pottery
   One jug (**6** from A604.40)
White-on-Dark Ware pottery
   Four shallow conical bowls or basins (**14** and **18** from A604.40; **17** from A601.5, A604.20; **19** from A604.10)
   Five conical bowls with or without spout (**23** from A604.34; **24** from A604.31; **25** from A604.20; **27** from A604.10; **36** from A601.1, A604.20)
   One conical bowl with frying pan handle (**44** from A604.20)
   Eight spouted conical jars (**47** from A604.22, A604.33; **48** from A604.32, 47; **49** from A604.17, 22, 40, 42, 43; **50** from A604.22, 40, 42, 46, 47, 53, 64–66; **51** from A604.20; **52** from A604.20, 45–46; **53** from A604.22; **58** from A604.40;)
One jar (**62** from A604.40)
One rounded cup with tiny handles (**67** from A604.40)

One rounded cup (**70** from A604.40)
One spouted kantharos (**71** from A604.41, 44)
Two bridge-spouted jars (**72** from A601.5, 604.13, 27, 62; **80** from A601.5, A604.13, 27, 28, 62)
One jar or jug (**86** from A604.24, 59, 62, 64, 69)
One jug (**87** from A604.35)
One closed vessel (**84** from A604.20)
Plain pottery and pottery with slight decoration
  One conical bowl (**112** from A604.20)
  Five basins with interior scoring (**114** from A604.7, 13, 17, 18, 21, 25, 26, 28, 39, 46, 47, 49, 50, 53, 59, 67; **115** from A604.40; **116** from A604.28; **117** from A604.13; **119** from A604.36)
  Four rounded cups (**125** from A604.32; **126** from A604.40, 46, 47, 53, 65; **129** from A604.27; **130** from A604.28)
  Eight jugs (**135** from A604.21, 27; **136** from A604.12, 22, 28, 40, 49, 53, 56, 62, 64; **137** from A604.17; **141** from A604.20; **143** and **152** from A604.27; **144** from A604.22; **148** from A604.22, 46, 47, 53, 56, 59, 62, 64)
  Four pithoi (**158** from A604.17; **163** and **166** from A604.40; **170** from A604.20)
  Two pithoi or jars (**184** from A604.27; **185** from A604.22)
  Two spouted jars (**192** from A604.13, 22, 27, 29, 40, 46, 47, 53, 55, 59, 67; **193** from A604.13)
  One collared jar (**202** from A604.20)
  Six hole-mouthed jars (**210** from A604.17, 22, 27, 28, 32, 37, 40, 44, 46, 48, 53, 59, 64, 65, 67; **212** from A604.20, 22, 40, 46, 64; **213** from A601.7, A604.13, 38, 56, 58, 59, 62, 64, 65, 67, 69; **214** from A600.1, A601.3, A604.2, 11, 17, 28; **222** from A604.13; **224** from A604.27)
  20 tripod cooking pots (**225** from A604.40, 46, 53, 59, 64, 65; **226** from A604.6, 7, 13, 17–19; **227** from A604.22, 59; **228** from A604.18, 20, 22, 28, 42, 46, 55, 64, 65; **230** from A604.14; **232** from A604.21, 29, 40, 53, 59, 61, 65, 67; **233** from A604.17, 22, 40, 46, 53, 59, 64, 65; **234** from A604.28, 40, 47, 53, 56, 59, 64, 65, 57; **236** from A604.13, 22, 28, 29, 40, 42, 47, 55–57, 59, 61; **252** from A604.7, 15, 16, 17, 19, 20, 21, 28, 40, 46, 55, 56, 57, 59, 61, 62, 64, 65, 67; **254** from A604.11; **242, 262, 274** from A604.40; **264** from A604.22; **267, 269, 283** from A604.13; **268** from A604.20; **275** from A604.27)
  13 closed vessels (jar, cooking pot, or jug: **342, 357, 378** from A604.17; **364** from A604.13; **371** and **372** from A604.27; **374** from A604.22; **376** and **346** from A604.20; **380** from A604.38, 62, 69, 85; **384** from A604.30; **385** from A604.8; **391** from A604.20, 46, 47, 53, 55, 56)
  12 sherds of cooking dishes (**291** from A604.20; **295, 299, 303, 317, 329** from A604.40; **300** from A604.13; **302** from A604.28; **305** from A604.27; **318** and **320** from A604.17; **327** from A604.10;)
  Two vats (**330** from A604.43; **332** from A604.40)

One flat lid (**333** from A604.20)
Ceramic objects
  One potter's disk (**511** from A604.42)
  One loomweight (**509** from A604.28)
Ground stone
  11 hand tools: one pounder (**413**) and one possible tool (**489**) from A604.10; one pounder (**416**) and one possible tool (**492**) from A604.20; one pounder-abrader (**434**) and one possible tool (**494**) from A604.40; one abrader (**446**) from A604.28; one pumice abrader/polisher (**458**) and one polisher (**460**) from A604.13; one possible tool (**491**) from A604.11; one grinder (**493**) from A604.22.
  Four querns (**471** from A604.20; **472** from A604.22, 59; **473** and **474** from A604.28)
  One stone weight (**469** from A604.40)
Chipped stone
  One blade (**394** from A604.13)
Marine invertebrates
  Two seashells: one *Monodonta* sp. (A604.17), one *Patella* sp. (A604.28)
Botanical remains
  4/2 *Olea europaea* stones (A604.22)
Soil samples
  Standard: A604.13 (2), A604.17 (1), A604.22 (1), A604.27 (1), A604.28 (1)
  Intensive: A604.3 (1), A604.13 (1), A604.17 (3)

Stratum 2 Lower Exterior (Layers 5–12): A604.6–7, A604.9, A604.11–12, A604.15–16, A604.19, A604.21, A604.26, A604.29, A604.39

Below Layer 4 it was possible to determine whether a locus lay inside or outside the rock shelter. The lower part of Stratum 2 outside the chamber consisted of eight layers (Layers 5–12). Several of these layers contained small and large stones that appear to have tumbled from wall A605 (Figs. 12, 14, 15). Although fewer ceramic objects lay outside the chamber (Pls. 20B–26B), the assemblage is interesting because of the noticeable differences it displays compared to the deposit inside the rock shelter, suggesting a specific intent for each deposit. The most striking example is the presence of shellfish and beach pebbles outside the chamber in contrast with their absence within the rock shelter.

Like on the interior, the excavation outside revealed two parts to the deposit. The upper part includes Layers 5–9 (Fig. 14). On the west side of this space, these layers (at an elevation of +36.31–36.12 m asl) were removed with Locus A604.7 and A604.12 (Pls. 20B–21B). The soil was soft and pale yellow (2.5Y 8/4), and two scatters of pottery were recovered next to a pair of large stones (Fig. 14; Pls. 20B–21B). To the south there were large pieces of a pithos with a drain (**189**); to the north lay parts of three vessels (Fig. 14; Pls. 21B, 22A), including a cooking dish (**316**), a closed vessel (**388**), and a pithos (**171**). Two loomweights were recovered on either side of the pottery clusters—**505** to the east (Fig. 14; Pls. 22B, 23A) and **506** to the west—and a whetstone (**455**; Fig. 14) lay just outside the enclosure wall.

On the east side of the stratum, Layers 5–9 (at an elevation of +36.29–36.11 m asl) were removed in sequence with Locus A604.6, A604.9, and A604.11 (Fig. 14; Pls. 23B, 24A). The soil was soft and reddish yellow (7.5YR 6/6), and it contained many stones ranging in size from 0.20–0.40 m, large fragments of pottery, and concentrations of seashells with small beach pebbles. The stones again appear to have tumbled from the enclosure wall. A loomweight (**501**), two obsidian blades (**392, 393**) and a worked flake, and several ground stone tools, including a whetstone (**454**), three pounders (**411, 412, 414**), one faceted tool (**445**), and two possible tools (**488, 490**), were recovered between the wall and two flat stones to the north (Fig. 14).

Layers 10–12 make up the lower part of this stratum (Pls. 25A, 26A), which lay at an elevation of +36.11–36.01 m asl. The soil on the west side was soft and pale yellow (2.5Y 8/4), and its removal exposed the sloping bedrock floor of a small corner of the rock shelter that lay outside the enclosure wall at this level (Fig. 15; Pls. 24B, 25A, 26A). A stone slab (**483**) was the only find recovered from the space. To the east the soil was darker (10YR 6/6, brown yellow), and it contained several large stones (Fig. 15) that probably fell from the enclosure wall, much more broken pottery, and small concentrations of seashells and pebbles (Pls. 24B, 25A). A spindle whorl (**510**; Fig. 15; Pl. 25B) was found to the north, while an obsidian blade (**395**) and a large number of stone tools lay between the enclosure wall and the tumble (Fig. 15). This group included two pounders (**417, 418**), one pounder-abrader (**423**), a grinder (**450**), a pumice-abrader (**458**), and a whetstone (**456**).

This stratum also produced pottery fragments that joined a small number of objects found in the fill below the chamber. One tripod cooking pot (**226**) was also mended with fragments from inside and outside the chamber in this same stratum. Fragments of three more vessels (basin **97**, basin with scoring **114**, and tripod cooking pot **232**) were scattered in both Stratum 2 and the fill of Stratum 3 (Figs. 15, 16, 20; Pls. 28B, 29A).

The project again cataloged a large selection of sherds from this stratum to demonstrate the range of shapes in the deposit and to look for the contents of the vessels through an intensive program of chemical analysis. The shapes with White-on-Dark decoration include four shallow conical bowls (or basins; **9–11, 20**), two conical bowl with or without a spout (**30, 42**), two rounded cups with tiny lugs or handles on the lower body (**65, 66**), five bridge-spouted jars (**73–75, 78, 81**), one jar or jug (**85**), one jug (**88**), and one closed vessel (**95**). The shapes among the plain or slightly decorated class include seven conical bowls and basins (**97, 99, 100, 105, 106, 108, 110**), one basin with scoring inside (**114**), one spinning bowl (**121**), one rounded cup (**128**), three jugs (**135, 142, 146**), nine pithoi (**153, 155, 161, 162, 165, 167, 171, 173, 174**), five pithoi or jars (**176–179, 182**), one pithos with drains (**189**), one spouted jar (**194**), four collared jars (**198–201**), one bridge-spouted jar (**206**), five hole-mouthed jars (**216–219, 223**), 20 cooking pots (**226, 232, 237, 239, 241, 244–246, 248, 249, 254, 256, 259, 261, 265, 270, 272, 273, 281, 282**), 19 closed vessels (jar, cooking pot, or jug; **341, 344, 345, 350–353, 355, 356, 358, 367, 369, 370, 373, 381, 387–390**), 15 cooking dish sherds (**286–288, 292, 296, 301, 304, 307, 308, 312, 314–316, 321, 322**), and one flat lid (**2**).

Small Finds by Layer and Pass within the Stratum

Layer 5: A604.6–7, 604.15–16
   A604.6
      White-on-Dark Ware pottery
         One sherd of a shallow conical bowl or basin (**9**)
      Plain pottery and pottery with slight decoration
         One sherd of a shallow conical bowl (**108**)
         One sherd of a rounded cup (**128**)
         One sherd of a collared jar (**198**)
         One sherd of a hole-mouthed jar (**217**)
         Three sherds of a tripod cooking pots (**226** from A604.6, 7, 13, 17–19; **241, 265**)
         One sherd of a closed vessel (**353**)
         Two sherds of cooking dishes (**292, 314**)

Ground stone
	Two hand tools: one faceted hand tool (**445**), possible tool (**488**)
Soil samples: one standard, three intensive
Marine invertebrates
	Two seashells: *Patella* sp., one *Pisania* sp.
A604.7
	White-on-Dark Ware pottery
		One sherd of a bridge-spouted jar (**73**)
	Plain pottery and pottery with slight decoration
		Three bowls or basins and sherds (**97** from A604.5, 7, 12, 23, 26, 29, 46, 53; **100, 105**)
		One basin with scoring inside (**114** from A604.7, 13, 17, 18, 21, 25, 26, 28, 39, 40, 46, 47, 49, 50, 53, 55, 59, 67)
		One pithos sherd (**174**)
		One sherd of a hole-mouthed jar (**218**)
		One fragmentary tripod cooking pot (**226** from A604.6, 7, 13, 17–19)
		One sherd of a closed vessel (**95**)
		Three sherds of cooking dishes (**286, 288, 296**)
	Ceramic objects
		Two discoid loomweights (**505, 506**)
A604.15
	Plain pottery and pottery with slight decoration
		One pithos (**189** from A604.13, 15, 17, 55, 57, 62)
		One sherd of hole-mouthed jar (**216**)
A604.16
	White-on-Dark Ware pottery
		One sherd of bridge-spouted jar (**81**)
	Plain pottery and pottery with slight decoration
		One pithos sherd (**171**)
		One sherd of a tripod cooking pot (**256**)
		Two sherds of closed vessels (jar, cooking pot, or jug: **387, 388**)
		One sherd of cooking dish (**316**)
	Ground stone
		One hand tool: grinder (**449**)
	Soil samples: two intensive
	Marine invertebrates
		One *Pisania* sp. seashell
Layer 6: A604.9 (and A604.6–7 and A604.15–16, which continue in it)
A604.9
	White-on-Dark Ware pottery
		One sherd of a shallow bowl (**10**)
		One jug sherd (**88**)
	Plain pottery and pottery with slight decoration
		Two pithos sherds (**165, 173**)
		One sherd of collared jar (**199**)
		One sherd of bridge-spouted jar (**206**)
		One sherd of tripod cooking pot (**259**)
		Two sherds of cooking dishes (**287, 301**)
		Two sherds of closed vessels (**355, 356**)
	Ceramic objects
		One discoid loomweight (**501**)
Ground stone
	Two pounders (**411, 412**)
Chipped stone
	One blade fragment (**392**)
	One worked flake (**398**)
Soil samples: one standard, one intensive
Marine invertebrates
	Two *Patella* sp., one *Pisania* sp. seashells
A604.15–16 (see Layer 5)
Layer 7: A604.11 (and A601.6–7, A604.9, A604.15–16, which continue in it)
A604.11
	White-on-Dark Ware pottery
		One sherd of conical bowl with or without spout (**42**)
		Two sherds of rounded cups with tiny lugs or handles on lower body (**65, 66**)
		One sherd of bridge-spouted jar (**74**)
		One sherd of jar or jug (**85**)
	Plain pottery and pottery with slight decoration
		One bowl sherd (**106**)
		One sherd of spinning bowl (**121**)
		One pithos sherd (**155**)
		One collared jar sherd (**200**)
		Three sherds of tripod cooking pots (**254, 281, 282**)
		Five sherds of closed vessels (jar, cooking pot, or jug: **344, 350–352, 358**)
		One sherd of a cooking dish (**308**)
	Ground stone
		Three hand tools: one pounder (**414**), one whetstone (**454**), and one possible tool (**490**)
	Chipped stone
		One blade fragment (**393**)
	Soil samples: one standard, two intensive
	Marine invertebrates
		One *Monodonta* sp. seashell
A604.15–16 (see Layer 5)
Layer 8: A604.12 (and A604.11, 15–16, which continue in it)
A604.12
	White-on-Dark Ware pottery
		One jug sherd (**92**)
	Plain pottery and pottery with slight decoration
		One basin (**97** from A604.5, 7, 12, 23, 26, 29, 46, 53)
		One jug (**136** from A604.12, 22, 28, 40, 49, 53, 56, 62, 64)
		One sherd of a closed vessel (**381** from A604.12, 19, 40, 46, 53, 55, 59, 62, 65)
	Ground stone
		One whetstone fragment (**455**)
A604.15–16 (see Layer 5)
Layer 9: no finds recorded
Layer 10: A604.19, 25–26
A604.19
	Early Minoan I

One flat lid (**2**)
White-on-Dark Ware pottery
    One sherd from shallow conical bowl or basin (**11**)
    One conical bowl with or without a spout (**30**)
Plain pottery and pottery with slight decoration
    Three sherds from conical bowls or basins (**99**, **100**, **110**)
    Two jug sherds (**142**, **146**)
    One pithos sherd (**167**)
    One pithos or jar sherd (**182**)
    Five sherds of tripod cooking pots (**226** from A604.6, 7, 13, 17, 18, 19; **237**, **245**, **248**, **249**)
    Five closed vessels (jar, cooking pot, or jug: **341**, **345**; **381** from A604.12, 19, 40, 46, 53, 55, 59, 62, 65; **389**, **390**)
    Four sherds of cooking dishes (**304**, **312**, **315**, **322**)
Ground stone
    One hand tool: hammer stone (**415**)
Soil samples: one standard
Marine invertebrate
    Seven seashells: one *Fasciolaria lignaria*, two *Pisania* sp., four *Patella* sp.
A604.25
Plain pottery and pottery with slight decoration
    One basin with scoring inside (**114** from A604.7, 13, 17, 18, 21, 25, 26, 28, 39, 40, 46, 47, 49, 50, 53, 55, 59, 67)
A604.26
Plain pottery and pottery with slight decoration
    One basin (**97** from A604.5, 7, 12, 23, 26, 29, 46, 53)
    One basin with scoring inside (**114** from A604.7, 13, 17, 18, 21, 25, 26, 28, 39, 40, 46, 47, 49, 50, 53, 55, 59, 67)
Layer 11: A604.21 (and A604.25–26, which continue in it)
A604.21
White-on-Dark Ware pottery
    One sherd from a shallow conical bowl or basin (**20**)
    One bridge-spouted jar sherd (**78**)
Plain pottery and pottery with slight decoration
    One basin with scoring inside (**114** from A604.7, 13, 17, 18, 21, 25, 26, 28, 39, 40, 46, 47, 49, 50, 53, 55, 59, 67)
    Two jugs (**135** from A604.21, 27; **239**)
    Three pithos sherds (**153**, **161**, **162**)
    Four sherds of pithoi or jars (**176–179**)
    One sherd of a spouted jar (**194**)
    One sherd of a collared jar (**201**)
    Six tripod cooking vessels and sherds (**232** from A604.21, 29, 40, 46, 53, 59, 61, 64, 65, 67; **239**, **244**, **246**, **270**, **272**)
    Two sherds of closed vessels (jar, cooking pot, or jug: **367**, **369**)
    One cooking dish sherd (**307**)
Ceramic objects
    One spindle whorl (**510**)
Ground stone
    Four hand tools: two pounders (**417**, **418**), one abrader (**433**), one grinder (**450**)
    One slab (**483**)
Soil samples: one standard
Marine invertebrates
    One *Patella* sp. seashell
A604.25 (see Layer 10)
    Soil samples: one intensive
A604.26 (see Layer 10)
    Soil samples: one intensive
Layer 12: A604.29, A604.39
A604.29
White-on-Dark Ware pottery
    One sherd of bridge-spouted jar (**75**)
Plain pottery and pottery with slight decoration
    One basin (**97** from A604.5, 7, 12, 23, 26, 29, 46, 53)
    One spouted jar (**192** from A604.13, 22, 27, 29, 40, 46, 47, 53, 55, 59, 67)
    Two sherds of hole-mouthed jars (**219**, **223**)
    Four tripod cooking pots and sherds (**232** from A604.21, 29, 40, 46, 53, 59, 61, 64, 65, 67; **236** from A604.13, 22, 28, 29, 40, 42, 47, 55–57, 59, 61, 62; **261**, **273**)
    Two sherds of closed vessels (**370**, **373**)
    One sherd of a cooking dish (**321**)
Ground stone
    One whetstone fragment (**456**)
    One pounder-abrader (**423**)
Chipped stone
    One blade fragment (**395**)
Soil samples: one standard, one intensive
Marine invertebrates
    One *Patella* sp., one *Patella caerulea*, one *Patella aspera*, three *Monodonta* sp. seashells
A604.39
Plain pottery and pottery with slight decoration
    One basin with scoring inside (**114** from A604.7, 13, 17, 18, 21, 25, 26, 28, 39, 40, 46, 47, 49, 50, 53, 55, 59, 67)

Summary of Stratum 2 Lower Exterior: Small Finds by Type Including Both Complete and Fragmentary Cataloged Objects

White-on-Dark Ware pottery
    Four shallow conical bowls or basins (**9** from A604.6; **10** from A604.9; **11** from A604.19; **20** from A604.21)

Two conical bowls with or without spout (**30** from A604.19; **42** from A604.11)

Two rounded cups with tiny lugs or handles on lower body (**65** and **66** from A604.11)

Five bridge-spouted jars (**73** from A604.7; **74** from A604.11; **75** from A604.29; **78** from A604.21; **81** from A604.16)

Two jars or jugs (**85** from A604.11; **92** from A604.12)

One jug (**88** from A604.9)

One closed vessel sherd (**95** from A604.7)

Plain pottery and pottery with slight decoration

Seven conical bowls or basins (**97** from A604.5, 7, 12, 23, 26, 29, 46, 53; **99**, **110** from A604.19; **100**, **105** from A604.7; **106** from A604.11; **108** from A604.6)

One basin with scoring inside (**114** from A604.7, 13, 17, 18, 21, 25, 26, 28, 39, 40, 46, 47, 49, 50, 53, 55, 59, 67)

One spinning bowl (**121** from A604.11)

One rounded cup (**128** from A604.6)

Three jugs (**135** from A604.21, 27; **142** and **146** from A604.19)

Nine pithoi (**153**, **161**, **162** from A604.21; **155**, **165**, and **173** from A604.9; **167** from A604.19; **171** from A604.16; **174** from A604.7)

Five pithoi or jars (**176**–**179** from A604.21; **182** from A604.19)

One pithos with drain (**189** from A604.13, 15, 17, 55, 57, 62)

One spouted jar (**194** from A604.21)

Four collared jars (**198** from A604.6; **199** from A604.9; **200** from A604.11; **201** from A604.21)

One bridge-spouted jar (**206** from A604.9)

Five hole-mouthed jars (**216** from A604.15; **217** from A604.6; **218** from A604.7; **219**, **223** from A604.29)

20 tripod cooking pots (**226** from A604.6, 7, 13, 17–19; **232** from A604.21, 29, 40, 46, 53, 59, 61, 64, 65, 67; **237**, **245**, **248**, **249**, from A604.19; **239**, **244**, **246**, **270**, **272** from A604.21; **241**, **265** from A604.6; **254**, **281**, **282** from A604.11; **256** from A604.16; **259** from A604.9; **261**, **273** from A604.29)

18 fragments of closed vessels (jar, cooking pot, or jug: **341**, **345**, **389**, **390** from A604.19; **344**, **350**–**352**, **358** from A604.11; **353** from A604.6; **387**, **388** from A604.16; **355**, **356** from A604.9; **367**, **369** from A604.21; **370**, **373** from A604.29)

15 fragments of cooking dishes (**286**, **288**, **296** from A604.7; **287**, **301** from A604.9; **292**, **314** from A604.6; **307** from A604.21; **308** from A604.11; **304**, **312**, **315**, **322** from A604.19; **316** from A604.16; **321** from A604.29)

One fragment of flat lid (**2** from A604.19)

Ceramic objects

Three loomweights (**501** from A604.9; **505**, **506** from A604.7)

One spindle whorl (**510** from A604.21)

Ground stone

16 hand tools and fragments: two pounders (**411**, **412**) from A604.9; one pounder (**414**), one whetstone (**454**), and one possible tool (**490**) from A604.11; one pounder (**415**) from A604.19; two pounders (**417**, **418**), one abrader (**433**), and one grinder (**450**) from A604.21; one pounder-abrader (**423**) and one whetstone (**456**) from A604.29; one facetted hand tool (**445**) and one possible tool (**488**) from A604.6; one grinder (**449**) from A604.16; one whetstone (**455**) from A604.12

One work slab (**483** from A604.21)

Chipped stone

Three blades (**392** from A604.9; **393** from A604.11; **395** from A604.29)

One worked flake (**398** from A604.9)

Marine invertebrates

20 seashells: one *Fasciolaria lignaria*, eight *Patella* sp., one *Patella caerulea*, one *Patella aspera*, five *Pisania* sp., four *Monodonta* sp.

Soil samples

Standard: A604.6 (1), A604.9 (1), A604.11 (1), A604.19 (1), A604.21 (1), A604.29 (1)

Intensive: A604.6 (3), A604.11 (2), A604.16 (2), A604.25 (1), A604.26 (1), A604.29 (1)

Stratum 3 (Layers 13–21)

The lowest stratum of the context consists of a fill that was removed in nine layers (Layers 13–21). It contained a substantial assemblage of broken pottery and stone tools that complement the finds from the main chamber of the rock shelter (Pls. 27–38). Many fragments from this level joined objects in Stratum 2, suggesting that all the material was deposited at the same time and that many objects were already broken before or during the process. A few finds in Stratum 3, however, distinguish the deposit from Strata 1 and 2, such as a handful of earlier EM IIB pottery sherds, larger numbers of ground and chipped stone tools, small amounts of olive stones and grape seeds, and the overwhelming majority of the shellfish remains, the last of which perhaps represents food consumed when the deposit was created.

The excavation exposed the fill in four levels: Stratum 3a (Layers 13–14; Fig. 16), Stratum 3b (Layers 15–16; Fig. 17), Stratum 3c (Layers 17–19; Fig. 18), and Stratum 3d (Layers 20–21;

Fig. 19). These artificial levels illustrate the position of finds and the sloping bedrock floor of the rock shelter, which was fully revealed after excavating Layer 21. A second part of the roof of the rock shelter, preserved in the southeastern corner of the trench, was also exposed in Stratum 3. Because this portion of the chamber appears to continue underneath the modern road, we were unable to excavate it completely for fear of weakening the pavement and endangering traffic.

The southern and northern halves of Stratum 3a were excavated separately. In the southern half, Loci A604.47 and A604.53 lay at an elevation of +36.05–36.00 m asl (Pl. 27). The soft soil on the west side of the chamber was the same pale yellow color (2.5Y 8/3) as the sloping bedrock floor (Pl. 27A). A stone pestle (**465**) and piercer (**467**) were found there (Fig. 16). The pale brown (10YR 7/4) soil in the center of the rock shelter probably represented part of the earth fill introduced to raise the floor. A basin with scoring inside (**114**) lay against the southern edge of the chamber (Fig. 16; Pl. 28B). To the north lay a stone polisher (**462**), fragments of a wide-mouthed jug (**150**), and a quern (**477**) near two burned stones (Fig. 16). There were no other traces of fire in any stratum, suggesting that these rocks came into contact with a fire lit outside the chamber. The presence of a small number of carbonized olive pits and grape seeds inside the rock shelter may also be connected to this fire. A pounder-abrader was recovered near the entrance of the chamber (**424**; Fig. 16), and further digging to the south revealed a thick deposit of broken pottery in the southeastern corner of the rock shelter where part of the sloping roof was preserved underneath the road.

The northern half of Stratum 3a was removed as Loci A604.46 and A604.49. It contained a small number of fragmentary vessels: a basin (**97**; Fig. 16; Pl. 29A), a vessel of unknown shape (**340**), a conical bowl (**26**; Pl. 28A), a tripod cooking pot (**271**), a complete rounded cup (**124**), and nine stone tools. Two pounder-abraders (**426**, **436**) and two pieces of querns (**475**, **478**) were found on the west side; to the north lay another pounder-abrader (**425**), a piece of a quern (**479**), a chopper (**439**), a polisher (**461**), and a pestle (**464**). The locations of a pounder-abrader (**435**), a discoid loomweight (**503**), and fragments of an olive were not recorded. A large number of seashells (mostly limpets) were recovered on the east side of the stratum in Locus A604.49 in an area that lay outside the chamber. These shells probably form part of a more substantial deposit recovered below in Stratum 3b (Fig. 17).

The second layer of the fill (Stratum 3b) lay at an elevation of +36.00–35.96 m asl. The soil matrix was similar to that noted in Stratum 3a pale yellow (2.5Y 8/4) on the perimeter where sloping bedrock was revealed and pale brown (10YR 7/4) in the middle. The southern and northern halves of the stratum (Fig. 17; Pls. 30A–31B) were again excavated separately as Loci A604.56 and A604.58 to the north and Loci A604.55 and A604.57 to the south. Two small clusters of pottery were collected in the center of the chamber (Pl. 31B), and several stone tools (Fig. 17) were recorded on the east side, including two abraders (**427**, **428**) and a quern fragment (**480**). Work to the north recovered a discoid loomweight (**504**; Fig. 17) and several stone tools (Fig. 17). From east to west they include two obsidian blades (**396**, **397**), a possible tool (**497**), a scraper (**466**), two grinders (**451**, **542**), two choppers (**440**, **441**), and two more possible tools (**495**, **496**). Finally, there were a large number of seashells in the deposit (114 shells), and the location of the largest concentration in A604.56 (76 shells) is marked on the plan (Fig. 17).

Stratum 3c contained Layers 17–19 and lay at an elevation of +35.96–35.83 m asl. The soil continued the same pattern observed in the upper levels with darker soil restricted to the center of the rock shelter and lighter soil on the periphery where bedrock was exposed. Digging in Layer 17 was restricted to Locus A604.59 in the southern half of the chamber (Fig. 17; Pls. 32A–33B) and Locus A604.61, which attempted to remove the thick deposit of broken pottery in the southeast corner that ran underneath the road (Fig. 17; Pl. 34A). The excavation of Layers 18–19 exposed sloping bedrock across most of the northern part of the trench (Pl. 32A) and cleaned the deposit of ceramics and stone tools in the center of the rock shelter. A moderate number of stone tools and seashells were found in most parts of the rock shelter, with clusters of pottery in the middle and the southeast corner. The central deposit contained a cooking dish fragment (**328**), a bridge-spouted jar or jug sherd

(**76**), and a fragment of a Vasiliki Ware jug (**7**). A closed vessel (**382**) lay to the east. The assemblage of stone tools (Fig. 18) included three limestone pounder-abraders (**430, 437, 438**), one pumice abrader (**459**), three pounders (**419–421**), one grinder (**453**), three choppers (**442–444**), and one quern fragment (**481**). Finally, a piece of a potter's disk (**513**), an obsidian blade (**399**), a stone weight (**470**), and three loomweights (**502, 507, 508**) were found together on the north side of the chamber.

Stratum 3d (Layers 20–21; Fig. 19) lay at an elevation of +35.83–35.79 m asl, and its excavation removed the final level of material from the rock shelter. The upper part was removed as Locus A604.65, which revealed clusters of pottery labeled A604.68–69 (Pls. 34B–38). The thick deposit in the southeast corner was also probed in Locus A604.66 (Pl. 36A). A second pass removed the remaining pottery and stone tools in the center of the chamber as Locus A604.67 (Pls. 36B–37B). The locations of four vases were recorded on the plan (Fig. 19): a complete wide-mouthed jug (**151**; Fig. 19; Pl. 38A) and a fragmentary tripod cooking pot (**227**), pieces of a jar (**380**; Fig. 19; Pl. 38B), and a jug (**137**; Fig. 19; Pl. 38B). The stone tools included four pounder-abraders (**429, 431, 432, 448**), one polisher (**463**), two possible tools (**138, 499**), and four obsidian flakes (**400, 406–408**).

A large selection of sherds was again cataloged to show the range of shapes in the deposit and provide samples for chemical residue analysis. Among the early material there was one cup of Final Neolithic or EM I date (**1**) and two Vasiliki Ware bowl sherds and a jug sherd of EM IIB date (**3, 4, 7**). Among the shapes in the EM III White-on-Dark Ware class were sherds of four shallow conical bowls or basins (**15, 16, 21, 22**), one conical bowl without a spout (**35**), 10 conical bowls with or without spouts (**26, 29, 31–35, 37, 39, 43**), one conical bowl with a frying pan handle (**45**), five spouted conical jars (**52, 56, 57, 59, 60**), one conical or rounded cup (**63**), one rounded cup (**69**), five bridge-spouted jars (**72, 77, 79, 80, 83**), one jar or jug (**76**), one jug (**91**), and one vessel with an elliptical base (**94**). Among the plain or slightly decorated open shapes were nine conical bowls and basins (**101, 103, 104, 107, 109, 111, 131–133**), one basin with scoring inside (**114**), one undecorated cup (**123**), and three rounded cups (**124, 126, 134**). The closed vessels, mostly in fragments, included eight jugs (**136–138, 140, 145, 147–149**), one wide-mouthed jug (**151**), four pithoi (**164, 168, 169, 172**), five pithoi or jars (**180, 181, 183, 186, 187**), four pithoi with drains (**188–191**), four spouted jars (**192, 195–197**), one collared jar (**203**), one small jar with knobs (**204**), two bridge-spouted jars (**205, 207**), four hole-mouthed jars (**210, 212, 213, 220**), 20 tripod cooking pots (**225, 227, 228, 232, 233, 235, 236, 238, 243, 250, 251, 255, 258, 266, 271, 276–279, 284**), six closed vessels (**366, 368, 375, 382, 383, 391**), 15 cooking dish sherds (**285, 290, 293, 294, 297, 298, 309–311, 313, 319, 323, 325, 326, 328**), one vat (**331**), four flat lids (**355–358**), and two unknown vessels (**339, 340**).

Small Finds by Layer and Pass within the Stratum: A604.46–A604.67

Layer 13: south A604.47, 50, 54; north A604.46, 48, 51, 52
  A604.46
    Vasiliki Ware pottery
      One sherd of a shallow bowl (**4**)
    White-on-Dark Ware pottery
      Two sherds of a spouted conical jar (**50** from A604.22, 40, 42, 46, 47, 53, 64–66; **52** from A604.20, 45–46)
      One bridge-spouted jar sherd (**77**)
    Plain pottery and pottery with slight decoration
      One basin (**97** from A604.5, 7, 12, 23, 26, 29, 46, 53)
      One basin with scoring inside (**114** from A604.7, 13, 17, 18, 21, 25, 26, 28, 39, 40, 46, 47, 49, 50, 53, 55, 59, 67)
      One rounded cup (**126** from A604.40, 46, 47, 53, 65)
      Two fragmentary jugs (**148** from A604.22, 46, 47, 53, 56, 59, 62, 64; **149**)
      One pithos (**172**)
      One spouted jar (**192** from A604.13, 22, 27, 29, 40, 46, 47, 53, 55, 59, 67)
      One sherd of hole-mouthed jar (**212** from A604.20, 22, 40, 46, 64)
      Five tripod cooking pots and fragments (**225** from A604.40, 46, 53, 59, 64, 65; **228** from A604.18, 20, 22, 28, 42, 46, 55, 64, 65; **232** from A604.21, 29, 40, 46, 53, 59, 61, 64, 65, 67; **233** from A604.17, 22, 40, 46, 53, 59, 64, 65; **235**)
      Two sherds of a closed vessel (**382** from A604.46, 47, 55, 59, 60, 67; **391** from A604.20, 46, 47, 53, 55, 56)
      One sherd of a cooking dish (**313**)
      One sherd of an unknown vessel (**340**)
      Two sherds of flat lid (**335, 337**)

Ceramic object
  One piece of a potter's wheel disk (**514**)
Ground stone
  One hand tool: pounder-abrader (**424**)
  Three quern fragments (**475, 476, 478**)
Chipped stone
  One flake (**401**)
Soil samples: one standard, three intensive
Botanical remains: 0/11 *Olea europaea* stones
Marine invertebrates
  Five seashells: one *Patella* sp., one *Monodonta* sp., one *Bolinus brandaris*, two *Pisania* sp.
A604.47
  White-on-Dark Ware pottery
    One sherd of a conical bowl with or without spout (**32**)
    Two spouted conical jars (**48** from A604.42, 47; **50** from A604.22, 40, 42, 46, 47, 53, 64–66)
  Plain pottery and pottery with slight decoration
    One basin with scoring inside (**114** from A604.7, 13, 17, 18, 21, 25, 26, 28, 39, 40, 46, 47, 49, 50, 53, 55, 59, 67)
    One rounded cup (**126** from A604.40, 46, 47, 53, 65)
    One jug fragment (**148** from A604.22, 46, 47, 53, 56, 59, 62, 64)
    One wide-mouthed jug fragment (**151** from A604.47, 60, 68)
    Two spouted jars (**192** from A604.13, 22, 27, 29, 40, 46, 47, 53, 55, 59, 67; **197**)
    Two fragments of tripod cooking pots (**234** from A604.28, 40, 47, 53, 56, 59, 64, 65, 67; **236** from A604.13, 22, 28, 29, 40, 42, 47, 55–57, 59, 61, 62)
    Two sherds of a closed vessel (**382** from A604.46, 47, 55, 59, 60, 67; **391** from A604.20, 46, 47, 53, 55, 56)
    Two sherds of cooking dishes (**293, 294**)
    One lid sherd (**338**)
  Ground Stone
    One quern (**477**)
A604.48
  Plain pottery and pottery with slight decoration
    One jug (**137** from A604.6, 17, 40, 48, 53, 56, 57, 62, 64, 65, 69)
    One hole-mouthed jar (**210** from A604.17, 22, 27, 28, 32, 37, 40, 44, 46, 48, 53, 59, 64, 65, 67)
A604.50
  Plain pottery and pottery with slight decoration
    One basin with scoring inside (**114** from A604.7, 13, 17, 18, 21, 25, 26, 28, 39, 40, 46, 47, 49, 50, 53, 55, 59, 67)
A604.51: no finds recorded
A604.52
  White-on-Dark Ware pottery
    One sherd of conical bowl (**26**)
  Plain pottery and pottery with slight decoration
    One rounded cup (**124**)
    One sherd of tripod cooking pot (**271**)
A604.54
  Plain pottery and pottery with slight decoration
    One fragment of a wide-mouthed jug (**150**)
Layer 14: south A604.53 (and A604.50, 54, which continue in it); north A604.49 (and A604.48, 51–52, which continue in it)
A604.49
  White-on-Dark Ware pottery
    One sherd of a closed elliptical vase (**94**)
  Plain pottery and pottery with slight decoration
    Three sherds of a shallow bowl or basin (**103, 107, 111**)
    One basin with scoring (**114** from A604.7, 13, 17, 18, 21, 25, 26, 28, 39, 40, 46, 47, 49, 50, 53, 55, 59, 67)
    One sherd of an undecorated cup (**123**)
    One sherd of a small jar with knobs (**204**)
    One jug (**136** from A604.12, 22, 28, 40, 49, 53, 56, 62, 64)
    One sherd of a hole-mouthed jar (**220**)
    One sherd of a tripod cooking pot (**284**)
    Two sherds of closed vessels (**366**; **383** from A604.49, 67)
    One sherd of cooking dish (**309**)
  Ceramic object
    One loomweight (**503**)
  Ground stone
    Six hand tools: four pounder-abraders (**425, 426, 436, 435**), one polisher (**461**), one pestle (**464**)
    One chopper (**439**),
    One quern fragment (**479**)
  Soil samples: one standard, four intensive
  Marine invertebrates
    54 seashells: 14 *Patella* sp., 20 *Patella caerulea*, one *Patella aspera*, eight *Buccinulum corneum*, 10 *Bolinus brandaris*
A604.53
  White-on-Dark Ware pottery
    One sherd of shallow conical bowl (**21**)
    Two sherds of conical bowls with or without spouts (**29, 37**)
    One spouted jar (**50** from A604.22, 40, 42, 46, 47, 53, 64–66)
    One sherd of bridge-spouted jar (**79**)
    One jug sherd (**91** from A604.13, 20, 53, 57, 65)
  Plain pottery and pottery with slight decoration
    Four conical bowls or basins and sherds (**97** from A604.5, 7, 12, 23, 26, 29, 46, 53; **131–133**)
    One basin with scoring inside (**114** from A604.07, 13, 17, 18, 21, 25, 26, 28, 39, 40, 46, 47, 49, 50, 53, 55, 59, 67)

One rounded cup (**126** from A604.40, 46, 47, 53, 65)
Four jugs (**136** from A604.12, 22, 28, 40, 49, 53, 56, 62, 64; **137** from A604.6, 17, 40, 48, 53, 56, 57, 62, 64, 65, 69; **138**; **148** from A604.22, 46, 47, 53, 56, 59, 62, 64)
One pithos (**164**)
One pithos or jar (**186**)
One spouted jar (**192** from A604.13, 22, 27, 29, 40, 46, 47, 53, 55, 59, 67)
One hole-mouthed jar (**210** from A604.17, 22, 27, 28, 32, 37, 40, 44, 46, 48, 53, 59, 64, 65, 67)
Six tripod cooking pots and sherds (**225** from A604.40, 46, 53, 59, 64, 65; **232** from A604.21, 29, 40, 46, 53, 59, 61, 64, 65, 67; **233** from A604.17, 22, 40, 46, 53, 59, 64, 65; **234** from A604.28, 40, 47, 53, 56, 59, 64, 65, 67; **238**, **278**)
One sherd of a closed vessel (**391** from A604.20, 46, 47, 53, 55, 56)
One sherd of a cooking dish (**325**)
One sherd of a flat lid (**336**)
Ground stone
Two hand tools: one polisher (**462**) and one pestle (**465**)
One piercer (**467**)
Soil samples: one standard
Botanical remains: 0/9 *Vitis vinifera*
Marine invertebrates: two *Patella* sp.

Layer 15: south part with A604.55 (and A604.50, A604.54, which continue in it)
A604.55
White-on-Dark Ware pottery
One sherd of a shallow conical bowl (**22**)
One sherd of a spouted conical bowl (**35**)
One sherd of a spouted conical jar (**57**)
Plain pottery and pottery with slight decoration
One basin with scoring inside (**114** from A604.7, 13, 17, 18, 21, 25, 26, 28, 39, 40, 46, 47, 49, 50, 53, 55, 59, 67)
One jug sherd (**145**)
One sherd of a pithos with drain (**189** from A604.13, 15, 17, 55, 57, 62)
One spouted jar (**192** from A604.13, 22, 27, 29, 40, 46, 47, 53, 55, 59, 67)
One sherd of a collared jar (**203**)
Two tripod cooking pots (**228** from A604.18, 20, 22, 28, 42, 46, 55, 64, 65; **236** from A604.13, 22, 28, 29, 40, 42, 47, 55–57, 59, 61, 62)
Three sherds of closed vessels (**381** from A604.12, 19, 40, 46, 53, 55, 59, 62, 65; **382** from A604.46, 47, 55, 59, 60, 67; **391** from A604.20, 46, 47, 53, 55, 56)
One sherd of a cooking dish (**297**)
Ground stone
One pounder-abrader (**427**)
One quern fragment (**480**)

Soil samples: one standard
Marine invertebrates
28 seashells: 10 *Patella* sp., 13 *Patella caerulea*, four *Monodonta* sp., one *Bolinus brandaris*

Layer 16: south part with A604.57 (and A604.50, 604.54, which continue in it); north part with A604.56 and A604.58
A604.56
White-on-Dark Ware pottery
Two sherds of conical bowls with or without spouts (**39**, **43**)
Two sherds of spouted conical jars (**56**, **59**)
One sherd of a rounded cup (**69**)
Plain pottery and pottery with slight decoration
Four jugs (**136** from A604.12, 22, 28, 40, 49, 53, 56, 62, 64; **137** from A604.6, 17, 40, 48, 53, 56, 57, 62, 64, 65, 69; **140**; **148** from A604.22, 46, 47, 53, 56, 59, 62, 64)
One hole-mouthed jar (**213** from A601.7, A604.13, 38, 56, 58, 59, 62, 64, 65, 67, 69)
Three sherds of tripod cooking pots (**234**; **236** from A604.13, 22, 28, 29, 40, 42, 47, 55, 56, 57, 59, 61, 62; **250** from A604.56)
Two sherds of closed vessels (**375**; **391** from A604.20, 46, 47, 53, 55, 56)
One vat sherd (**331**)
One sherd of an unknown vessel (**339**)
Ground stone
Seven hand tools: one abrader (**447**), two grinders (**451**, **452**), one scraper (**466**), three possible tools (**495–497**)
Two chopper fragments (**440**, **441**)
One mortar fragment (**482**)
Chipped stone
One blade fragment (**396**)
Soil samples: two intensive
Marine invertebrates
76 seashells: one *Spondylus gaederopus*, 14 *Patella* sp., 31 *Patella caerulea*, 17 *Patella aspera*, five *Monodonta* sp., one *Bolinus brandaris*, one *Hexaplex trunculus*, five *Buccinulum corneum*, one *Pisania* sp.

A604.57
White-on-Dark Ware pottery
One jug fragment (**91** from A604.13, 20, 53, 57, 65)
Plain pottery and pottery with slight decoration
One sherd of a shallow bowl (**109**)
One jug (**137** from A604.6, 17, 40, 48, 53, 56, 57, 62, 64, 65, 69)
One sherd of a pithos with drain (**189** from A604.13, 15, 17, 55, 57, 62)
One sherd of bridge-spouted jar (**205**)
One tripod cooking pot (**236** from A604.13, 22, 28, 29, 40, 42, 47, 55–57, 59, 61, 62)
One sherd of a cooking dish (**311**)
Ground stone

One pounder-abrader (**428**)
Marine invertebrates
11 seashells: one *Arca noae*, nine *Patella caerulea*, one *Monodonta* sp.
A604.58
White-on-Dark Ware pottery
One sherd of a conical bowl (**33**)
Plain pottery and pottery with slight decoration
One pithos or jar sherd (**180**)
One hole-mouthed jar (**213** from A601.7, A604.13, 38, 56, 58, 59, 62, 64, 65, 67)
One sherd of a closed vessel (**368**)
Two sherds of cooking dishes (**290**, **310**)
Ceramic object
One loomweight (**504**)
Chipped stone
One blade fragment (**397**)
Soil samples: one standard
Marine invertebrates
Four seashells: one *Patella* sp., one *Patella caerulea*, one *Monodonta* sp., one *Bolinus brandaris* seashells
Layer 17: south part with A604.59–61
A604.59
Plain pottery and pottery with slight decoration
One basin with scoring inside (**114** from A604.7, 13, 17, 18, 21, 25, 26, 28, 39, 40, 46, 47, 49, 50, 53, 55, 59, 67)
One jug fragment (**148** from A604.22, 46, 47, 53, 56, 59, 62, 64)
One sherd of a pithos with drain (**188**)
One spouted jar (**192** from A604.13, 22, 27, 29, 40, 46, 47, 53, 55, 59, 67)
One bridge-spouted jar (**207**)
Two hole-mouthed jars (**210** from A604.17, 22, 27, 28, 32, 37, 40, 44, 46, 48, 53, 59, 64, 65, 67; **213** from A601.7, A604.13, 38, 56, 58, 59, 62, 64, 65, 67, 69)
Seven tripod cooking pots and sherds (**225** from A604.40, 46, 53, 59, 64, 65; **227** from A604.22, 59; **232** from A604.21, 29, 40, 46, 53, 59, 61, 64, 65, 67; **233** from A604.17, 22, 40, 46, 53, 59, 64, 65; **234** from A604.28, 40, 47, 53, 56, 59, 64, 65, 67; **236** from A604.13, 22, 28, 29, 40, 42, 47, 55–57, 59, 61, 62; **277**)
One sherd of a closed vessel (**382** from A604.46, 47, 55, 59, 60, 67)
Two sherds of cooking dishes (**285**, **298**)
Ceramic object
One sherd of a potter's disk (**512**)
Ground stone
Three hand tools: one pounder (**419**), one pounder-abrader (**437**), and one grinder (**453**)
Three choppers (**442**–**444**)
One quern fragment (**481**)

One weight (**470**)
Chipped stone
One blade (**399**)
Soil samples: three intensive
Marine invertebrates
39 seashells: seven *Patella* sp., 20 *Patella caerulea*, one *Patella aspera*, three *Bolinus brandaris*, two *Buccinulum corneum*, six *Pisania* sp.
A604.60
Plain pottery and pottery with slight decoration
One wide-mouthed jug (**151** from 604.47, 60, 68)
One sherd of closed vessel (**382** from A604.46, 47, 55, 59, 60, 67)
A604.61
White-on-Dark Ware pottery
One sherd of a tall spouted jar (**60**)
Plain pottery and pottery with slight decoration
One sherd of a rounded cup (**134**)
One jug sherd (**147**)
One sherd of a pithos with drain (**191**)
One sherd of a pithos or jar (**187** from A604.1, 61, 62)
Three tripod cooking pots and sherds (**232** from A604.21, 29, 40, 46, 53, 59, 61, 64, 65, 67; **236** from A604.13, 22, 28, 29, 40, 42, 47, 55–57, 59, 61, 62; **279**)
One cooking dish (**326**)
Ground stone
Three hand tools: one pounder (**420**), one pounder-abrader (**430**), and one pumice abrader/polisher (**459**)
Layer 18: A604.62
Vasiliki Ware pottery
One bowl sherd (**3**)
White-on-Dark Ware pottery
One conical or rounded cup sherd (**63**)
Two sherds of conical bowls (**31**, **34**)
Two bridge-spouted jars (**72** from A601.5, 604.13, 27, 62; **80** from A601.5, A604.13, 27, 28, 62)
Plain pottery and pottery with slight decoration
Three jugs (**136** from A604.12, 22, 28, 40, 49, 53, 56, 62, 64; **137** from A604.6, 17, 40, 48, 53, 56, 57, 62, 64, 65, 69; **148** from A604.22, 46, 47, 53, 56, 59, 62, 64)
One pithos with drain (**189** from A604.13, 15, 17, 55, 57, 62)
Two pithos or jar sherds (**181**; **187** from A604.1, 61, 62)
One sherd of a spouted jar (**196**)
Two hole-mouthed jars (**213** from A601.7, A604.13, 38, 56, 58, 59, 62, 64, 65, 67, 69; **380** from A604.38, 62, 69)
One tripod cooking pot and a sherd (**236** from A604.13, 22, 28, 29, 40, 42, 47, 55–57, 59, 61, 62; **251**)

Ground stone
: Two hand tools: one pounder (**421**) and one possible tool (**498**)
Chipped stone
: One primary chunk/flake (**402**)
Marine invertebrates
: Five seashells: four *Patella* sp., one *Pisania* sp.

Layer 19: A604.63–64
 A604.63
  Plain pottery and pottery with slight decoration
   One sherd of conical bowl (**104**)
  Ceramic objects
   Three loomweights (**502, 507, 508**)
  Ground stone
   One pounder-abrader (**438**)
  Chipped stone
   One flake (**403**)
  Soil samples: two intensive
  Marine invertebrates
   10 seashells: one *Patella* sp., four *Patella caerulea*, two *Patella aspera*, two *Bolinus brandaris*, one *Pisania* sp.
 A604.64
  Vasiliki Ware pottery
   One jug sherd (**7**)
  White-on-Dark Ware pottery
   One spouted conical jar (**50** from A604.22, 40, 42, 46, 47, 53, 64–66)
   One jar or jug fragment (**76** from A604.24, 59, 62, 64, 69)
   One fragmentary jug (**148** from A604.22, 46, 47, 53, 56, 59, 62, 64)
  Plain pottery and pottery with slight decoration
   Two jugs (**136** from A604.12, 22, 28, 40, 49, 53, 56, 62, 64; **137** from A604.6, 17, 40, 48, 53, 56, 57, 62, 64, 65, 69)
   Two hole-mouthed jars and a fragment (**210** from A604.17, 22, 27, 28, 32, 37, 40, 44, 46, 48, 53, 59, 64, 65, 67; **212** from A604.20, 22, 40, 46, 64; **213** from A601.7, A604.13, 38, 56, 58, 59, 62, 64, 65, 67, 69)
   Five tripod cooking pots and sherds (**225** from A604.40, 46, 53, 59, 64, 65; **228** from A604.18, 20, 22, 28, 42, 46, 55, 64, 65; **232** from A604.21, 29, 40, 46, 53, 59, 61, 64, 65, 67; **233** from A604.17, 22, 40, 46, 53, 59, 64, 65; **234** from A604.28, 40, 47, 53, 56, 59, 64, 65, 67)
   One sherd of a closed vessel (**243**)
   Two sherds of a cooking dish (**319, 328**)
  Marine invertebrates
   30 seashells: 22 *Patella caerulea*, six *Patella aspera*, one *Pisania* sp. one *Buccinum undatum*

Layer 20: A604.65–66, A604.68–69
 A604.65
  FN–EM I pottery
   One cup fragment (**1** from A604.65, 67)
  White-on-Dark Ware pottery
   One sherd of a conical bowl with frying pan handle (**45**)
   One sherd of spouted conical jar (**50** from A604.22, 40, 42, 46, 47, 53, 64–66)
   One sherd of bridge-spouted jar (**83**)
   One jug sherd (**91** from A604.13, 20, 53, 57, 65)
  Plain pottery and pottery with slight decoration
   One rounded cup (**126** from A604.40, 46, 47, 53, 65)
   One sherd of spouted jar (**195**)
   One jug (**137** from A604.6, 17, 40, 48, 53, 56, 57, 62, 64, 65, 69)
   Two hole-mouthed jars (**210** from A604.17, 22, 27, 28, 32, 37, 40, 44, 46, 48, 53, 59, 64, 65, 67; **213** from A601.7, A604.13, 38, 56, 58, 59, 62, 64, 65, 67, 69)
   One spouted jar (**195**)
   Five tripod cooking pots (**225** from A604.40, 46, 53, 59, 64, 65; **228** from A604.18, 20, 22, 28, 42, 46, 55, 64, 65; **232** from A604.21, 29, 40, 46, 53, 59, 61, 64, 65, 67; **233** from A604.17, 22, 40, 46, 53, 59, 64, 65; **234** from A604.28, 40, 47, 53, 56, 59, 64, 65, 67)
  Ground stone
   Three hand tools: two possible tools (**499, 500**) and one pounder-abrader (**429**)
  Soil samples: two intensive
 A604.66
  White-on-Dark Ware pottery
   One spouted conical jar (**50** from A604.22, 40, 42, 46, 47, 53, 64–66)
  Plain pottery and pottery with slight decoration
   Two pithos sherds (**168, 169**)
   One pithos or jar sherd (**183**)
   One sherd of pithos with drain (**190**)
   One sherd of a tripod cooking pot (**258**)
  Chipped stone
   Two flakes (**404, 405**)
 A604.68
  Plain pottery and pottery with slight decoration
   One wide-mouthed jug (**151** from A604.47, 60, 68)
 A604.69
  White-on-Dark Ware pottery
   One jar or jug sherd (**86** from A604.24, 59, 62, 64, 69)
   One jug (**137** from A604.6, 17, 40, 48, 53, 56, 57, 62, 64, 65, 69)
   One hole-mouthed jar (**213** from A601.7, A604.13, 38, 56, 58, 59, 62, 64, 65, 67, 69)
   One closed vessel (**380** from A604.38, 62, 69)

Layer 21: A604.67 (and A604.68–69, which continue in it)
 A604.67
  FN–EM I pottery

One cup fragment (**1** from A604.65, 67)
White-on-Dark Ware pottery
Two shallow bowl sherds (**15, 16**)
Plain pottery and pottery with slight decoration
One conical bowl sherd (**101**)
One basin with scoring inside (**114** from A604.7, 13, 17, 18, 21, 25, 26, 28, 39, 40, 46, 47, 49, 50, 53, 55, 59, 67)
One spouted jar (**192** from A604.13, 22, 27, 29, 40, 46, 47, 53, 55, 59, 67)
Two hole-mouthed jars (**213** from A601.7, A604.13, 38, 56, 58, 59, 62, 64, 65, 67, 69; **210** from A604.17, 22, 27, 28, 32, 37, 40, 44, 46, 48, 53, 59, 64, 65, 67)
Five tripod cooking pots and sherds (**232** from A604.21, 29, 40, 46, 53, 59, 61, 64, 65, 67; **234** from A604.28, 40, 47, 53, 56, 59, 64, 65, 67; **255, 266, 276**)
Two fragments of closed vessels (**382** from A604.46, 47, 55, 59, 60, 67; **383** from A604.49, 67)
One cooking dish (**324**)
Ground stone
Four hand tools: two pounder-abraders (**431, 432**), one abrader (**448**), and one polisher (**463**)
Chipped stone
Four flakes (**400, 406–408**)
Soil samples: one standard, two intensive
Marine invertebrates
19 seashells: one *Arca noae*, one *Pinna nobilis*, 11 *Patella* sp., two *Monodonta* sp., two *Bolinus brandaris*, one *Fasciolaria lignaria*, one *Pisania* sp. The *Pinna* fragment is perforated.

Summary of Stratum 3: Small Finds by Type Including Both Complete and Fragmentary Cataloged Objects

FN–EM I pottery
One cup (**1** from A604.65, A604.67)
EM IIB Vasiliki Ware pottery
Two bowls (**3** from A604.62; **4** from A604.46)
One jug (**7** from A604.64)
White-on-Dark Ware pottery
Four shallow conical bowls or basins (**15** and **16** from A604.67; **21** from 2604.53; **22** from A604.55)
Ten conical bowls with or without spouts (**26** from A604.52; **29, 37** from A604.53; **31, 34** from A604.62; **32** from A604.47; **33** from A604.58; **35** from A604.55; **39, 43** from A604.56)
One conical bowl with frying pan handle (**45** from A604.65)
Seven spouted conical jars (**48** and **50** from A604.47; **52** from A604.20, 45–46; **56** and **59** from A604.56; **57** from A604.55; **60** from A604.61)
One conical or rounded cup (**63** from A604.62)
One rounded cup, type of handle unknown (**69** from A604.56)
Five bridge-spouted jars (**72** from A601.5, 604.13, 27, 62; **77** from A604.46; **79** from A604.53; **80** from A601.5, A604.13, 27, 28, 62; **83** from A604.65)
One jar or jug (**76** from A604.24, 59, 62, 64, 69)
One jug (**91** from A604.13, 20, 53, 57, 65)
One vessel with elliptical base (**94** from A604.49)
Plain pottery and pottery with slight decoration
10 conical bowls or basins (**97** from A604.5, 7, 12, 23, 26, 29, 46, 53; **101** from A604.67; **103, 107, 111** from A604.49; **104** from A604.63; **109** from A604.57; **131–133** from A604.53)
One basin with scoring inside (**114** from A604.7, 13, 17, 18, 21, 25, 26, 28, 39, 40, 46, 47, 49, 50, 53, 55, 59, 67)
One undecorated cup (**123** from A604.49)
Three rounded cups (**124** from A604.52; **126** from A604.40, 46, 47, 53, 65; **134** from A604.61)
Eight jugs (**136** from A604.12, 22, 28, 40, 49, 53, 56, 62, 64; **137** from A604.6, 17, 40, 48, 53, 56, 57, 62, 64, 65, 69; **138** from A604.53; **140** from A604.56; **145** from A604.55; **147** from A604.61; **148** from A604.22, 46, 47, 53, 56, 59, 62, 64; **149** from A604.46)
Two wide-mouthed jugs (**150** from A604.54; **151** from A604.47, 60, 68)
Four pithoi (**164** from A604.53; **168** and **169** from A604.66; **172** from A604.46)
Five pithoi or jars (**180** from A604.58; **181** from A604.62; **183** from A604.66; **186** from A604.53; **187** from A604.1, 61, and 62)
Four pithoi with drains (**188** from A604.59; **189** from A604.13, 15, 17, 55, 57, 62; **190** from A604.66; **191** from A604.61)
Three spouted jars (**192** from A604.13, 22, 27, 29, 40, 46, 47, 53, 55, 59, 67; **195** from A604.65; **196** from A604.62; **197** from A604.47)
One collared jar (**203** from A604.55)
One small jar with knobs (**204** from A604.49)
Two bridge-spouted jars (**205** from A604.57; **207** from A604.59)
Four hole-mouthed jars (**210** from A604.17, 22, 27, 28, 32, 37, 40, 44, 46, 48, 53, 59, 64, 65, 67; **212** from A604.20, 22, 40, 46, 64; **213** from A601.7, A604.13, 38, 56, 58, 59, 62, 64, 65, 67, 69; **220** from A604.49)
21 tripod cooking pots and fragments (**225** from A604.40, 46, 53, 59, 64, 65; **227** from A604.22, 59; **228** from A604.18, 20, 22, 28, 42, 46, 55, 64, 65; **232** from A604.21, 29, 40, 46, 53, 59, 61, 64, 65, 67; **233** from A604.17, 22, 40, 46, 53, 59, 64, 65; **234** from A604.28, 40, 47, 53, 56, 59, 64, 65, 67; **235** from A604.46; **236** from A604.13, 22, 28, 29, 40, 42, 47, 55–57, 59, 61, 62; **238, 278** from A604.53; **243** from A604.64; **250** from

A604.56; **251** from A604.62; **255**, **266**, and **276** from A604.67; **258** from A604.66; **271** from A604.52; **277** from A604.59; **279** from A604.61; **284** from A604.49)

Six closed vessels (jar, cooking pot, or jug) **366** from A604.49; **368** from A604.58; **375** from A604.56; **382** from A604.46, 47, 55, 59, 60, 67; **383** from A604.49, 67; **391** from A604.20, 46, 47, 53, 55, 56)

15 fragments of cooking dishes (**285** and **298** from A604.59; **290** and **310** from A604.58; **293** and **294** from A604.47; **297** from A604.55; **309** from A604.49; **311** from A604.57; **313** from A604.46; **319** and **328** from A604.64; **324** from A604.67; **325** from A604.53; **326** from A604.61)

One vat (**331** from A604.56)

Four flat lids (**335** and **337** from A604.46; **336** from A604.53; **338** from A604.47)

Two unknown vessels (**339** from A604.56; **340** from A604.46)

Ceramic objects

Five loomweights (**502**, **507**, and **508** from A604.63; **503** from A604.49; **504** from A604.58)

Two potter's disk fragments (**512** from A604.59; **514** from A604.46)

Ground stone

34 hand tools: one pounder (**419**), one pounder-abrader (**437**), and one grinder (**453**) from A604.59; one hammer stone (**420**), one pounder-abrader (**430**), and one pumice abrader/polisher (**459**) from A604.61; one pounder (**421**) and one possible tool (**498**) from A604.62; one pounder-abrader (**424**) from A604.46; four pounder-abraders (**425**, **426**, **435**, **436**), one polisher (**461**), and one pestle (**464**) from A604.49; one pounder-abrader (**427**) from A604.55; one pounder-abrader (**428**) from A604.57; one pounder-abrader (**429**) and two possible tools (**499**, **500**) from A604.65; three pounder-abraders (**431**, **432**, **448**) and one polisher (**463**) from A604.67; one pounder-abrader (**438**) from A604.63; one abrader (**447**), two grinders (**451**, **452**), one scraper (**466**), and three possible tools (**495**–**497**) from A604.56; one polisher (**462**) and one pestle (**465**) from A604.53

Six choppers (**439** from A604.49; **440** and **441** from A604.56; **442**–**444** from A604.59)

One mortar (**482** from A604.56)

One piercer (**467** from A604.53)

Seven querns (**475**, **476**, **478** from A604.46; **477** from A604.47; **479** from A604.49; **480** from A604.55; **481** from A604.59)

One weight (**470** from A604.59)

Chipped stone

Three blades (**396** from A604.56; **397** from A604.58; **399** from A604.59)

Eight flakes (**400**, **406**–**408** from A604.67; **401** from A604.46; **403** from A604.63; **404**, **405** from A604.66)

One primary chunk/flake (**402** from A604.62)

Botanical remains

0/11 *Olea europaea* stones (A604.46), 0/9 *Vitis vinifera* (A604.53)

Marine invertebrates

282 seashells: two *Arca noae*, one *Pinna nobilis*, one *Spondylus gaederopus*, 65 *Patella* sp., 96 *Patella caerulea*, 27 *Patella aspera*, 14 *Monodonta* sp., 21 *Bolinus brandaris*, one *Hexaplex trunculus*, 16 *Buccinulum corneum*, one *Buccinum undatum*, one *Fasciolaria lignaria*, 13 *Pisania* sp.

Soil samples

Standard, one from each locus: A604.46, 49, 53, 55, 58, 67

Intensive: A604.46 (3), 49 (4), 56 (2), 59 (3), 63 (2), 65 (2), 67 (2)

Concluding Remarks

Several threads of evidence help us reconstruct the sequence of events involved in building and filling the rock shelter in EM III. First, it appears that the builders found a man-made cavity in the soft marl slope with an opening on the northeast side that served their purpose. That this involved very little additional digging to modify the interior plan of the rock shelter is suggested by the absence of pale yellow soil and chips that would have resulted from such work in the earth packing later used to raise the earth floor inside the chamber.

The space within the chamber was roughly elliptical in section (Fig. 20). The lowest point of the bedrock floor lay in the center of the rock shelter at an elevation of +35.75 m asl. The bedrock floor sloped upward on all sides to reach a maximum interior space at approximately +36.25 m asl on the north, west, and south sides. It was impossible to define the eastern limit that ran underneath the highway. At this level the roughly rectangular plan of the chamber measures more than 4.0 m on a southeast to northwest line and 2.75 m from inside the blocking wall to the back of the chamber on the south side. The original line and height of the roof of the rock shelter is lost. The best-preserved section on the west side reaches an elevation of +36.95 m asl, and its slope mirrors that of the floor below (Fig. 19). A similar, but lower section of the roof is preserved on the east side at an elevation of +36.55 m asl.

Instead of depositing the ceramic vases and stone tools on the bedrock floor of the chamber, the builders chose to introduce an earth fill to create a level surface at an elevation of +36.00 m asl, which was retained with the lower courses of blocking wall A605. This construction allowed them to spread the objects out on the largest possible flat surface within the rock shelter and perhaps created a smaller opening on the northeast side, which needed to be blocked or closed after the objects had been placed inside the chamber. A small amount of broken EM IIB pottery was found in the lowest levels of the fill, but many of the mended EM III vases had joins from both the fill and main chamber, suggesting that much of the material was brought to the shelter in a broken state and then carelessly put inside with the earth fill. Even the best-preserved objects from the main deposit of fill were usually broken or missing small pieces (e.g., conical jar **46** and kantharos **71**).

At the same time, the distribution of the pots and stone tools in Stratum 2 suggests that there was some effort to organize the deposit. Several of the largest and best-preserved vessels (e.g., hole-mouthed jars **208** and **209**) were placed in the back of the chamber on the west side while cooking pots, pouring vessels, and cups were concentrated in the center and east sides. This may also reflect the fact that more space for larger objects existed on the west side. Stone tools were found all over the floor and fill, but they also were concentrated on either side of the blocking wall where at least some of the stones may have been used secondarily as building material.

There was very little organic material in the deposit. Seashells were consistently recovered on the east side, and the largest concentrations were in the fill of Stratum 3 where limpets may have been the remains of an offering made or a meal eaten during the construction. The absence of animal bone is noteworthy, but this pattern may be a reflection of taphonomic conditions like soil chemistry, which adversely affected preservation. The botanical record is nearly as poor: only a few burned olive stones came from inside the chamber, and the fill yielded a few grape pips. The absence of any signs of fire inside the chamber may again point to poor conditions for organic preservation (an exception is two burned stones on the lowest floor surface of Stratum 2 inside the chamber). The olives and stones may have been burned in a cooking fire lit outside during construction of the chamber or in another context of the small unknown nearby settlement where this assemblage is likely to have originated.

In the Cretan Early and Middle Bronze Age rock shelters of this type were typically used for burial; however, the actual function of the rock shelter at Alatzomouri is not straightforward. No human remains were found inside the chamber (although a skull fragment was found elsewhere in the track of the bulldozer), and the assemblage does not include the stone vases and jewelry that are typically found in Early Minoan burials on Crete. Instead, the pottery and stone tools appear to represent a domestic assemblage that was deposited after its destruction, perhaps in the nearby village recorded during the Gournia survey (Watrous 2012, 111). Several possible interpretations are explored in the conclusion to this volume.

After the rock shelter was closed, the deposit appears to have survived undisturbed for more than four millennia. Parts of the blocking wall appear to have fallen in a scatter in front of the entrance to the chamber, but it was only with the arrival of the bulldozer in January of 2007 that things literally began to fall apart. We are fortunate that the area was neglected in later periods because the excavated results have provided a small but unique window into the material culture of the EM III period on the isthmus of Ierapetra, which will be important for understanding the period just before the construction of the Middle Bronze Age palaces in this part of Crete (Betancourt 2006, 71–73; Apostolakou, Betancourt, and Brogan 2007–2008; Watrous and Schultz 2012b; Brogan 2013, 559–560; Watrous et al. 2015).

4

Pottery

by

Philip P. Betancourt and Thomas M. Brogan

The pottery from the Alatzomouri Rock Shelter consisted of three contexts. Objects from inside a small man-made cavity carved into the soft marl (called a rock shelter) were recorded as two different layers. A large assemblage of almost complete vessels and sherds filled the cave, and a lower earth fill created the rough floor inside the cavern (see Ch. 3, p. 35). The third context, found outside the cave, had only a little pottery, including a burial with a human skull fragment and two vases.

The deposit is important for several reasons. First, it provides secure evidence for the stylistic identification of EM III, the final pottery stage before the beginning of the Middle Bronze Age. This stage in the development of Minoan ceramic styles has been a problem for over a century because most previous deposits have been mixed with material from the MM IA style (Betancourt 1984). The main deposit in the rock shelter is all in the EM III style, and it includes a large amount of both finely decorated and plainer material from this Minoan ceramic phase (discussed in detail in Ch. 12). In addition, the deposit contains a good selection of the surviving artifacts in other classes from a single period of early Minoan history, with loomweights, stone tools, potter's disks, a vat, and other objects from a brief period of time. Because these humble objects cannot be dated stylistically by themselves, the closed context allows them to be placed within their proper chronological period. The deposit seems to belong to a single household that includes both domestic items as well as objects used for industrial purposes, like a vat for pressing grapes or olives and clay disks that have been regarded in the past as potter's batts. If the identification of the deposit as the remains of a single household is correct, the assemblage suggests that domestic buildings in EM III housed some industrial activity as well as food preparation and storage.

The objects are mostly local to this part of Crete, and they may provide information on the nearby but still unexcavated town on the hill of Alatzomouri (Watrous 2012, 111, survey site 17, map 2). The town is known only from survey. Alternative associations include Gournia (Hawes et al. 1908; Watrous et al. 2015, with additional bibliography) and Pera Alatzomouri/Sphoungaras (Watrous 2012, 109–110, survey site 10). Wherever the people who buried this deposit came from, they had access to the mainstream of pottery manufacture and other craft productions from the region. The fabric of the majority of the pottery (Mirabello Fabric) indicates that the manufacture location is local to the region. Its granodiorite to diorite inclusions occur geologically in the landscape near Gournia and Alatzomouri (Dierckx and Tsikouras 2007). The presence of Mirabello Fabric temper grains adhering to an occasional base where the vessel was set on a surface that had temper on it while it was not dry yet (an example is in the University of Pennsylvania Museum number MS 4615-37) proves that temper existed as a separate material in a workshop making the ware. Natural deposits of such sand-sized fragments occur geologically both in and near Gournia, and extensive deposits were visible during the early 1980s in the erosion features near Sphoungaras, between Gournia and the sea (Betancourt 1984, 159). Wasters and kilns found at Gournia indicate local production at this site (Watrous et al. 2015, 420).

In addition to the positive information provided by this deposit, the enigmatic context raises several new questions. The pottery and the other objects come from a small sealed rock shelter that looks like a burial place, but it contained no human bones. A burial was outside the cavity, but placing the grave goods in a separate chamber is not usual in Minoan funerary customs. The explanation for the breaking of the objects and their subsequent placement in a closed underground depository will be debated for a long time.

Although almost all of the items date to EM III, a few earlier pieces of pottery can be recognized by their fabric and style. They are almost all small fragments. The early classes include both decorated vessels made in fine fabrics and coarser undecorated and slightly decorated containers. Probable functions include both serving and shipping or storage.

The pottery is organized following the order established for Myrtos (Warren 1972) with open vessels before closed ones, simple forms before more complex items, and shallow shapes before deeper vases. Colors are given in the Munsell system (Kollmorgen 1992). Nomenclature mainly follows the suggestions of Betancourt (1985). Dimensions are in centimeters. Groups of nearby sherds or vases were assigned locus numbers as they were excavated, and individual objects were assigned object numbers; these numbers are listed in the catalog entries for convenient reference. Residue numbers were assigned to sherds that were sampled for potential analysis by gas chromatography (the methodology is given in Koh and Betancourt 2010, 16–17). Not all collected samples have been analyzed; the vials of solvent containing microscopic residue are in storage at the INSTAP Study Center for East Crete. The pottery and other objects are stored either at the INSTAP Study Center or in the local archaeological museums of East Crete.

The words "ware" and "style" are distinguished in this publication. A style is defined as a pottery class that is manufactured with a specific surface treatment but whose fabric has either not been defined or has been found to be variable. A ware is a ceramic class manufactured from a specific clay fabric with a technology that includes a definable surface treatment. Both Vasiliki Ware and East Cretan White-on-Dark Ware are defined as Wares (capitalized) because both their distinctive surface decorations and their fabric (Mirabello Fabric) are unique to East Cretan production. The styles of their ornament have been discussed in detail by Betancourt (1979, 1984). For the fabric, the production (probably centered at Gournia, Pera Alatzomouri, and/or Alatzomouri unexcavated site 17) used clays with tiny grains of igneous rocks consisting of pale colored (often white) plagioclase feldspar, dark mica (once called biotite), and a black amphibole. The igneous rock formations in this class come only from the region between Alatzomouri and Priniatikos Pyrgos on the coast of the Gulf of Mirabello (Dierckx and Tsikouras 2007). The petrography of Mirabello Fabric has been discussed many times (first recognized by George Myer: see Betancourt 1979, 5, table 1; see also Myer 1984; Day 1991, 92–94; 1997, 225; Myer, McIntosh, and Betancourt 1995, 144–145; Whitelaw et al. 1997, 268; Vaughan 2002, 153–154; Barnard 2003, 7; Day, Joyner, and Relaki 2003, 17–18; Day et al. 2005, 183–187; Nodarou 2013b, 168). Without using ceramic petrography, it has been identified by macroscopic means in a system devised for Kavousi and called Types 2 and 3 (Haggis and Mook 1993). Where macroscopic examination seemed secure, the fabric is noted in the following catalog. Detailed study of the fabric by petrography is presented in Chapter 11.

Objects from the Grave

Three objects were recovered from the grave near the rock shelter. They consisted of two ceramic vases and a fragment of a human skull. Both of the vessels, a jar and a jug with a low spout, can be assigned to EM III, which is the same period as the main assemblage of objects inside the cave. The similar date and the nearby locations suggest that the two deposits are related, but one cannot be absolutely certain. The three objects were displaced by the bulldozer that discovered the site, and no completely positive association can be proved.

The jar is a hole-mouthed type with a slightly constricted mouth and horizontal handles on the shoulder. Several similar examples were found in the assemblage (**208–224**). The jug belongs to a class of pouring vessel with a low spout that developed in the EM III ceramic phase. The presence of knobs on the sides of the spout and neck, a characteristic that was common in the previous period (Betancourt 1979, 46), almost disappeared by MM I. Dark bands on the rim and handle like those on this jug are characteristic of plain vessels from EM III (Betancourt 2006, 85, fig. 5.6:78–81).

GRAVE 1A (PAR 447; Fig. 21; Pl. 5B). Hole-mouthed jar, almost complete. H. 44, d. of rim 32, d. of base 16 cm. Mirabello Fabric (light reddish brown, 5YR 6/4). Thickened rim. Dark slip on rim, inside and out. From outside the rock shelter. *Date*: EM III.

GRAVE 1B (PAR 448; Fig. 21; Pl. 5B). Jug, complete. H. 16, d. of neck 4, d. of base 6.3 cm. Mirabello Fabric (between pink 5YR 7/3 and 7/4). Globular body. Dark band on rim and on handle. From outside the rock shelter. *Date*: EM III.

Antiques or Heirlooms from the Rock Shelter

Only a few sherds from the Alatzomouri Rock Shelter were manufactured before EM III. This paucity of early pieces is an important characteristic of the deposit. Almost all of the material in the assemblage comes from a very narrow time period, which firmly demonstrates that the body of material is not a gradual accumulation that was deposited in the cavity over many years, but a selection of broken vessels and fragments that were mostly new or almost new when they were placed in the deposit. Except for one large jug (**8**), the early sherds are all small fragments that do not join any other pieces from the assemblage.

Final Neolithic to Early Minoan I

Cup

The earliest fragment in the deposit is a piece of a burnished cup from FN or EM I. It is tempered with white calcite and organic matter, heavily burnished, and fired to a relatively low temperature. This class is common in the settlements along the northern coast of this part of Crete (for discussion of the class, see Betancourt 2008, 14, fig. 2.1).

1 (PAR 46, HN 14625; Fig. 22). Cup, base sherd. D. of base 8 cm. A coarse fabric (reddish brown, 5YR 5/4), with calcite and organic matter temper, well burnished. From levels 20, 21; loci A604.65, 67; object 431; residue 1258. *Date*: FN–EM I. *Bibl.*: Betancourt 2008, 14, fig. 2.1.

Early Minoan I

Pyxis Lid

Lids like this one with a cylindrical shape and a pair of pierced lugs placed opposite each other on the upper part were made to fit globular pyxides with rounded bodies and vertical necks (shaped like a collared jar). The date is established by the presence of the shape in stratified EM I levels at Hagia Triada (for the bibliography, see Todaro 2001, 15 n. 24, figs. 1, 2) and in an EM I well at Knossos (Hood 1990, fig. 2:16). Similar lids from the Cyclades and from the northeastern Aegean suggest the origin of the shape is from some region north of Crete (Sotirakopoulou 1997, 527, fig. 2:2). For examples from Crete, see the following:

Hagia Triada: Todaro 2001, figs. 1, 2
Hagios Nikolaos near Palaikastro: Tod 1902–1903, 341, figs. 1, 2; Mortzos 1972, figs. 37, 38
Knossos: Hood 1990a, fig. 2:16
Koutsokeras: Xanthoudides 1924, 74–75, pl. 40:a, lower row, 2nd to 5th from right

Lebena: Warren 2004, figs. 22–24, nos. 150–211, pls. 54, 55

Partira: Mortzos 1972, pls. 1, 2

Pyrgos cave: Xanthoudides 1918, 154, fig. 9:61

The lid is pattern-burnished, a style called the Pyrgos Style after the context in the Pyrgos cave. The Pyrgos Style is discussed by Betancourt (2008, 56–63, with a list of 37 find-spots for this style on p. 59).

2 (PAR 134, HN 14709; Fig. 22). Pyxis, lid sherd. D. of lower rim ca. 16 cm. A pale fabric (light brown, 7.5YR 6/4, on both interior and exterior surface, with red areas, 10R 5/8), pattern burnished on top. Small lug handle at rim. From level 10, locus A604.19, object 224, residue 1002. Pyrgos Ware. *Date*: EM I.

Early Minoan IIB

Shallow Conical Bowl in Vasiliki Ware

Vasiliki Ware is a handmade pottery tradition using burnished surfaces and overall coats of slip that are mottled during the firing process to create pale brown to medium brown to red to black mottled effects (Betancourt 1979). The surface treatment combine casual, random decoration that varies in color with an overall ornamental scheme that is planned ahead of time. The effects are achieved at the end of the firing process with a method of selective burning that creates the mottling by the careful oxidation and reduction of different parts of the vase. The tradition is one of the most distinctive and easily recognizable of the specialized productions of EM Crete. The class can be termed a Ware because both the decoration and the fabric are unique to the tradition. The fabric is a type of Mirabello Fabric with small fragments of rocks in the granodiorite-diorite class, in some cases probably not added deliberately by the potters but present naturally in the clay used by the workshop.

The shallow bowl is a common shape during EM II in eastern Crete, although it is seldom given mottled effects. Examples that confirm the date come from Myrtos (Warren 1972, 166–169, figs. 50–53) and Vasiliki (Seager 1906–1907, 116). A list of examples is published by Betancourt (1979, 32–33).

3 (PAR 490, HN 14534; Fig. 22). Shallow bowl, rim sherd. D. of rim ca. 27 cm. Mirabello Fabric (dark gray, 5YR 4/1). Slipped on both upper and lower surfaces, mottled. From locus A604.62. Vasiliki Ware. *Date*: EM IIB.

4 (PAR 100, HN 14722; Fig. 22). Shallow bowl, rim sherd. D. of rim ca. 30 cm. A fine fabric (pink, 7.5YR 7/4). Slipped on interior and exterior and burnished; mottled, red to black. From level 13, locus A604.46. Vasiliki Ware. *Date*: EM IIB.

Jugs in Vasiliki Ware

Jugs with raised spouts were first used in Crete during EM I (Betancourt 2008). By EM IIB they were firmly established for both storage and serving. Vasiliki Ware vessels include some of the finest examples of the typical mottled decoration. They usually have a piriform shape, a small ring base, a single vertical handle, and a raised spout that can sometimes be elevated to produce a very tall form. Only a few jugs (mostly small examples) lack the ring base. Over a hundred examples of EM IIB jugs have been published (for a list with references, see Betancourt 1979, 46–48).

5 (PAR 99, HN 14720; Fig. 22). Jug, body sherd. Max. dim. 6.2 cm. A fine Mirabello Fabric (pink, 7.5YR 7/4). Slipped on exterior and burnished; mottled red to brown to black. From level 12, locus A604.28. Vasiliki Ware. *Date*: EM IIB.

6 (PAR 126, HN 14834; Fig. 22). Jug, base sherd. D. of base ca. 10 cm. A fine Mirabello Fabric (pink, 5YR 7/4). Slipped on exterior and burnished; mottled red to brown to black. From level 12, locus A604.40, object 383, residue 1207. Petrography sample PAR 09/01. Vasiliki Ware. *Date*: EM IIB.

7 (PAR 499, HN 14525; Fig. 22). Jug, body sherd. Max. dim. 3.8 cm. A fine Mirabello Fabric (gray, 5YR 3/1). Slipped on exterior and burnished; mottled. Incised nicks on the body. From level 19, locus A604.64. Vasiliki Ware. *Date*: EM IIB.

Jug, South Coast Class

The South Coast tradition for EM IIB pottery was discussed in detail by Whitelaw and colleagues (1997). It is characterized by a South Coast Fabric recipe and several distinctive forming techniques including thrusting the lower end of the handle through the wall of the upper shoulder. Examples like the one from Alatzomouri were excavated at Myrtos (Warren 1972, pl. 53:P465, P466).

8 (PAR 116, HN 14752; Fig. 22; Pls. 10A, 10B, 39). Jug, almost complete. Pres. h. 32.5, d. of base 13 cm. South Coast Fabric (reddish yellow, 5YR 6/6, with a slightly darker core). Raised spout, handle with rectangular section attached on the neck and thrust through the shoulder, small knobs on shoulder. Dark slip on exterior in imitation of Vasiliki Ware. From levels 2–7, 3–7, 12; loci A601.5, A604.5, 36; object 118; residues 1242, 1368. *Date*: EM IIB.

The Main Assemblage from the Rock Shelter, Early Minoan III

East Cretan White-on-Dark Ware

Shallow Conical Bowls and Basins

The shallow conical bowl has straight sides and a straight or slightly outturned rim. Its form is open and shallow. Like all Minoan vases from before MM IB, it was handmade. It was a common vessel during EM II (Warren 1972, 166–169, figs. 50–53), and it continued to be popular in EM III and MM IA. Mirabello Fabric was used as the raw material. Parallels from the Gulf of Mirabello region made of East Cretan White-on-Dark Ware include the following:

Chrysokamino: Betancourt 2006, 90, fig. 5.7:87, 89
Gournia: Hall 1904–1905, pl. 32:3; Betancourt and Silverman 1991, fig. 1:306, 307
Malia: Chapouthier, Demargne, and Dessenne 1962, pls. 5:below; 36:center left
Palaikastro: Forsdyke 1925, 78
Pera Alatzomouri or Sphoungaras: Watrous 2012, fig. 12:B151
Vasiliki: Blinkenberg and Johansen 1924, pl. 30:4

For the same shape made in a different White-on-Dark Ware production tradition (the Lasithi Red Fabric Group), see Langford-Verstegen 2015, 56 (Hagios Charalambos Cave). The shallow shape survived in eastern Crete into MM IB (Tsipopoulou 2012a, 186, fig. 12; 2012b, 121, fig. 4 [Petras]).

9 (PAR 2, HN 14641; Fig. 22). Shallow bowl, base sherd. D. of base 10 cm. A pale fabric (light reddish brown, 5YR 6/3). Dark slip on exterior and interior. Added white not preserved. From level 5, locus A604.6, object 84, residue 1197. *Date*: EM III.

10 (PAR 36, HN 14616; Fig. 22). Shallow bowl, base sherd. D. of base 12 cm. Mirabello Fabric (light red, 2.5YR 6/6). Dark slip on exterior and interior. Added white not preserved. From level 6, locus A604.9, object 109, residue 1031. *Date*: EM III.

11 (PAR 63, HN 14829; Fig. 22). Shallow bowl, rim sherd. D. of rim 21 cm. Mirabello Fabric (light reddish brown, 5YR 6/4). Dark slip on interior of rim and bands on exterior. Added white not preserved. From level 10, locus A604.19, object 264, residue 1051. Petrography sample PAR 09/02. *Date*: EM III.

12 (PAR 75, HN 14761; Fig. 22). Shallow bowl, base sherd. D. of base 18 cm. Mirabello Fabric (light reddish brown, 5YR 6/6). Dark slip on interior. Added white bands on interior. From level 13, locus A601.7, object 13, residue 948. *Date*: EM III.

13 (PAR 81, HN 14790; Fig. 22). Shallow bowl, rim sherd. D. of rim ca. 32 cm. Mirabello Fabric (pink, 5YR 7/4). Horizontal handles rising from rim. Dark slip on inside of rim and on exterior. Added white traces (bands?). From level 3, locus A601.7, object 28, residue 959. *Date*: EM III.

14 (PAR 172, HN 14563; Fig. 22). Shallow basin, rim sherd. D. of rim over 38 cm. Mirabello Fabric (light reddish brown, 2.5YR 6/4). Slightly splayed rim. Dark slip on inside of rim and possibly on exterior (surface not preserved). Added white not preserved. From level 12, locus A604.40, object 378, residue 1116. *Date*: EM III.

15 (PAR 215, HN 14438; Fig. 22). Shallow bowl, rim sherd. D. of rim 20 cm. Mirabello Fabric (light red, 2.5YR 6/6). Dark slip on interior of rim and on exterior. Added white not preserved. From level 21, locus A604.67, object 430, residue 1046. *Date*: EM III.

16 (PAR 306, HN 14426; Fig. 22). Shallow basin, rim sherd. D. of rim over 42–44 cm. Mirabello Fabric (pink, 7.5YR 7/4). Dark slip on interior and on rim. Added white not preserved. From level 21, locus A604.67, object 429, residue 1047. *Date*: EM III.

17 (PAR 399, HN 14497; Fig. 22). Shallow bowl, base sherd. D. of base 12 cm. Mirabello Fabric (light brown, 7.5YR 6/4, with the core light brownish gray, 10YR 6/3). Dark slip on interior. Added white not preserved. From levels 2–7, 10; loci A601.5, A604.20; object 117; residue 1295. *Date*: EM III.

18 (PAR 187, HN 14575; Fig. 23). Shallow basin, base sherd. D. of base ca. 24–25 cm. Mirabello Fabric (reddish brown, 5YR 5/4). Dark slip on interior and exterior. Added white not preserved. From level 12, locus A604.40, object 325, residue 1105. *Date*: EM III.

19 (PAR 458, HN 14559; Fig. 23). Shallow basin, rim sherd. D. of rim 33 cm. Mirabello Fabric (reddish yellow, 5YR 7/6). Horizontal handles rising from rim. Dark band (red) on rim; wide band below rim band. Exterior surface missing. From level 6, locus A604.10. *Date*: EM III.

20 (PAR 460, HN 14501; Fig. 23). Shallow basin, rim sherd. D. of rim 40 cm. Mirabello Fabric (light reddish brown, 2.5YR 6/4). Thickened rim; small lug handle attached on rim. Dark slip on exterior. Added white diagonal bands (design mostly missing). From level 11, locus A604.21. *Date*: EM III.

21 (PAR 476, HN 14520; Fig. 23). Shallow bowl, rim sherd. D. of rim not measurable, max. dim. 11.1 cm. Mirabello Fabric (pinkish gray, 5YR 6/2). Horizontal handles rising from rim. Dark slip on interior of rim and top of handle, exterior missing. Added white not preserved. From level 14, locus A604.53. *Date*: EM III.

22 (PAR 482, HN 14529; Fig. 23). Shallow bowl, fragmentary. D. of rim ca. 30–40 cm. Mirabello Fabric (light reddish brown, 2.5YR 6/4). Thickened rim. Horizontal handles at rim. Dark slip on interior of rim. Added white not preserved. From level 15, locus A604.55. *Date*: EM III.

Conical Bowl without Spout

The conical bowl without a spout is uncommon, and most examples have a small open spout at the rim (i.e., Hawes et al. 1908, color pl. A4 [Hagia Photia Ierapetras]; Betancourt 1983, no. 240 [Vasiliki]). Unspouted examples with no handle come from Palaikastro (Bosanquet and Dawkins 1923, pl. 2a; MacGillivray et al. 1992, fig. 13:2) and Chrysokamino (Betancourt 2006, 85, fig. 5.6:75), but the last piece in not complete, so the presence or absence of a spout is not known.

One conical bowl with no spout comes from the rock shelter. It has two lugs at the rim, and it is decorated in the typical White-on-Dark style with the ware's most common motif, the pendent hatched triangles, extending from rim to base. For discussion of the shape, see Betancourt 1984, 39, shape 2A (usually with a spout); for the motif, see Betancourt 1984, 24, motif 2, nos. 3–5. The vessel is complete. This shape is very rare; it exists at Mochlos (Soles 1992, 97, fig. 41:M XI-2).

23 (PAR 103, HN 14650; Fig. 23; Pls. 19B, 39). Conical bowl, complete. H. 7.7, d. of rim 13.2, d. of base 5.4 cm. A fine Mirabello Fabric (pink, 7.5YR 7/4). Conical bowl with two opposed lugs at the rim. Dark slip on interior of rim and exterior. Added white pendent hatched triangles on exterior. From levels 11–12, locus A604.34, object 385, residue 1248. *Date*: EM III.

Conical Bowl with or without Spout

The spouted conical bowl is more common than the unspouted type. In most cases, the EM III bowl consists of an open container with a straight rim and wall, a flat base, and a conical shape with an open spout at the rim and a small lug opposite the spout. In many cases, a slight groove sets off the base. The shape is already present in Central Crete in EM I (from the Pyrgos cave, see Zois 1968b, pl. 25:7522). Many East Cretan examples come from EM IIB (Warren 1972, 170–171, figs. 54, 55; Betancourt 1979, 34–36), and except for the decoration, the EM III bowls are similar to their EM I and EM II predecessors. Many EM III–MM IA examples are known, with the main distribution in eastern Crete from Malia to Palaikastro:

Gournia: Hall 1904–1905, pl. 26A, B; Hawes et al. 1908, 37, fig. 15:8; Betancourt and Silverman 1991, fig. 1; Soles 1992, 11, fig. 4:G I-6

Hagia Photia Ierapetras: Hawes et al. 1908, color pl. A4

Malia: Demargne 1945, pl. 29:8610, 8656

Mochlos: Seager 1912, fig. 49:M61–M64

Palaikastro: Bosanquet and Dawkins 1923, 8, fig. 5b; Forsdyke 1925, 78, fig. 96, no. A 445-2;

Pera Alatzomouri or Sphoungaras: Watrous 2012, 109, fig. 11:137

Pseira: Banou 1995, fig. 43:AP 1; Betancourt 1999, fig. 20:BR 65–BR 67; Floyd 2009a, figs. 2:AF 41, 3:AF 60

Vasiliki: Seager 1906–1907, 120, 121, fig. 3:c; Betancourt 1983, fig. 19:240–242

Several conical bowls were present in the rock shelter. The following examples are all incomplete.

24 (PAR 105, HN 14647; Fig. 23; Pls. 20A, 39). Spouted conical bowl, almost complete. H. 10.5, d. of rim 18.5 x 20–22 cm (oval). Mirabello Fabric (between reddish brown, 5YR 5/4, and red, 2.5YR 4/6). Small lug at the rim and either another lug or an opposed open rim-spout (missing). Dark slip on interior of rim and on exterior. Added white pendent triangles. From levels 11–12, locus A604.31, object 386, residue 1252. *Date*: EM III.

25 (PAR 359, HN 14648; Fig. 24; Pl. 39). Conical bowl, complete profile. H. 6.2, d. of rim 12, d. of base 8 cm. A fine Mirabello Fabric (pinkish gray, 7.5YR 7/2). Dark slip on interior of rim and exterior. Added white pendent hatched triangles. From level 10; locus A604.20; objects 335, 336; residues 1145, 1220. *Date*: EM III.

26 (PAR 165, HN 14690; Fig. 24; Pl. 28A). Conical bowl, complete profile. D. of base ca. 7 cm. Mirabello Fabric (yellowish red, 5YR between 5/6 and 5/8). Dark paint on exterior. Added white not preserved. From level 14, locus A604.52, object 514, residue 1146. *Date*: EM III.

27 (PAR 457, HN 14489; Fig. 24). Conical bowl, rim sherd. D. of rim 12 cm. Mirabello Fabric (reddish yellow, 5YR 7/6). Conical shape, with concave profile. Dark slip on interior of rim and on exterior. Added white pendent hatched triangles. From locus A604.10. *Date*: EM III.

28 (PAR 59, HN 14600; Fig. 24). Conical bowl, rim sherd. D. of rim 12 cm. A pale fabric (pale brown, 10YR 6/3). Conical shape, with almost straight profile. Dark slip on interior of rim and on exterior. Added white traces. From level 4, locus A604.1, object 12, residue 967. *Date*: EM III.

29 (PAR 474, HN 14516; Fig. 24). Conical bowl, rim sherd. D. of rim 12 cm. Mirabello Fabric (pinkish gray, 7.5YR 6/2). Small lug handle at rim. Dark slip on exterior. Added white pendent hatched triangles. From locus A604.53. *Date*: EM III.

30 (PAR 133, HN 14781; Fig. 24). Conical bowl or cup, rim sherd. D. of rim ca. 9–12 cm. A pale fabric (pink, 7.5YR 7/4). Very slightly outturned rim. Dark slip on interior of rim and on exterior. Added white not preserved. From level 10, locus A604.19, object 255, residue 1042. *Date*: EM III.

31 (PAR 513, HN 14536; Fig. 24). Conical bowl or cup, base sherd. D. of base 8 cm. Mirabello Fabric (pink, 5YR 7/4). Conical shape, with almost straight profile. Dark slip (red) on interior of rim and on exterior. Added white not preserved. From level 18, locus A604.62. *Date*: EM III.

32 (PAR 469, HN 14517; Fig. 24). Conical bowl, rim sherd. D. of rim 14 cm. Mirabello Fabric (pinkish gray, 7.5YR 7/2), burnished. Lug handle attached at rim. Dark slip on interior and exterior. Added white not preserved. From level 13, locus A604.47. *Date*: EM III.

33 (PAR 304, HN 14428; Fig. 24). Conical bowl, rim sherd. D. of rim ca. 12 cm. Mirabello Fabric (pinkish gray, 7.5YR 6/2). Thickened rim. Traces of dark paint on inside of rim and on exterior. Added white not preserved. From level 16, locus A604.58, object 527, residue 1094. *Date*: EM III.

34 (PAR 489, HN 14521; Fig. 24). Conical bowl, rim sherd. D. of rim 12 cm. Mirabello Fabric (light reddish brown, 7.5YR 7/6, to light brown, 7.5YR 6/4). Double lug handle at rim. Dark slip (black) on inside of rim and on exterior. Added white not preserved. From level 18, locus A604.62. *Date*: EM III.

35 (PAR 481, HN 14531; Fig. 24). Spouted conical bowl, rim sherd with spout. D. of rim ca. 14–18 cm. Mirabello Fabric (light reddish brown, 5YR 6/4). Open spout. Small lug handle at rim. Dark slip on inside of rim and on exterior. Added white pendent hatched triangles. From locus A604.55. *Date*: EM III.

36 (PAR 449, HN 14513; Fig. 24). Conical bowl, rim sherd. D. of rim 12 cm. Mirabello Fabric (light brown, 7.5YR 6/4). Dark slip (black) on inside of rim and on exterior. Added white pendent hatched triangles on exterior. From levels 2, 10; loci A601.1, A604.20. *Date*: EM III.

37 (PAR 473, HN 14523; Fig. 24). Conical bowl, rim sherd. D. of rim 12 cm. Mirabello Fabric (light brown, 7.5YR 6/4). Dark slip (black) on exterior and interior of rim. Added white pendent hatched triangles on exterior. From level 14, locus A604.53. *Date*: EM III.

38 (PAR 93, HN 14836; Fig. 24). Conical bowl or cup, base sherd. D. of base 8 cm. Mirabello Fabric (light reddish brown, 2.5YR 6/4). Almost straight profile. Dark slip on exterior. Added white pendent hatched triangles on exterior. From level 3, locus A601.7, object 17, residue 962. Petrography sample PAR 09/04. *Date*: EM III.

39 (PAR 219, HN 14457; Fig. 24). Conical bowl, base sherd. D. of base ca. 8 cm. Mirabello Fabric (pink, 5YR 7/3). Dark paint on exterior. Added white pendent hatched triangles on exterior. From level 16, locus A604.56, object 442, residue 1218. *Date*: EM III.

40 (PAR 119, HN 14717; Fig. 24). Conical bowl, base sherd. D. of base ca. 6–7 cm. Mirabello Fabric (pink, 5YR 7/4). Dark paint on exterior. Added white traces on exterior. From level 3, locus A601.7, object 10, residue 951. *Date*: EM III.

41 (PAR 139, HN 14707; Fig. 24). Conical bowl, base sherd. D. of base ca. 8 cm. Mirabello Fabric (light reddish brown, 5YR 6/3). Dark slip on exterior. Added white not preserved. From level 3, locus A601.7, object 535, residue 951. *Date*: EM III.

42 (PAR 11, HN 14335; Fig. 25). Conical bowl, base sherd. D. of base 8 cm. A pale fabric (light reddish yellow, 5YR 7/6). Dark slip on exterior. Added white not preserved. From level 7, locus A604.11, object 186, residue 1076. *Date*: EM III.

43 (PAR 202, HN 14570; Fig. 25). Conical bowl, rim sherd. D. of rim ca. 16 cm. Mirabello Fabric (dark reddish gray, 5YR 4/2). Dark slip on interior of rim and on exterior. Added white not preserved. From level 16, locus A604.56, object 443, residue 1108. *Date*: EM III.

Conical Bowl with Frying Pan Handle

Although the bowl with a frying pan handle that extends horizontally or at a slight upward inclination from the rim is an uncommon shape, several parallels exist in East Cretan White-on-Dark Ware:

Malia: Demargne 1945, pl. 28:8499; Demargne and Gallet de Santerre 1953, pl. 46:c
Mochlos: Forsdyke 1925, 79, fig. 96:A 446
Palaikastro: Betancourt 1984, 40; MacGillivray et al. 1992, fig. 9:1

Pseira: Banou 1995, fig. 36:AA 7
Vasiliki: Seager 1906–1907, 120, fig. 3:a; Blinkenberg and Johansen 1924, pl. 30:5

Two bowls of this type are represented in the assemblage. They are both known only from handle fragments. The shape of their handles is typical for the class, suggesting they are ordinary members of the ware.

44 (PAR 454, HN 14499; Fig. 25). Conical bowl with frying pan handle, rim and handle sherd. D. of rim 16 cm. Mirabello Fabric (light reddish brown, 5YR 6/4). Raised tab handle with elliptical section. Dark slip on interior of rim and on exterior, underside of handle, and upper edges of handle. Added white not preserved. From level 10, locus A604.20. *Date*: EM III.

45 (PAR 493, HN 14548; Fig. 25). Conical bowl with frying pan handle, rim and handle sherd. D. of rim 16 cm. Mirabello Fabric (light reddish brown 5YR 6/4). Tab handle with elliptical section. Dark slip on interior of rim and on exterior, underside of handle, and upper edges of handle. Added white not preserved. From level 20, locus A604.65. *Date*: EM III.

Spouted Conical Jar

Conical jars with two horizontal handles placed below the rim and an open spout on the rim are known from several sites within the main distribution region of East Cretan White-on-Dark Ware. Their shape is related to the lower spouted conical bowls, differing mainly in the height. Examples include the following:

Malia: Demargne 1945, pls. 27:8519, 30:2
Mochlos: Seager 1909, fig. 13:center row right
Pseira: Banou 1995, fig. 43:AM 1; Betancourt 1999, fig. 20:BR 78
Vasiliki: Seager 1906–1907, 123, fig. 6; Foster 1978, pl. 1:33

46 (PAR 107, HN 14748; Fig. 25; Pls. 12B, 39). Spouted conical jar, complete. H. 25.8, d. of rim 27.5, d. of base 15.0 cm. Mirabello Fabric (red, 2.5YR 5/6). Everted rim, conical shape, open rim-spout opposite a small solid lug at the rim, two opposed horizontal handles with circular sections on upper body. Dark slip on interior of rim and on exterior. Added white groups of horizontal bands above and below a "diagonal(?)" frieze of chevrons. From levels 2–7; locus A601.4; objects 112, 113; residues 1269, 1291. *Date*: EM III.

47 (PAR 115, HN 14668; Fig. 25; Pls. 19B, 39). Spouted conical jar, two-thirds complete. H. 14.8, d. of rim 15.8, d. of base 8.4 cm. Mirabello Fabric (light brown, 7.5YR 6/6). Everted rim, conical shape, open rim-spout (missing) opposite a small solid double lug at the rim, two opposed horizontal handles with circular sections on upper body. Dark slip on interior of rim and on exterior. Added white groups of horizontal bands above and below frieze of large chevrons. From levels 11–12, loci A604.22, 33; objects 353, 393; residues 1249, 1287. *Date*: EM III.

48 (PAR 136, HN 14803; Fig. 25; Pl. 39). Spouted conical jar, complete profile. H. 22.1, d. of rim 23, d. of base 12.1 cm. Mirabello Fabric (pink, 5YR 7/4). Everted rim, conical shape, open rim-spout (missing), two opposed horizontal handles with circular sections on upper body. Dark slip on interior of rim and on exterior. Added white groups of horizontal bands above and below frieze of alternating groups of four diagonal lines. From levels 12–13; loci A604.32, 47; object 144; residues 1162, 1246. *Date*: EM III.

49 (PAR 394, HN 14432; Fig. 25; Pl. 39). Spouted conical jar, complete profile. H. 16.5, d. of rim 15, d. of base 8.2 cm. Mirabello Fabric (light reddish brown, 5YR 6/4). Everted rim, conical shape, open rim-spout (missing), two opposed horizontal handles with circular sections on upper body. Dark slip on interior of rim and on exterior. Added white groups of horizontal bands above base. From levels 9, 11–12; loci A604.17, 22, 40, 42, 43; objects 326, 381, 438; residues 1138, 1206, 1322. *Date*: EM III.

50 (PAR 405, HN 14798; Fig. 25; Pl. 39). Spouted conical jar, complete profile. H. 24.1, d. of rim 26, d. of base 13 cm. Mirabello Fabric (reddish yellow, 5YR 6/6). Everted rim, conical shape, open rim-spout (missing), opposite a small double lug at rim, two opposed horizontal handles with circular sections on upper body. Dark slip on interior of rim and on exterior. Added white groups of horizontal bands above missing main motif. From levels 11–14, 19, 20; loci A604.22, 40, 42, 46, 47, 53, 64–66; objects 323, 397, 439, 500; residues 1270, 1303, 1323, 1344. *Date*: EM III.

51 (PAR 204, HN 14441; Fig. 26). Spouted conical jar, base sherd. D. of base 14 cm. Mirabello Fabric (pink, 5YR 7/4). Dark slip on exterior. Added white band at base and other traces on exterior. From level 10, locus A604.20, object 348, residue 1131. *Date*: EM III.

52 (PAR 401, HN 14753; Fig. 26; Pl. 40). Spouted conical jar, base and lower part. D. of base ca. 12 cm. Mirabello Fabric (light reddish brown, 5YR 6/3). Dark slip on exterior. Added white pendent hatched triangles on exterior. From levels 10, 12, 13; loci A604.20, 45, 46; objects 395, 468; residues 1101, 1223. *Date*: EM III.

53 (PAR 206, HN 14805; Fig. 26). Spouted conical jar, base sherd. D. of base 14 cm. Mirabello Fabric (light reddish brown, 2.5YR 6/4). Dark slip on exterior. Added white bands on exterior. From level 11, locus A604.22, residue 1204. Petrography sample PAR 09/05. *Date*: EM III.

54 (PAR 45, HN 14623; Fig. 26). Spouted conical jar, rim sherd with spout. D. of rim not measurable, max. dim. 9.3 cm. Mirabello Fabric (pink, 5YR 7/4, with the core light reddish brown, 5YR 6/4). Conical jar with open rim-spout. Dark slip on interior of rim and on exterior. Traces of added white on exterior. From level 4, locus A604.1, object 11, residue 972. *Date*: EM III.

55 (PAR 91, HN 14814; Fig. 26). Spouted conical jar, rim sherd. D. of rim ca. 24 cm. Mirabello Fabric (light brown, 7.5YR 6/4). Probably a spouted conical jar; thickened, outturned rim. Dark paint on inside of rim and bands on exterior. Added white traces. From levels 2–7, locus A601.5, object 19, residue 1050. Petrography sample PAR 09/03. *Date*: EM III.

56 (PAR 135, HN 14706; Fig. 26). Spouted conical jar, body sherd. Max. dim. 8.3 cm. Mirabello Fabric (light reddish brown, 5YR 6/4). Probably a spouted conical jar. Dark slip on exterior. Added white quirks bounded by bands on exterior. From level 16, locus A604.56, object 165. No residue number. *Date*: EM III.

57 (PAR 195, HN 14675; Fig. 26). Spouted conical jar(?), base sherd. D. of base ca. 18 cm. Mirabello Fabric (reddish brown, 5YR 5/3, to reddish yellow, 5YR 6/6). Traces of dark slip (red) on exterior. Added white not preserved. From level 15, locus A604.55, object 485, residue 1135. *Date*: EM III.

58 (PAR 166, HN 14552; Fig. 26). Spouted conical jar(?), base sherd. D. of base ca. 12.2 cm. Mirabello Fabric (pink, 7.5YR 8/4). Dark slip traces at base of exterior. Added white not preserved. From level 12, locus A604.40, object 364, residue 1356. *Date*: EM III.

59 (PAR 484, HN 14738; Fig. 26). Spouted conical jar, body sherd with handle. D. of body not measurable, max. dim. of sherd 11.1 cm. Mirabello Fabric (light reddish brown, 5YR 6/4). Raised horizontal handle with circular section. Dark slip (black) on exterior and handle. Added white chevrons and bands. From level 15, locus A604.56. *Date*: EM III.

60 (PAR 487, HN 14526; Fig. 26). Spouted conical jar, base sherd. D. of base 10 cm. Mirabello Fabric (pink, 5YR 7/4). Slightly convex profile. Dark slip on exterior and handle. Added white wide diagonal bands. From level 17, locus A604.61. *Date*: EM III.

Jars

Sometimes jars of undetermined shape cannot be identified in detail because too little survives. Two bases of East Cretan White-on-Dark Ware vessels in this class are cataloged from the deposit.

61 (PAR 495, HN 14754; Fig. 26). Jar, lower part. Pres. h. 24, d. of base 18.2 cm. Mirabello Fabric (light yellowish brown, 10YR 6/4). Dark slip on exterior. Added white chevrons with two horizontal bands at base, on exterior. From level 3, locus A604.5. No object or residue number. *Date*: EM III.

62 (PAR 505, HN 14765; Fig. 26). Jar, base and body sherd. D. of base 18 cm. Mirabello Fabric (reddish yellow, 5YR 6/6). Conical lower part of vessel. Dark slip on exterior. Added white vertical band, three horizontal bands, and band of chevrons on exterior. From levels 12, 15; loci A604.40, 55. No object or residue number. *Date*: EM III.

Straight-sided or Rounded Cup with Vertical Handle

Cups with vertical handles were common at the end of the third millennium B.C. They were made in many different styles, and they were used throughout Crete. One of the most common classes has walls that vary from straight or almost straight to a conical overall form to a vessel with a rounded form. Parallels in East Cretan White-on-Dark Ware from EM III–MM IA (among many others) include the following:

Chrysokamino: Betancourt 2006, 74, fig. 5.3:21
Gournia: Hall 1904–1905, fig. 3:a, pl. 26:1; Betancourt and Silverman 1991, fig. 1:319
Malia: Demargne 1945, pls. 4, 29:8674
Mochlos: Seager 1912, figs. 28:XI.13, 50:88; Soles 1992, 92, fig. 37 with reference to two other cups.
Palaikastro: Bosanquet and Dawkins 1923, 8, fig. 5:d
Priniatikos Pyrgos: Betancourt 1983, fig. 6:19
Pseira: Banou 1995, figs. 34:AG 1, 36:AA 6; Floyd 1998, figs. 2:BS/BV 6–8, 16, 5:BS/BV 69, 9:BS/BV 147; 2009a, figs. 2:AF 42, 3:AF 61, 5:AF 107; Betancourt 1999, fig. 20:BR 65, BR 68
Sphoungaras: Hall 1912, 51, fig. 23:b

For the same shape in Lasithi White-on-Dark Ware, see Langford-Verstegen 2015, 57–59 (Hagios Charalambos Cave) and Pendlebury, Pendlebury, and Money-Coutts 1935–1936, 57, fig. 13:519 (Trapeza Cave).

63 (PAR 514, HN 14543; Fig. 26). Conical or rounded cup, rim sherd. D. of rim 16 cm. Mirabello Fabric (reddish yellow, 5YR 6/8). Conical shape, with rounded profile; vertical handle with almost circular section. Dark slip (red) on interior of rim and on exterior. Added white not preserved. From level 18, locus A604.62. No object or residue number. *Date*: EM III.

Cylindrical Cup

The shape of this container is uncertain, as too little survives for its reconstruction, but it may be a cylindrical cup, which is a shape with a flat base, a cylindrical shape, and a single vertical handle. The shape is discussed by Betancourt (1984, 42–43). This shape of cup is not common in this period. Parallels come from Mochlos (Seager 1909, fig. 13, top row no. 2) and Myrtos Pyrgos (Cadogan 1978, fig. 7, left).

64 (PAR 41, HN 14618; Fig. 27). Cylindrical cup(?), rim sherd. D. of rim ca. 20 cm. Mirabello Fabric (reddish yellow, 5YR 6/6). Almost vertical wall. Dark paint on inside of rim and on exterior. Added white bands below rim and angular motif on body, on exterior. From levels 2–7. locus A601.2, object 25, residue 958. *Date*: EM III.

Rounded Cup with Tiny Lugs or Handles

This is a highly specialized shape. It consists of a cup with a rounded profile like the cup with vertical handle, but it has four small lugs or miniature horizontal handles placed low on the body. The handles serve no practical purpose. The shape is not as common as the rounded cup with a vertical handle. All known examples are in the White-on-Dark Ware tradition. Parallels come from the Hagios Charalambos Cave (HNM 12,433, unpublished) and Vasiliki (Seager 1906–1907, 121, fig. 4c, d, f).

65 (PAR 503, HN 14544; Fig. 27). Rounded cup, body sherd with small handles. Max. dim. 3.7 cm. Mirabello Fabric (dark gray, 10YR 4/1). Very small lug handles low on the body. Dark slip on exterior. Added white not preserved. From level 7, locus A604.11. No object or residue number. Found south of the rock shelter. *Date*: EM III.

66 (PAR 512, HN 14546; Fig. 27). Rounded cup, body sherd with small handles. D. of rim 8 cm. Mirabello Fabric (between pale brown, 10YR 6/3, and light reddish brown, 10YR 6/4). Very small lug handles low on the body. Dark slip on interior of rim and exterior. Added white not preserved. From level 7, locus A604.11. No object or residue number. *Date*: EM III.

67 (PAR 198, HN 14777; Fig. 27). Rounded cup with knobs low on the body, base sherd. D. of base 7.8 cm. Mirabello Fabric (pink, 7.5YR 7/4). Small lug handles near the base. Dark slip on exterior. Added white not preserved. From level 12, locus A604.40, object 369, residue 1129. *Date*: EM III.

Rounded Cups

Most rounded cups have one vertical handle with a circular section. A rare class with tiny lugs or handles low on the body is known from a few examples, so one cannot be certain of the complete shape when only fragments survive. Some of the vessels had no handles. Parallels in East Cretan White-on-Dark Ware include the following examples:

Gournia: Hall 1904–1905, fig. 1:c
Malia: Demargne 1945, pl. 29:8529, 8603, 8604
Mochlos: Soles 1992, 92, fig. 37
Palaikastro: Bosanquet and Dawkins 1923, 8, fig. 5c; MacGillivray et al. 1992, fig. 9:5
Pseira: Floyd 1998, fig. 2:BS/BV 22, 31; Betancourt 1999, fig. 20:69; Betancourt, Reese, and Schoch 2003, fig. 28:2.42
Sphoungaras: Hall 1912, 51, fig. 23:a, c, e
Vasiliki: Seager 1906–1907, 121, fig. 4:a, b, e

The following cup fragments are all small, and their original class cannot be determined with certainty.

68 (PAR 88, HN 14730; Fig. 27). Rounded cup, body sherd. Max. dim. 6.6 cm. Mirabello Fabric (light reddish brown, 5YR 6/4). Dark slip on exterior. Added white not preserved. From level 4, locus A604.1, object 18, residue 974. *Date*: EM III.

69 (PAR 188, HN 14678; Fig. 27). Rounded cup, base sherd. D. of base 5.3 cm. Mirabello Fabric (light reddish brown, 2.5YR 6/4). Dark slip on exterior. Added white not preserved. From level 16, locus A604.56, object 441, residue 1020. *Date*: EM III.

70 (PAR 502, HN 14545; Fig. 27). Rounded cup, body sherd. D. of body 8.5 cm. Mirabello Fabric (pink, 5YR 7/4). Dark slip on exterior. Added white diagonal bands on exterior. From locus A604.40. No object or residue number. *Date*: EM III.

Spouted Two-handled Cup

The spouted two-handled cup is a new shape not previously recognized from East Cretan White-on-Dark Ware. It consists of an open container with a rounded profile and two opposed vertical handles. An open spout is placed at the rim midway between the handles on one side. The decoration, a horizontal frieze on the body bounded by horizontal bands above and below, suggests that it is conceptually related to the spouted conical jars, which have straight walls, horizontal handles, and a similar open rim-spout. Only a single example of this rare shape comes from the deposit.

71 (PAR 102, HN 14664; Fig. 27; Pl. 40). Spouted two-handled cup almost complete. H. 8.6, d. of rim 9.3, d. of base 7.6 cm. A fine Mirabello Fabric (pink, 7.5YR 7/4). Wide-mouthed two-handled cup with open rim-spout and two opposed vertical handles. Dark slip on inside of rim and on exterior. Added white bands below rim, low on body, and flanking a frieze of diamonds containing crossed lines on exterior; vertical lines on handles. From level 12, loci A604.41, 44, object 394, residue 1273. *Date*: EM III.

Bridge-spouted Jar

After the jug, the bridge-spouted jar was the most popular pouring vessel during EM III and MM IA. The most common shape had a rounded body and a thickened, inturned rim with two horizontal handles attached below the rim. The shape was already present with cursory white decoration in EM II (Warren 1972, 152, pl. 64:P675 [Myrtos]). East Cretan White-on-Dark Ware examples dating from EM III to MM IA come from several sites in the region of the Gulf of Mirabello:

Chrysokamino: Betancourt 2006, 89, no. 83
Evraika Field 298: Watrous 2012, fig. 13:B169
Gournia: Hall 1904–1905, pls. 29–30; Hawes et
    al. 1908, 37, fig. 15:9, 10; Betancourt and
    Silverman 1991, figs. 2:327–337, 3:338–346
Gournia Moon: Watrous 2012, fig. 13:B168
Gournia Survey Site 81: Watrous 2012, fig.
    13:B169
Malia: Demargne 1945, pl. 29:8492; Chapouthier, Demargne, and Dessenne 1962, pls. 5,
    36:9142
Palaikastro: MacGillivray et al. 1992, fig. 9:6
Pera Alatzomouri or Sphoungaras: Watrous
    2012, fig. 13:B161, B162, B164, B165, B173
Pseira: Floyd 1998, fig. 5:BS/BV 79; Betancourt
    1999, figs. 16:BR 13, 20:BR 74, BR 77
Vasiliki: Seager 1906–1907, pl. 30:a; Betancourt 1983, fig. 19:243

Bridge-spouted jars were also sometimes traded. An example made in East Cretan White-on-Dark Ware comes from the Psychro Cave in Central Crete where it was apparently considered appropriate for a dedication (Watrous 1996, pls. 9:a, 13:a, b).

72 (PAR 121, HN 14835; Fig. 27). Bridge-spouted jar, rim sherd and other sherds. H. 23, d. of rim 17, d. of base 15 cm. Mirabello Fabric (red, 2.5YR 5/6). Thickened, inturned rim. Dark slip on exterior. Added white bands below the rim, delineating a frieze of dots, on exterior. From levels 2–8, 12, 18; loci A601.5, A604.13, 27, 62; object 140; residues 1009, 1125, 1150, 1215, 1271, 1317. Petrography sample 09/06. *Date*: EM III.

73 (PAR 18, HN 14811; Fig. 27). Bridge-spouted jar, rim sherd. D. of rim 32 cm. Mirabello Fabric (reddish yellow, 5YR 6/6). Dark slip on interior of rim and on exterior. Added white not preserved. From level 5, locus A604.7, object 89, residue 1155. Petrography sample PAR 09/08. *Date*: EM III.

74 (PAR 16, HN 14637; Fig. 27). Hole-mouthed jar or bridge-spouted jar, rim sherd. D. of rim 22 cm. Mirabello Fabric (light reddish brown, 5YR 6/4). Dark slip on interior of rim and on exterior. Added white not preserved. From level 7, locus A604.11, object 181, residue 1070. *Date*: EM III.

75 (PAR 201, HN 14568; Fig. 27). Bridge-spouted jar(?), base sherd. D. of base ca. 10 cm. Mirabello Fabric (light reddish brown, 5YR 6/4). Dark paint on exterior. Added white not preserved. From level 12, locus A604.29, object 404, residue 1325. *Date*: EM III.

76 (PAR 391, HN 14800; Fig. 27). Bridge-spouted jar(?), base sherd. D. of base ca. 15 cm. Mirabello Fabric (light gray, 10YR 7/1). Traces of dark slip. Added white not preserved. From level 19, locus A604.64, object 498, residue 1290. *Date*: EM III.

77 (PAR 411, HN 14740; Fig. 27). Bridge-spouted jar(?), base sherd. D. of base ca. 14 cm. Mirabello Fabric (between pink, 5YR 7/3, and reddish yellow, 5YR 7/3). Traces of dark red slip. Added white not preserved. From level 13, locus A604.46, object 469, residue 1186. *Date*: EM III.

78 (PAR 159, HN 14759; Fig. 28). Closed vessel (bridge-spouted jar?), base sherd. D. of base ca. 22–26 cm. Mirabello Fabric (light brown, 7.5YR 6/4, with the core pinkish gray, 7.5YR 6/2). Dark slip on exterior. Added white not preserved. From level 11, locus A604.21, object 252, residue 1074. *Date*: EM III.

79 (PAR 224, HN 14757; Fig. 28). Bridge-spouted jar (or jug?), base sherd. D. of base 8–10 cm. Mirabello Fabric (pink, 5YR 7/3). Dark slip on exterior. Added white not preserved. From level 14, locus A604.53, object 477, residue 1073. *Date*: EM III.

80 (PAR 387, HN 14676; Fig. 28; Pl. 40). Bridge-spouted jar, three-quarters complete. H. 22.7, d. of rim 17, d. of base 15.3 cm. Mirabello Fabric (red, 2.5YR 7/6). Two horizontal handles below the rim. Dark slip on exterior. Added white not preserved. From levels 8, 12, 18; loci A601.5, A604.13, 24, 27, 28, 59, 62, 64, 69; objects 115, 140, 212, 351, 359, 363, 520; residues 1009, 1150, 1215, 1221, 1271, 1317. *Date*: EM III.

81 (PAR 453, HN 14787; Fig. 28). Bridge-spouted jar, rim sherd. D. of rim 24 cm. Mirabello Fabric (reddish yellow, 5YR 7/6). Dark slip on interior of rim and on exterior. Added white bands delineating a frieze of dots on exterior. From locus A604.16. No object or residue number. *Date*: EM III.

82 (PAR 456, HN 14736; Fig. 28). Bridge-spouted jar, rim sherd with spout. D. of rim 46–54 cm. Mirabello Fabric (pink, 5YR 7/4). Short spout (4.6 cm). Dark slip on exterior. Added white traces on exterior. From level 6, locus A604.10. No object or residue number. *Date*: EM III.

83 (PAR 501, HN 14535; Fig. 29). Bridge-spouted jar, body sherd with handle. Max. dim. 15.1 cm. Mirabello Fabric (reddish yellow, 5YR 7/6). Horizontal handle, circular section. Dark slip (reddish brown) on exterior. Added white not preserved. From level 20, locus A604.65. No object or residue number. *Date*: EM III.

Jug or Jar

Because much of the pottery in the assemblage is fragmentary, some of the sherds cannot be assigned to specific shapes. Many base fragments could belong to either jugs or jars.

84 (PAR 218, HN 14461; Fig. 29). Jug or jar, base sherd. D. of base ca. 15 cm. Mirabello Fabric (reddish yellow, 5YR 7/6, with the core light gray, 5YR 6/1). Dark slip on exterior. Added white band on lower body. From level 10, locus A604.20, object 347, residue 1161. *Date*: EM III.

85 (PAR 9, HN 14774; Fig. 29). Jar or jug, base sherd. D. of base 19–20 cm. A pale fabric (pink, 5YR 7/4). Pronounced base. Dark slip on exterior. Added white traces of band on lower body. From level 7, locus A604.11, object 183, residue 1025. *Date*: EM III.

86 (PAR 378, HN 14649; Fig. 29). Jar or jug, base sherd. D. of base 13.9 cm. Mirabello Fabric (yellowish red, 5YR 5/6). Dark slip on exterior. Added white not preserved. From levels 11, 17–19, 21; loci A604.24, 59, 62, 64, 69; objects 349, 501; residues 1289, 1363. *Date*: EM III.

Jugs

The jug with a raised spout and a single vertical handle was popular throughout the Early Bronze Age. The bodies varied from rounded to piriform, and the spout could be short or taller. Early Minoan III examples were often given rounded bodies and short spouts, contrasting with the more piriform shape with a taller spout preferred for Vasiliki Ware during EM IIB (see above). For jugs in East Cretan White-on-Dark Ware, see the following (among many others):

Gournia: Hall 1904–1905, pl. 31
Gournia Survey: Watrous et al. 2012, fig. 12:B155
Mochlos: Seager 1912, figs. 18, 19:V.a
Palaikastro: Dawkins 1902–1903, 305, fig. 5:g
Sphoungaras: Hall 1912, 51, fig. 23:d, f
Vasiliki: Seager 1906–1907, 123, fig. 7

87 (PAR 360, HN 14661; Fig. 29; Pl. 19B). Jug, lower part. D. of base 9.8 cm. Mirabello Fabric (between pink, 5YR 7/3 and 7/4). Globular body. Traces of dark slip on exterior. Added white not preserved. From levels 11–12, locus A604.35, object 390, residue 1315. *Date*: EM III.

88 (PAR 34, HN 14614; Fig. 29). Jug, neck and shoulder sherd. Max. dim. 7.6 cm. Mirabello Fabric (reddish yellow, 5YR 6/6). Dark slip on exterior. Added white traces on exterior. From level 6, locus A604.9, object 104, residue 1054. *Date*: EM III.

89 (PAR 47, HN 14608; Fig. 29). Jug, body sherd with spout. Max. dim. 11.2 cm. A pale fabric (light reddish brown, 5YR 6/4). Dark slip on inside of rim and on exterior. Added white not preserved. From level 4, locus A604.1, object 8, residue 981. *Date*: EM III.

90 (PAR 398, HN 14496; Fig. 29). Jug, body sherd from base of neck, with handle. Max. dim. 5.8 cm. Mirabello Fabric (reddish brown, 5YR 6/4). Dark slip on exterior. Added white not preserved. From level 4, locus A604.1, object 22, residue 980. Petrography sample PAR 09/07. *Date*: EM III.

91 (PAR 396, HN 14665; Fig. 30). Jug, base sherd. D. of base 8 cm. Mirabello Fabric (yellowish red, 5YR 5/6). Globular body. Traces of dark slip. Added white not preserved. From levels 8, 10, 14, 16, 20; loci A604.13, 20, 53, 57, 65; object 414; residue 1120. *Date*: EM III.

92 (PAR 123, HN 14768; Fig. 30). Jug (or jar?), base sherd. D. of base ca. 16 cm. Mirabello Fabric (light reddish brown, 2.5YR 5/4). Dark slip on exterior. Added white traces of a band on exterior. From level 8, locus A604.12, no object number, residue 1049. *Date*: EM III.

Collared Jar

The collared jar is a closed vessel with a rounded profile on the body and a vertical neck with no spout. The shape has a long earlier history in Crete as well as in the Cyclades (Betancourt 2008, 38), but it is very rare in East Cretan White-on-Dark Ware.

93 (PAR 42, HN 14619; Fig. 30). Collared jar, body sherd from base of neck. D. of neck ca. 12–14 cm. Mirabello Fabric (surface between pink, 5YR 7/4, and reddish yellow, 5YR 7/6). Dark slip on exterior. Added white traces of hatched triangles on exterior. From level 3, locus A601.7, object 30, residue 957. *Date*: EM III.

Vessel with Elliptical Base

An unique elliptical base from the assemblage cannot be associated with certainty with any specific shape because of the absence of the upper part. In earlier times, elliptical shapes were most often used for small covered boxes called pyxides (Betancourt 1980, 54, fig. 5.10, from Hagios Onouphrios).

94 (PAR 98, HN 14791; Fig. 30). Closed elliptical vessel, base sherd. Dims. of elliptical base 4 x 7.5 cm. Mirabello Fabric (light red, 2.5YR 6/6). Small knob feet. Dark slip on exterior. Added white horizontal band above base and the beginning of another motif on exterior. From level 14, locus A604.49, object 460, residue 1104. It is possible that the vessel was a pyxis. *Date*: EM III.

Closed Vessel

A base from a closed vessel of unknown shape was covered with dark red slip. The decoration in white was no longer preserved.

95 (PAR 24, HN 14827; Fig. 30). Closed vessel, base sherd. D. of base 16 cm. Mirabello Fabric (reddish yellow, 5YR 6/6). Traces of dark slip (red) on exterior. Added white not preserved. From level 5, locus A604.7, object 90, residue 1043. Petrography sample 09/19. *Date*: EM III.

Teapot

The teapot, one of the characteristic EM III shapes that began to be produced in EM IIB, has a rounded body, a vertical handle, and an elongated spout set at the end of a cylindrical extension that joins the spout to the body below the rim. Illustrations of EM IIB teapots with elongated spouts include the following: Amouretti 1970, pl. 24, no. 1 (Malia); Warren 1972, 201–202, figs. 85, 86, pls. 62, 63 (Myrtos); Betancourt 1979, 41, fig. 10 (Vasiliki). Other EM IIB examples are listed by Betancourt (1979, 49–51). An EM IIB example with white bands on the body comes from Myrtos (Warren 1972, 151, no. P665, not illustrated). The EM III examples continue the shape of the earlier teapots, both with and without an added base. By the early years of the Middle Minoan period, the elongated spout was replaced with a shorter one. White-on-Dark Ware teapots with long spouts from EM III to MM IA are published from the following sites:

Archanes: Sakellarakis and Sapouna-Sakellaraki 1997, 386, fig. 339: lower row number 3
Gournia: Betancourt and Silverman 1991, 13, no. 347
Mochlos: Seager 1912, figs. 18:IV.2 (incorrectly restored as a cup); fig. 49:74; Maraghiannis and Karo 1907–1921, II, pl. 10; Pendlebury 1939, pl. 13:4b
Palaikastro: Forsdyke 1925, 79, no. A 450-2
Pseira: Maraghiannis and Karo 1907–1921, II, pl. 22:6; Banou 1995, fig. 36:AA 3; Betancourt 1999, fig. 20:BR 75, BR 76
Vasiliki: Seager 1906–1907, 129, fig. 13; Hawes et al. 1908, pl. 12:24; Maraghiannis and Karo 1907–1921, II, pls. 25:10, 12

The single teapot from this deposit has the long spout inherited from EM IIB, but it lacks the conical base that is often applied to the earlier versions. The decoration is simpler than on most of the published parallels, suggesting that they may be a little later than this EM III vase. The two tiny handles on the side of the spout would have helped the user hold the vessel steady while pouring.

96 (PAR 101, HN 14674; Fig. 30; Pls. 10A, 11A, 14A, 40). Teapot, almost complete. H. 27–28, d. of rim 18 cm. A fine Mirabello Fabric (pink, 5YR 7/4). Everted rim, long raised spout, handle with circular section opposite spout, two tiny handles on top of spout. Dark slip on inside of rim and on exterior. Added white pendent band extending from spout to below handle on the exterior, splitting to double band at the boundary between spout and body; pendent arcs with zigzags on lower body. From levels 3–7, locus A604.3, object 129, residue 1268. *Date*: EM III.

Plain Pottery and Pottery with Slight Decoration

In addition to the fine decorated style known as East Cretan White-on-Dark Ware, the pottery from the assemblage in the rock shelter had a large number of vessels that were either plain or only slightly decorated. The difference in decorative systems was not casual or random. The shapes in the two classes were different, and the decorated and plainer pieces were clearly treated very differently, a distinction that must have extended to how and when they were used. Cooking vessels, for example, were never decorated. The largest storage jars (pithoi and other large vessels) were given only cursory ornament. Industrial vessels and vessels used in preparing food, like transport jars, vats, and basins, were also given slight attention.

Open Vessels

Conical Bowls and Basins

An open vessel whose height is less than the rim's diameter is called a bowl or a basin. Bowls have diameters of 30 cm or less, and basins have diameters greater than 30 cm. Most examples of bowls and basins in this period are designed to be lifted with two horizontal handles. The shape varies from shallow to deep. Rim bands in dark paint are common. The EM III form is inherited from EM II (see, e.g., Pendlebury, Pendlebury, and Money-Coutts 1935–1936, 51, nos. 303–305 [Trapeza Cave]). Parallels from EM III to MM IA include the following:

Chrysokamino: Betancourt 2006, 78, 90–93, nos. 22, 88–94

Sphoungaras: Betancourt 1983, fig. 13:114, 115

Vasiliki: Betancourt 1983, figs. 14, 15:142, 148, 149

97 (PAR 137, HN 14673; Fig. 31; Pls. 29A, 40). Basin, three-quarters complete. H. 18.1, d. of rim 33.0, d. of base 18.9 cm. Mirabello Fabric (reddish yellow, 5YR 6/6). Everted rim, conical shape, two opposed horizontal handles with circular sections at rim. Dark band on rim. From levels 3, 5, 6, 11–14; loci A604.5, 7, 12, 23, 26, 29, 46, 53; objects 211, 213, 332, 334, 449, 476; residues 1103, 1251, 1254, 1274, 1279, 1306, 1345. *Date*: EM III.

98 (PAR 66, HN 14583; Fig. 31). Shallow bowl or basin with out-turned rim, rim sherd. D. of rim not measurable, max. dim. 5.6 cm. Mirabello Fabric (light reddish brown, 5YR 6/4). Dark band on rim. From level 4, locus A604.1, object 19, residue 973. *Date*: EM III.

99 (PAR 148, HN 14816; Fig. 31). Basin, rim sherd with handle. D. of rim over 40 cm. Mirabello Fabric (light reddish brown, 5YR 6/4). Everted rim, horizontal handle with circular sections at rim. Dark band on rim. From level 10, locus A604.19, object 256, residue 1301. Petrography sample PAR 09/09. *Date*: EM III.

100 (PAR 21, HN 14755; Fig. 31). Basin, base sherd. D. of base 24 cm. Mirabello Fabric (reddish yellow, 5YR 6/6, with the core light brown, 7.5YR 6/4). Traces of dark slip (red) on interior. From level 5, locus A604.7, object 95, residues 1098, 1181. *Date*: EM III.

101 (PAR 305, HN 14427; Fig. 31). Shallow basin, rim sherd. D. of rim 56 cm. Mirabello Fabric (pink, 5YR 7/4). Dark band on rim. From level 21, locus A604.67, object 428, residue 1085. *Date*: EM III.

102 (PAR 83, HN 14728; Fig. 31). Bowl, base sherd. D. of base ca. 15 cm. Mirabello Fabric (light red, 2.5YR 6/6). From level 3, locus A601.7, object 12, residue 947. *Date*: EM III.

103 (PAR 95, HN 14726; Fig. 31). Shallow bowl or basin, base sherd. D. of base 12 cm. Mirabello Fabric (reddish yellow, 5YR 6/6). Dark paint (red) on the lower part of the interior (perhaps a wide band). From level 14, locus A604.49, object 459, residue 1173. *Date*: EM III.

104 (PAR 343, HN 14782; Fig. 31). Shallow bowl or basin, base sherd. D. of base ca. 12 cm. Mirabello Fabric (light brown, 7.5YR 6/4). From level 19, locus A604.63, object 531, residue 1191. *Date*: EM III.

105 (PAR 82, HN 13729; Fig. 31). Basin (or jar?), rim sherd. D. of rim ca. 30 cm. Mirabello Fabric (reddish yellow, 7.5YR 6/6). Thickened and flat rim. Dark band on rim. From level 5, locus A604.7, object 88, residue 1039. *Date*: EM III.

106 (PAR 452, HN 14512; Fig. 31). Shallow, conical bowl, rim sherd. D. of rim ca. 46 cm. Mirabello Fabric (pink, 5YR 7/3). Horizontal handle with circular section attached on rim. Dark band (red) on rim and handle and drips on interior. From level 7, locus A604.11. No object or residue number. *Date*: EM III.

107 (PAR 509, HN 14540; Fig. 31). Shallow, conical bowl, rim sherd. D. of rim not measurable, max. dim 6.8 cm. A fine fabric (pale brown, 10YR 6/3). Raised handle with oval section attached at rim. Red slip on exterior. From level 14, locus A604.49. No object or residue number. *Date*: EM II–III.

108 (PAR 455, HN 14762; Fig. 31). Shallow, conical bowl, rim sherd. D. of rim 31.5 cm. Mirabello Fabric (grayish brown, 10YR 5/1). Thickened rim. Dark band (black) on interior of rim. From level 5, locus A604.6. No object or residue number. *Date*: EM III.

109 (PAR 486, HN 14524; Fig. 31). Shallow bowl, rim sherd. D. of rim ca. 44 cm. Mirabello Fabric (pink, 7.5YR 7/4). Small raised lug on rim. Dark band on interior of rim. From level 16, locus A604.57. No object or residue number. *Date*: EM III.

110 (PAR 412, HN 14491; Fig. 31). Shallow bowl, rim sherd. D. of rim ca. 10 cm. Mirabello Fabric (light reddish brown, 2.5YR 6/4). Dark band on interior of rim. From level 10, locus A604.19, object 261, residue 1122. *Date*: EM III.

111 (PAR 472, HN 14515; Fig. 31). Basin, rim sherd. D. of rim ca. 34–38 cm. Mirabello Fabric (reddish yellow, 7.5YR 7/6). Thick ledge rim; raised horizontal handle with oval section attached at rim. From level 14, locus A604.49, no object number or residue number. *Date*: EM III.

112 (PAR 200, HN 14569; Fig. 31). Bowl, base sherd. D. of base ca. 14 cm. Mirabello Fabric (pinkish gray, 5YR 6/2). Dark traces on interior. From level 10, locus A604.20, object 341, residue 1017. *Date*: EM III.

113 (PAR 71, HN 14772; Fig. 31). Basin, base sherd. D. of base 42 cm. Mirabello Fabric (reddish yellow, 5YR 7/6). Traces of slip (pink) on interior and exterior. From level 3, locus A604.7, object 24, residue 965. *Date*: EM III.

Basin with Interior Scoring

Large basins with scoring inside them on the walls and base are common from many Minoan periods. They have occasionally been regarded as beehives, but no analysis has ever confirmed the use (for discussion of beehives, see Melas 1999). As an alternative, they may have been used for washing clothing, for grating cheese, or for some other purpose. Parallels exist from many Minoan sites, occurring over a long period of time. Parallels include the following (among many others):

Chrysokamino: Ferrence and Shank 2006
Karpathos island, seven survey sites near Aphiartos: Melas 1985, 270, 271, 276, 277, 282, figs. 79:421–427, 80:485, 85:665–667, 86:707, 757, 778, 91:967–970, 983; 1999, 388, pl. 107:a
Karpathos island, two survey sites near Leukos: Melas 1985, 296, 297, figs. 105:1125–1127, 106:1246–1248; 1999, 488
Kasos island, four survey sites near Chelatros: Melas 1985, 300–302, figs. 109:1350, 1351, 110:1433–1435, 111:1484, 1485; 1999, 488, pl. 107:a
Kato Syme: Melas 1999, pl. 108:a
Kommos: Betancourt 1990, 98, 168, nos. 467, 1483; Watrous 1992, 25, no. 439
Nerokourou: Kanta and Rocchetti 1989, 193, fig. 64:482
Pseira: Floyd 1998, 75, no. 263

114 (PAR 414, HN 14645; Fig. 32; Pl. 28B). Basin with scoring inside, three-quarters complete. H. 22, d. of rim 39.8, d. of base 28.8 cm. Mirabello Fabric (pink, 5YR 7/4, to reddish yellow, 5YR 7/6, with the core pinkish gray, 5YR 6/2). Thickened rim, groove above base, two horizontal handles with circular sections below rim, lug opposite the handles. Vertical grooves on interior. Dark slip on rim and dabs on handles. From levels 5, 8–15, 21; loci A604.7, 13, 17, 18, 21, 25, 26, 28, 39, 40, 46, 47, 49, 50, 53, 55, 59, 67; objects 26, 139, 233, 333, 398, 402, 446, 458, 518; residue 1336. *Date*: EM III.

115 (PAR 408, HN 14495; Fig. 32). Basin with scoring inside, base sherd. D. of base ca. 36 cm. Mirabello Fabric (light reddish brown, 5YR 6/4). Both horizontal and vertical grooves in the interior. From levels 12, 14; loci A604.40, 49; objects 402, 458; residues 1297, 1335. *Date*: EM III.

116 (PAR 409, HN 14828; Fig. 32). Basin with scoring inside, base sherd. D. of base ca. 34 cm. Mirabello Fabric (light reddish brown, 5YR 6/4, with the core between pinkish gray, 5YR 6/2, and gray, 5YR 5/1). Both horizontal and vertical grooves in the interior. From level 12, locus A604.28, object 518, residue 1320. Petrography sample PAR 09/11. *Date*: EM III.

117 (PAR 410, HN 14493; Fig. 32). Basin with scoring inside, base sherd. D. of base ca. 32 cm. Mirabello Fabric (light reddish brown, 2.5YR 6/4, with the core gray, 2.5YR N/6). Both horizontal and vertical grooves in the interior. From level 8, locus A604.13, object 261, residue 1008. *Date*: EM III.

118 (PAR 415, HN 14490; Fig. 32). Basin with scoring inside, base sherd. D. of base ca. 32 cm. Mirabello Fabric (light reddish brown, 2.5YR 6/4, and the exterior surface slipped[?], very pale brown, 10YR 7/4). Both horizontal and vertical grooves in the interior. From level 3, locus A601.7, object 26, residue 961. *Date*: EM III.

119 (PAR 120, HN 14831; Fig. 32). Basin with scoring inside, base sherd. D. of base ca. 40 cm. Mirabello Fabric (mostly light red, 2.5YR 6/6, to light reddish brown, 2.5YR 6/4). Almost vertical wall. From level 12, locus A604.36, object 396, residue 1255. Burned. Petrography sample PAR 09/10. *Date*: EM III.

120 (PAR 156, HN 14695; Fig. 32). Basin with scoring inside, base sherd. D. of base ca. 24 cm. Mirabello Fabric (light brown, 7.5YR 6/4). From level 3, locus A601.7, object 21, residue 977. *Date*: EM III.

Spinning Bowl

Spinning bowls (or fiber-wetting bowls) are recognizable from their depiction in Egyptian wall paintings (Dothan 1963, 105; Barber 1991, 45, 48, fig. 2:5, 10). They consist of shallow bowls with one or more interior handles. The bowls are used to hold a ball of thread in position so that it will unwind as the thread is pulled upward through the interior handle for use in weaving (if used for flax, they would also contain water to wet the thread). Several examples of the bowls are known from Minoan Crete:

Drakones: Xanthoudides 1924, pl. 42:5033
Kommos: Betancourt 1990, 171, 182, nos. 1563, 1846
Myrtos: Warren 1972, 207, 209, fig. 91:P 701
Phaistos: Levi 1976, vol. I*, pl. 213

121 (PAR 382, HN 14799; Fig. 33). Shallow spinning bowl, mostly complete. D. of rim 25, d. of base 17 cm. Mirabello Fabric (pale brown, 10YR 6/3, with the core light brownish gray, 10YR 6/2). Two horizontal handles with circular section at rim; scar from the interior handle. Dark band on rim, inside and out, and on handle, with drips on interior. From level 7, locus A604.11, object 192, residue 1048. *Date*: EM III.

Cups with Straight Sides

Cups with straight sides are common in all periods of Minoan Crete. The shape is much less frequent in this assemblage than is usual.

122 (PAR 44, HN 14621; Fig. 33). Cup, complete profile. H. 5.6, d. of rim 10, d. of base 8 cm. A pale fabric (between pink, 7.5YR 7/4, and light brown, 7.5YR 6/4). From level 4, locus A604.1, object 14, residue 976. *Date*: EM III.

123 (PAR 94, HN 14724; Fig. 33). Cup, base sherd. D. of base 4 cm. Mirabello Fabric (light reddish brown, 2.5YR 6/4). From level 14, locus A604.49, object 453, residue 1061. *Date*: EM III.

Rounded Cups

Cups with rounded bodies, vertical handles, and bands on the rims are common in eastern Crete during EM III. This deposit includes examples of various sizes. Parallels from other EM III to MM IA deposits include the following:

Chrysokamino: Betancourt 2006, 85, fig. 5.6:78–81
Mochlos: Seager 1912, figs. 23:VI.6, 49:58–60, 50:89
Priniatikos Pyrgos: Betancourt 1983, fig. 6:19
Sphoungaras: Hall 1912, 55, fig. 28:a, b, e

An unusual aspect of the group from Pacheia Ammos is the presence of several very large examples. One cup has a rim diameter of 16 cm, which is very rare in this period.

124 (PAR 104, HN 14652; Fig. 33; Pl. 40). Rounded cup, complete. H. 8.5 to top of handle, d. of rim 10.6 x 11.7, d. of base 8 cm. Mirabello Fabric (reddish yellow, 5YR 6/6). Dark band on rim, inside and out, and on outer side of handle. From levels 13–14, which are the lowest levels, locus A604.52, object 447, residue 1284. *Date*: EM III.

125 (PAR 113, HN 14651; Fig. 33; Pls. 19B, 41). Rounded cup, almost complete. H. ca. 14 to top of restored handle, d. of rim 13.6, d. of base 9.2 cm. Mirabello Fabric (reddish yellow, 7.5YR 7/4). Dark band on rim, inside and out, with drips on the exterior. From levels 3–7, locus A604.32, object 128, residue 1247. *Date*: EM III.

126 (PAR 110, HN 14663; Fig. 33; Pl. 41). Rounded cup, complete profile. Rest. h. ca. 12 (including missing handle), d. of rim ca. 13, d. of base ca. 8 cm. Mirabello Fabric (pink, 5YR 7/4, with the core grayish pink, 5YR 7/2). Dark band on rim, inside and out, with a drip on the exterior. From levels 12, 13, 14, 20; loci A604.40, 46, 47, 53, 65; objects 401, 415, 440; residues 1064, 1130. *Date*: EM III.

127 (PAR 1, HN 14642; Fig. 33). Rounded cup, base sherd. D. of base 4.5 cm. A pale fabric (light reddish brown, 5YR 6/3). From level 4, locus A604.1, object 10, residue 970. *Date*: EM III.

128 (PAR 26, HN 14630; Fig. 33). Rounded cup, base sherd. D. of base 6 cm. Mirabello Fabric (light reddish brown, 5YR 6/4). From level 5, locus A604.6, object 79, residue 1052. *Date*: EM III.

129 (PAR 461, HN 14504; Fig. 33). Rounded cup, rim sherd. D. of rim 14.8 cm. Mirabello Fabric (reddish yellow, 5YR 7/6). Almost straight, thickened rim. Dark band on rim, inside and out. From level 12, locus A604.27. No object or residue number. *Date*: EM III.

130 (PAR 462, HN 14779; Fig. 33). Rounded cup, rim sherd. D. of rim 11.5 cm. Mirabello Fabric (pink, 5YR 7/4). Almost straight rim. Dark band on rim, inside and out. From level 12, locus A604.28. No object or residue number. *Date*: EM III.

131 (PAR 480, HN 14532; Fig. 34). Rounded cup, rim sherd. D. of rim 15 cm. Mirabello Fabric (pink, 5YR 7/4). Dark band on rim, inside and out, with drips on interior and exterior. From level 14, locus A604.53. No object or residue number. *Date*: EM III.

132 (PAR 479, HN 14530; Fig. 34). Rounded cup or bowl, rim sherd. D. of rim 15 cm. Mirabello Fabric (pink, 7.5YR 7/4, and pinkish gray core, 7.5YR 6/2). Dark band on rim, inside and out. From level 14, locus A604.53. No object or residue number. *Date*: EM III.

133 (PAR 478, HN 14528; Fig. 34). Rounded cup or bowl, rim sherd. D. of rim ca. 9.9 cm. Mirabello Fabric (light reddish brown, 5YR 6/4). Dark band on rim with drips inside and out. From locus A604.53, no object or residue number. *Date*: EM III.

134 (PAR 491, HN 14510; Fig. 34). Rounded or conical cup, rim and handle sherd. Max. dim. 7.3 cm. Mirabello Fabric (pinkish gray, 7.5YR 5/3). Vertical handle with almost circular section. Dark band on rim, inside and out. From level 17, locus A604.61. No object or residue number. *Date*: EM III.

CLOSED VESSELS

Jugs

Jugs from EM III have elevated spouts, but they are not as exaggerated as in EM II. The shape of the body of the EM III pouring vessels varies considerably, with the form ranging from globular to piriform. They occur in several sizes, closely following the shapes that had been used in EM IIB (for the EM IIB examples, see Warren 1972, pls. 49–51 [Myrtos]; Betancourt 1979, 37, fig. 11:15, 16 [Vasiliki]). Examples from EM III to MM IA come from Gournia (Betancourt and Silverman 1991, fig. 3:449) and Mochlos (Seager 1912, figs. 18:V, 19:Va).

135 (PAR 379, HN 14685; Fig. 34). Jug, almost complete. D. of base ca. 14 cm. Mirabello Fabric (grayish brown, 10YR 5/2, to yellowish red, 5YR 5/6). From levels 11, 12; loci A604.21, 27; objects 234, 362; residues 1136, 1316. *Date*: EM III.

136 (PAR 395, HN 14686; Fig. 34; Pl. 41). Jug, almost complete. D. of base ca. 15 cm. Mirabello Fabric (pink, 5YR 7/4, with the core gray, 5YR 5/1). Dark band on rim and handle. From levels 8, 11, 12, 14, 16; loci A604.12, 22, 28, 40, 49, 53, 56, 62, 64; object 327; residue 1364. *Date*: EM III.

137 (PAR 402, HN 14667; Fig. 34; Pl. 38B, 41). Jug: spout, handle, lower body, and other sherds. D. of base ca. 14.5 cm. Mirabello Fabric (light brown, 7.5YR 6/4). Slightly raised open spout. Dark band on rim, both inside and out. From levels 5, 9, 12–14, 16, 19, 20; loci A604.6, 17, 40, 48, 53, 56, 57, 62, 64, 65, 69; objects 207, 450; residues 1250, 1282. *Date*: EM III.

138 (PAR 336, HN 14424; Fig. 35). Jug, rim sherds with spout and handle. D. of rim not measurable, max. dim. 9.3 cm. Mirabello Fabric (pink, 5YR 7/2). Dark bands on rim and on handle, with drips and a line on the body. From level 14, locus A604.53, object 479, residue 1331. *Date*: EM III.

139 (PAR 30, HN 14627; Fig. 35). Jug, spout sherd. D. of rim not measurable, max. dim. 4.7 cm. Mirabello Fabric (light red, 2.5YR 6/6). Short shallow spout. From level 4, locus A604.1, object 23, residue 982. *Date*: EM III.

140 (PAR 189, HN 14577; Fig. 35). Jug, spout sherd. D. of rim not measurable, max. dim. 6.8 cm. Mirabello Fabric (red, 2.5YR 5/6). From level 16, locus A604.56, object 445, residue 1065. *Date*: EM III.

141 (PAR 199, HN 14572; Fig. 35). Jug, spout sherd. D. of rim not measurable, max. dim. 10.6 cm. Mirabello Fabric (light brown, 7.5YR 6/4). Dark band on rim of spout, inside and out. From level 10, locus A604.20, object 338, residue 1286. *Date*: EM III.

142 (PAR 131, HN 14715; Fig. 36). Jug, neck and shoulder sherd. Neck d. not measurable, max. dim. 8.3 cm. A pale fabric (pinkish gray, 7.5YR 6/2). From level 10, locus A604.19, object 440, residue 1062. *Date*: EM III.

143 (PAR 174, HN 14564; Fig. 36). Jug, neck sherd. Neck d. not measurable, max. dim. 7.7 cm. Mirabello Fabric (light red, 2.5YR 6/6). From level 12, locus A604.27, object 356, residue 1143. *Date*: EM III.

144 (PAR 180, HN 14560; Fig. 36). Jug, base sherd. D. of base ca. 9 cm. Mirabello Fabric (reddish yellow, 7.5YR 6/6). From level 11, locus A604.22, object 324, residue 1154. *Date*: EM III.

145 (PAR 179, HN 14808; Fig. 36). Jug, spout and neck sherd. W. of upper part 6.6 cm. Mirabello Fabric (pink, 5YR 7/3). Dark band on rim and top of handle. From level 15, locus A604.55, object 483, residue 1330. Petrography sample PAR 09/12. *Date*: EM III.

146 (PAR 132, HN 14710; Fig. 36). Jug, neck sherd. Neck d. not measurable, max. dim. 8.3 cm. Mirabello Fabric (pinkish gray, 7.5YR 6/2). Handle attached below the rim. From level 10, locus A604.19, object 440, residue 1062. *Date*: EM III.

147 (PAR 232, HN 14446; Fig. 36). Jug, rim sherd with spout. Max. dim. 11 cm. Mirabello Fabric (light reddish brown, 5YR 6/4). Open spout. Dark band on rim of spout. From level 17, locus A604.61, object 533, residue 1211. *Date*: EM III.

148 (PAR 384, HN 14739; Fig. 36). Jug, base and body sherd with handle. D. of base ca. 15 cm. Mirabello Fabric (red, 2.5YR 5/6). From levels 11, 13, 14, 16–19; loci A604.22, 46, 47, 53, 56, 59, 62, 64; objects 320, 510; residues 1203, 1353. *Date*: EM III.

149 (PAR 508, HN 14549; Fig. 36). Jug, rim sherd with scar for handle. D. of rim ca. 10 cm. A pale fabric (gray, 5YR 6/1). Outturned rim, handle with thin oval section attached just below the rim. From level 13, locus A604.46. No object or residue number. *Date*: EM II–III.

Wide-mouthed Jugs

Many of the handmade, wide-mouthed jugs from EM III to MM I had bridged spouts (Seager 1912, 67, fig. 37:XVI.8 [Mochlos]). If they are undecorated, they cannot be assigned a close date except by reference to their context.

150 (PAR 346, HN 14802; Fig. 36; Pl. 41). Wide-mouthed jug, almost half complete. Restored h. ca. 20, d. of rim 12, d. of base 10 cm. Mirabello Fabric (reddish brown, 5R 5/4). One vertical handle with circular section opposite missing spout. From level 13, locus A604.54, object 448. No residue number. Burned. *Date*: EM III.

151 (PAR 108, HN 14656; Fig. 37; Pl. 41). Wide-mouthed jug, complete. H. 12.9, d. of rim 14.5, d. of base 14.5 cm. Mirabello Fabric (weak red, 10R 4/3). Open, almost horizontal spout; one vertical handle with circular section opposite spout. From levels 13, 17, 20, 21; loci A604.47, 60, 68; objects 432, 451; residues 1021, 1338. *Date*: EM III.

152 (PAR 212, HN 14655; Fig. 37). Wide-mouthed jug, spout sherd. Max. dim. 4.2 cm. Mirabello Fabric (brown, 7.5YR 5/4, with the core dark reddish brown, 5YR 3/3). From level 12, locus A604.27, object 355, residue 1075. *Date*: EM III.

Pithoi

The tradition of making large ceramic vessels in East Crete during EM III is very little understood because of the scarcity of complete examples (for discussion, see Christakis 2005, 79). Large clay storage containers called pithoi are an important aspect of Minoan storage strategies, and they have been discussed in detail (Christakis 2005, 2008). The complete absence of any sherds with applied decorative moldings or other plastic additions to the exterior from the eastern region of the Gulf of Mirabello during EM III and MM I indicates that the vessels from this deposit were all plain, in spite of the fact that raised decorative moldings began to be used on Cretan pithoi in other regions as early as EM I (Betancourt 2008, 79–81, figs. 5.47–5.50), and they continued to be present on a regular basis into the Late Bronze Age (Christakis 2005). The presence of a ridge below the rim on the examples from this deposit should be regarded not as decoration but as a useful detail intended to facilitate grasping the vessel while moving it. Few firmly dated complete pithoi are known from EM III contexts (the examples from the Pacheia Ammos cemetery published by Seager [1916] are all later). Several fragments of pithoi come from the small rock shelter, but no complete examples were present in the assemblage. Parallels come from Gournia (Watrous et al. 2012, fig. 14:B186–B190).

153 (PAR 144, HN 14704; Fig. 37). Pithos, rim sherd. D. of rim ca. 50 cm. Mirabello Fabric (between pink, 5YR 7/4, and light reddish brown, 5YR 6/3, with the core pinkish gray, 5YR 7/2). Thickened rim, ridge below the rim. Dark slip on rim. From level 11, locus A604.21, object 238, residue 1240. *Date*: EM III.

154 (PAR 39, HN 14624; Fig. 37). Pithos, rim sherd. D. of rim 46 cm. Mirabello Fabric (reddish yellow, 5YR 6/6, with the core, light gray, 5YR 7/1). Thickened rim, ridge below the rim. Dark slip on rim. From level 4, locus A604.1, object 6, residue 986. *Date*: EM III.

155 (PAR 13, HN 14626; Fig. 37). Pithos, rim sherd. D. of rim 26 cm. Mirabello Fabric (light reddish brown, 5YR 6/3). Groove on inside of rim. Thickened rim, ridge below the rim. Dark slip on rim, inside and out. From level 7, locus A604.11, object 190, residue 1010. *Date*: EM III.

156 (PAR 43, HN 14620; Fig. 37). Pithos, rim sherd. D. of rim 42 cm. Mirabello Fabric (light reddish brown, 5YR 6/3, with the core light gray, 5YR 7/1). Thickened rim, ridge below the rim. Dark slip on rim, inside and out. From level 4, locus A604.1, object 7, residue 984. *Date*: EM III.

157 (PAR 50, HN 14605; Fig. 37). Pithos, rim sherd. D. of rim 26 cm. Mirabello Fabric (reddish yellow, 5YR 6/6). Thickened rim, ridge below the rim. From level 4, locus A604.1, object 5, residue 983. *Date*: EM III.

158 (PAR 56, HN 14603; Fig. 37). Pithos, rim sherd. D. of rim 36 cm. Mirabello Fabric (very pale brown, 10YR 7/3, with the core light reddish brown, 2.5YR 6/4). Thickened rim, ridge not preserved. Dark band on rim, inside and out. From level 9, locus A604.17, object 209, residue 1214. *Date*: EM III.

159 (PAR 92, HN 14721; Fig. 37). Pithos, rim sherd. D. of rim ca. 46 cm. Mirabello Fabric (light brown, 7.5YR 6/4). Thickened rim, ridge below the rim. Dark band on rim and below rim. From level 3, locus A601.7, object 15, residue 968. *Date*: EM III.

160 (PAR 86, HN 14732; Fig. 37). Pithos, rim sherd. D. of rim ca. 44 cm. Mirabello Fabric (pink, 5YR 7/4). Thickened rim, ridge below the rim. Dark band on rim. From level 3, locus A601.7, object 16, residue 959. *Date*: EM III.

161 (PAR 125, HN 14712; Fig. 37). Pithos, rim sherd. D. of rim ca. 30 cm. Mirabello Fabric (pink, 5YR 7/4). Thickened rim, ridge below the rim. From level 11, locus A604.21, object 232, residue 1296. *Date*: EM III.

162 (PAR 151, HN 14699; Fig. 38). Pithos, rim sherd. D. of rim ca. 30 cm. Mirabello Fabric (light reddish brown, 5YR 6/4). Thickened rim, ridge below the rim. From level 11, locus A604.21, object 236, residue 1305. *Date*: EM III.

163 (PAR 177, HN 14562; Fig. 38). Pithos, rim sherd. D. of rim ca. 32 cm. Mirabello Fabric (reddish yellow, 5YR 6/6, with the core gray, 5YR 6/1). Thickened rim, ridge not preserved. Dark band (red) on rim. From level 12, locus A604.40, object 380, residue 1196. *Date*: EM III.

164 (PAR 178, HN 14567; Fig. 38). Pithos, rim sherd. D. of rim ca. 52 cm. Mirabello Fabric (light reddish brown, 5YR 6/4, with the core weak red, 2.5YR 5/2). Thickened rim, ridge below the rim. Dark band on rim and below rim. From level 14, locus A604.53, object 473, residue 1308. *Date*: EM III.

165 (PAR 214, HN 14439; Fig. 38). Pithos, rim sherd. D. of rim ca. 36 cm. Mirabello Fabric (light reddish brown, 2.5YR 6/4). Thickened rim, ridge below the rim. Dark band on rim, inside and out. From level 6, locus A604.9, object 100, residue 1340. *Date*: EM III.

166 (PAR 261, HN 14487; Fig. 38). Pithos, rim sherd. D. of rim ca. 40 cm. Mirabello Fabric (light brown, 5YR 6/4, with the core light gray, 10YR 6/2). Thickened rim, ridge below the rim. Dark band on rim. From level 12, locus A604.40, object 377, residue 1198. *Date*: EM III.

167 (PAR 341, HN 14423; Fig. 38). Pithos, rim sherd. D. of rim ca. 32 cm. Mirabello Fabric (light brown, 7.5YR 6/4). Thickened rim, ridge below the rim. Dark band (red) on rim. From level 10, locus A604.19, object 266, residue 1175. *Date*: EM III.

168 (PAR 357, HN 14422; Fig. 38). Pithos, rim sherd. D. of rim ca. 28 cm. Mirabello Fabric (reddish yellow, 5YR 7/6). Thickened rim, ridge below the rim. From level 20, locus A604.66, object 506, residue 1276. *Date*: EM III.

169 (PAR 389, HN 14435; Fig. 38). Pithos, rim sherd. D. of rim ca. 32 cm. Mirabello Fabric (pink, 5YR 7/4). Thickened rim, ridge below the rim. From level 20, locus A604.66, object 504, residue 1293. *Date*: EM III.

170 (PAR 170, HN 14793; Fig. 38). Pithos, rim sherd. D. of rim ca. 27–28 cm. Mirabello Fabric (reddish brown, 5YR 4/4). Thickened rim, ridge below rim. Dark band on rim, inside and out. From level 10, locus A604.20, object 343, residue 1060. *Date*: EM III.

171 (PAR 393, HN 14819; Fig. 38; Pl. 22A). Pithos, rim sherd with handle. D. of rim ca. 56 cm. Mirabello Fabric (light reddish brown, 5YR 6/3). Thickened rim, ridge below the rim, handle with circular section. Dark band on rim and handle. From level 5, locus A604.16, object 197, residue 1366. Petrography sample PAR 09/13. *Date*: EM III.

172 (PAR 413, HN 14492; Fig. 39). Pithos, rim sherd. D. of rim ca. 52–54 cm. Mirabello Fabric (light reddish brown, 2.5YR 6/4). Thickened rim, ridge below the rim. Dark band on rim, inside and out. From level 13, locus A604.46, object 463, residue 1312. *Date*: EM III.

173 (PAR 20, HN 14634; Fig. 39). Pithos, rim sherd. D. of rim ca. 34 cm. Mirabello Fabric (light reddish brown, 5YR 6/4). Thickened rim, ridge not preserved. From level 6, locus A604.9, object 110, residue 1041. *Date*: EM III.

174 (PAR 31, HN 14611; Fig. 39). Pithos, rim sherd. D. of rim 42 cm. Mirabello Fabric (reddish yellow, 5YR 6/6, with a dark core, dark reddish gray, 5YR 4/2). Thickened rim, ridge below the rim. From level 5, locus A604.7, object 93, residue 1056. *Date*: EM III.

Pithoi or Jars

This class of large storage vessel is distinguished from the pithos rim sherds cataloged above by the absence of a ridge below the rim, but the overall shape is uncertain. The rim configuration is clearly related to that of the hole-mouthed jars, but whether all of these pieces were parts of that shape is uncertain.

175 (PAR 90, HN 14809; Fig. 39). Pithos or jar, rim sherd. D. of rim ca. 44 cm. Mirabello Fabric (light reddish brown, 5YR 6/4). Thickened rim. Dark band on rim, inside and out. From level 3, locus A601.7, object 22, residue 966. Petrography sample PAR 09/16. *Date*: EM III.

176 (PAR 153, HN 14698; Fig. 39). Pithos or jar, rim sherd. D. of rim ca. 26 cm. Mirabello Fabric (reddish yellow, 7.5YR 6/6). Thickened rim. Dark band on rim, inside and out. From level 11, locus A604.21, object 240, residue 1029. *Date*: EM III.

177 (PAR 160, HN 14771; Fig. 39). Pithos or jar, rim sherd. D. of rim ca. 28–36 cm. Mirabello Fabric (reddish yellow, 7.5YR 6/6, with the core pinkish gray, 7.5YR 6/2). Thickened rim. From level 11, locus A604.21, object 246, residue 1081. *Date*: EM III.

178 (PAR 163, HN 14769; Fig. 39). Pithos or jar, rim sherd. D. of rim ca. 40 cm. Mirabello Fabric (light reddish brown, 2.5YR 6/4, with a pale surface). Thickened rim. From level 11, locus A604.21, object 245, residue 1036. *Date*: EM III.

179 (PAR 167, HN 14550; Fig. 39). Pithos or jar, rim sherd. D. of rim ca. 40 cm. Mirabello Fabric (light brown, 7.5YR 6/4, with the core pinkish gray, 7.5YR 7/2). Thickened rim. From level 11, locus A604.21, object 235, residue 1128. *Date*: EM III.

180 (PAR 216, HN 14437; Fig. 39). Pithos or jar, rim sherd. D. of rim ca. 28 cm. Mirabello Fabric (light reddish brown, 5YR between 6/3 and 6/4). Thickened rim. From level 16, locus A604.58, object 524, residue 1277. *Date*: EM III.

181 (PAR 236, HN 14760; Fig. 39). Pithos or jar, rim sherd. D. of rim ca. 28 cm. Mirabello Fabric (reddish yellow, 5YR 6/6, with the core light gray, 5YR 7/1).

Thickened rim. From level 18, locus A604.62, object 509, residue 1354. *Date*: EM III.

182 (PAR 390, HN 14434; Fig. 39). Pithos or jar, rim sherd. D. of rim ca. 38 cm. Mirabello Fabric (light yellowish brown, 10YR 6/4, with the core light red, 2.5YR 6/6, and light brownish gray, 10YR 6/2). Thickened rim. From level 10, locus A604.19, object 260, residue 1299. *Date*: EM III.

183 (PAR 392, HN 14433; Fig. 39). Pithos or jar, rim sherd. D. of rim ca. 34 cm. Mirabello Fabric (pink, 7.5YR 7/2). Thickened rim. From level 20, locus A604.66, object 505, residue 1310. Petrography sample PAR 09/14. *Date*: EM III.

184 (PAR 182, HN 14566; Fig. 39). Pithos or jar, rim sherd. D. of rim ca. 12 cm. Mirabello Fabric (between pink, 7.5YR 7/4, and light brown, 7.5YR 6/4). Thickened rim. Dark slip on rim. From level 12, locus A604.27, object 358, residue 1055. *Date*: EM III.

185 (PAR 356, HN 14763; Fig. 39). Pithos or jar, rim sherd. D. of rim ca. 27 cm. Mirabello Fabric (reddish yellow, 7.5YR 7/6, with the core gray, 7.5YR N6/). Thickened rim. From level 11, locus A604.22, object 357, residue 1115. *Date*: EM III.

186 (PAR 477, HN 14527; Fig. 39). Pithos or jar, rim sherd. D. of rim ca. 32 cm. Mirabello Fabric (light reddish brown, 5YR 6/4). Thickened outturned rim. Dark band on rim. From level 14, locus A604.53, no object or residue number. Lid number PAR 467 fits this vessel. *Date*: EM III.

187 (PAR 504, HN 14747; Fig. 39). Pithos or jar, rim and body sherds with handle. D. of rim ca. 28 cm. Mirabello Fabric (reddish yellow, 5YR 6/6 and 7/6). Horizontal handle with circular section. Dark band on rim and top of handle. From levels 4, 17, 18; loci A604.1, 61, 62. No object or residue number. *Date*: EM III.

Pithoi or Jars with Drains

Several of the fragments from near the bases of pithoi or large jars had drains. This detail is not very common in eastern Crete.

188 (PAR 169, HN 14551; Fig. 40). Pithos or jar with drain, base sherd. D. of base ca. 24 cm. Mirabello Fabric (light reddish brown, 2.5YR 6/4). Groove above base, hole (d. 2.1 cm) above base. Dark bands above base. From level 17, locus A604.59, object 493, residue 1350. *Date*: EM III.

189 (PAR 358, HN 14657; Fig. 40). Pithos or jar with drain, base sherd. D. of base ca. 24 cm. Mirabello Fabric (reddish yellow, 5YR 6/6, and light reddish brown, 5YR 6/4). Hole above base. Dark drips on exterior. From levels 5–9, 18; loci A604.13, 15, 17, 55, 57, 62; objects 196, 205, 218; residues 1139, 1362, 1367. *Date*: EM III.

190 (PAR 342, HN 14807; Fig. 40). Pithos or jar with drain, base sherd. D. of base ca. 24 cm. Mirabello Fabric (reddish yellow, 5YR 6/6). Hole above base. From level 20, locus A604.19, object 507, residue 1195. Petrography sample PAR 09/20. *Date*: EM III.

191 (PAR 488, HN 14522; Fig. 40). Pithos with drain, base sherd. D. of base ca. 30–32 cm. Mirabello Fabric (reddish yellow, 5YR 6/6). Hole (d. ca. 3–4 cm) above base. From level 17, locus A604.61. No object or residue number. *Date*: EM III.

Spouted Jars

Six examples of jars with spouts are cataloged. Rim diameters vary from 22–30 cm. The shape is a storage vessel suitable for a number of different commodities, including liquids.

192 (PAR 400, HN 14669; Fig. 40; Pl. 41). Spouted jar, complete. H. 25.1, d. of rim 25.8, d. of base 16.1 cm. Mirabello Fabric (pink, 7.5YR 7/4). Inturned rim, open spout, two horizontal handles with circular sections. Dark band on rim, inside and out, and on handles. From levels 11, 12, 14, 15, 17; loci A604.13, 22, 27, 29, 40, 46, 47, 53, 55, 59, 67; objects 319, 326, 464, 479, 486, 494; residues 1131, 1193, 1261, 1280, 1311, 1342. *Date*: EM III.

193 (PAR 53, HN 14597; Fig. 40). Spouted jar, spout sherd. Max. dim. 8.2 cm. A pale fabric (between pink, 5YR 7/4, and reddish yellow, 5YR 7/6). Conical jar with open spout at rim. Dark band on rim and spout. From level 4, locus A604.13, object 9, residue 969. *Date*: EM III.

194 (PAR 157, HN 14692; Fig. 40). Spouted jar, spout sherd. Max. dim. 5 cm. Mirabello Fabric (pink, 5YR 7/3, with the core pinkish gray, 7.5YR 7/2). From level 11, locus A604.21, object 250, residue 1093. *Date*: EM III.

195 (PAR 203, HN 14830; Fig. 40). Spouted jar, rim sherd with spout. D. of rim ca. 30 cm. Mirabello Fabric (reddish yellow, 5YR 7/6). Thickened rim, conical shape, open spout. Dark band on rim and spout, inside and out. From level 20, locus A604.65, object 421, residue 1321. Petrography sample PAR 09/15. *Date*: EM III.

196 (PAR 234, HN 14445; Fig. 40). Spouted jar, rim sherd with spout. Max. dim. 8.2 cm. Mirabello Fabric (strong brown, 7.5YR 4/6, and partly light reddish brown, 5YR 6/3). Open spout. From level 18, locus A604.62, object 512, residue 1241. *Date*: EM III.

197 (PAR 471, HN 14735; Fig. 40). Spouted jar, rim sherd with spout. D. of rim 22 cm. Mirabello Fabric (reddish yellow, 7.5YR 6/6). Thickened, outturned rim. Dark slip on rim, inside and out. From level 13, locus A604.47. No object or residue number. *Date*: EM III.

Collared Jars

A collared jar is a storage vessel with a vertical or flaring neck/rim and no spout. Six examples of this vessel are cataloged.

198 (PAR 4, HN 14792; Fig. 40). Collared jar, rim sherd. D. of rim 26, h. of rim 1.06 cm. A medium-coarse fabric (light reddish brown, 5YR 6/4). Low collar. From level 4, locus A604.6, object 83, residue 1030. *Date*: EM III.

199 (PAR 6, HN 14640; Fig. 40). Collared jar, rim sherd. D. of rim not measurable, max. dim. 8.6 cm. A pale fabric (reddish yellow, between 5YR 7/6 and 5YR 6/6). Closed vessel, possibly a jar with outturned rim. From level 6, locus A604.9, object 105, residue 1053. *Date*: EM III.

200 (PAR 17, HN 14821; Fig. 40). Collared jar, rim sherd. D. of rim 10.7, h. of neck 3.6 cm. Mirabello Fabric (light reddish brown, 5YR 6/3). From level 7, locus A604.11, object 193, residue 1243. Petrography sample PAR 09/17. *Date*: EM III.

201 (PAR 143, HN 14806; Fig. 41). Collared jar, rim sherd. D. of rim 12 cm. Mirabello Fabric (light reddish brown, 5YR 6/4). From level 11, locus A604.21, object 248, residue 1028. Petrography sample PAR 09/18. *Date*: EM III.

202 (PAR 150, HN 14693; Fig. 41). Collared jar, rim sherd. D. of rim 12 cm. Mirabello Fabric (light reddish brown, 5YR 6/4). From level 10, locus A604.20, object 337, residue 1229. *Date*: EM III.

203 (PAR 226, HN 14450; Fig. 41). Collared jar, rim sherd. D. of rim 12 cm. Mirabello Fabric (light reddish brown, 5YR 6/3). Dark paint on exterior. From level 15, locus A604.55, object 484, residue 1068. *Date*: EM III.

Small Vessel with Knobs

The addition of knobs to the side of a small vessel is unusual, but an example exists from Malia from a context with pottery from EM III and MM IA (Demargne 1945, pl. 28:8607).

204 (PAR 97, HN 14734; Fig. 41). Small jar with knobs on the body, base sherd. D. of base 8 cm. Mirabello Fabric (light red, 2.5YR 6/6). From level 14, locus A604.49, object 452, residue 1059. *Date*: EM III.

Bridge-spouted Jars

Most bridge-spouted jars that have been published from this period are decorated in White-on-Dark Ware (see above). The undecorated bridge-spouted jar exists in EM III, but previously published contexts at Chrysokamino and Gournia have not furnished many examples for comparison (Betancourt 2006, 82, fig. 5.5:45, 46; Watrous et al. 2012, fig. 15:B192).

205 (PAR 485, HN 14741; Fig. 41). Bridge-spouted jar, rim sherd with spout. D. of rim 32 cm. Mirabello Fabric (pink, 5YR 7/4). Thickened inturned rim. Dark slip on rim, inside and out. Traces of added white bands. From level 16, locus A604.57, no object or residue number. *Date*: EM III.

206 (PAR 19, HN 13636; Fig. 41). Bridge-spouted jar, spout sherd. Max. dim. 5.2 cm. Mirabello Fabric (reddish yellow, 5YR 7/6) with a dark core (pinkish gray, 7.5YR 6/2). Shallow spout. From level 6, locus A604.9, object 107, residue 1019. *Date*: EM III.

207 (PAR 208, HN 14833; Fig. 41). Bridge-spouted jar, rim sherd with spout. D. of rim 16 cm. Mirabello Fabric (brown, 7.5YR 5/4). Thickened, inturned rim, wide shallow spout. From level 17, locus A604.59, object 491, residue 1086. Petrography sample PAR 09/29. *Date*: EM III.

Hole-mouthed Jar

The hole-mouthed jar is a vessel with a thickened rim, two horizontal handles, and a wide but slightly constricted mouth. During EM III, this form and the tripod cooking pot are made from similar fabrics in a closely related potting and firing tradition. As a result, it is impossible to distinguish between the two shapes when diagnostic features are not preserved. The shape has not been recorded previously from this period very often because complete vessels are rarely preserved. The only other known complete example from EM III to MM IA comes from the cemetery at Pacheia Ammos (Seager 1916, pls. 2:I-b, 11:XI-e).

208 (PAR 406, HN 14671; Fig. 41; Pls. 15A, 42). Hole-mouthed jar, complete. H. 40.7, d. of rim 34, d. of base 18.5 cm. Mirabello Fabric (red, 5YR 6/4). Thickened inturned rim, two horizontal handles with circular sections on the shoulder. From levels 2–8; loci A601.2, 5; object 230; residue 1263. Burned. *Date*: EM III.

209 (PAR 407, HN 14683; Fig. 41; Pls. 15B, 42). Hole-mouthed jar, almost complete. H. 39.5, d. of rim 33.7, d. of base 18.6 cm. Mirabello Fabric (reddish brown, 2.5YR 4/4). Thickened inturned rim, two horizontal handles with circular sections on the shoulder. From levels 2–8; loci A601.3, A603.3; object 227; residue 1257. Burned. *Date*: EM III.

210 (PAR 436, no HN number; Fig. 42; Pl. 19A). Hole-mouthed jar almost complete. H. 44, d. of rim ca. 38, d. of base 17.6 cm. Mirabello Fabric (red, 2.5YR 5/6, with the core reddish brown, 5YR 5/4). Thickened inturned rim, horizontal handles with circular sections on the shoulder. From levels 9, 11–14, 17, 19–21; loci A604.17, 22, 27, 28, 32, 37, 40, 44, 46, 48, 53, 59, 64, 65, 67; objects 221, 325, 384; residues 1166, 1256, 1260. Body extremely warped from the firing. *Date*: EM III.

211 (PAR 40, HN 14684; Fig. 42). Hole-mouthed jar, upper part. H. 22 cm. Mirabello Fabric (between very dark gray, 5YR 3/1, and reddish brown, 5YR 4/4, with the surface red, 2.5YR 4/6). Thickened inturned rim, horizontal handles with circular sections on the shoulder. From level 3, locus A601.7, object 11, residue 957. *Date*: EM III.

212 (PAR 404; no HN number; Fig. 42). Hole-mouthed jar, almost complete. D. of rim ca. 44 cm. Mirabello Fabric (dark reddish gray, 5YR 4/2). Thickened inturned rim, horizontal handles with circular section on the shoulder. From levels 10–13, 19; loci A604.20, 22, 40, 46, 64; object 367; residue 1337. *Date*: EM III.

213 (PAR 89, HN 14681; Fig. 42). Hole-mouthed jar, almost complete. H. 41, d. of rim ca. 34, d. of base 19 cm. Mirabello Fabric (red, 2.5YR 5/6, to strong brown, 7.5YR 4/6, to yellowish red, 5YR 5/8). From levels 3, 8, 12, 16–21; loci A601.7, A604.13, 38, 56, 58, 59, 62, 64, 65, 67, 69; objects 14, 502, 522; residues 954, 1245, 1365. *Date*: EM III.

214 (PAR 433, HN 14670; Fig. 43; Pl. 15B). Hole-mouthed jar, almost complete. H. 35, d. of rim ca. 34, d. of base 20 cm. Mirabello Fabric (light reddish brown, 5YR 6/4, with the core reddish gray, 5YR 5/2). Thickened rim, horizontal handles with circular sections on the shoulder. From levels 1, 2–9, 12; loci A600.1, A601.3, 8, A604.2, 11, 17, 28; objects 228, 231; residues 1259, 1264. The small size suggests the possibility that the vessel is a cooking pot rather than a hole-mouthed jar, but all the certain cooking pots from this context have a single vertical handle, not the horizontal type preserved on this fragment. *Date*: EM III.

215 (PAR 515, HN 14691; Fig. 43). Hole-mouthed jar, rim sherd with handle. D. of rim 31 cm. Mirabello Fabric (red, 2.5YR 4/6). Thickened inturned rim, horizontal handles with circular sections on the shoulder. No find context recorded. No object or residue number. *Date*: EM III.

216 (PAR 496, HN 14644; Fig. 43). Hole-mouthed jar, rim sherd. D. of rim 30 cm. Mirabello Fabric (red, between 2.5YR 4/6 and 5/6, with the core brown, 7.5YR 4/6). Thickened rim, horizontal handles with circular sections on the shoulder. From level 5, locus A604.15. No object or residue number. *Date*: EM III.

217 (PAR 35, HN 14610; Fig. 43). Hole-mouthed jar, rim sherd. D. of rim 54 cm. Mirabello Fabric (red, 2.5YR 5/8). Thickened rim. From level 5, locus A604.6, object 81, residue 1033. *Date*: EM III.

218 (PAR 79, HN 14826; Fig. 43). Hole-mouthed jar, rim sherd. D. of rim 26–28 cm. Mirabello Fabric (red, 2.5YR 5/6, with the core weak red, 5YR 4/2). Thickened rim. From level 5, locus A604.7, object 92, residue 1171. Petrography sample PAR 09/27. *Date*: EM III.

219 (PAR 192, HN 14815; Fig. 43). Hole-mouthed jar, rim sherd. D. of rim ca. 32 cm. Mirabello Fabric (dark reddish brown, 5YR 3/4). Thickened rim. From level 12, locus A604.29, object 406, residue 1127. Burned. Petrography sample 09/31. *Date*: EM III.

220 (PAR 112, HN 14820; Fig. 43). Hole-mouthed jar, rim sherd. D. of rim ca. 30 cm. Mirabello Fabric (reddish yellow, 5YR 6/8). Thickened rim. From level 14, locus A604.49, object 14, residue 1231. Petrography sample PAR 09/33. *Date*: EM III.

221 (PAR 72, HN 14590; Fig. 43). Hole-mouthed jar, rim sherd. D. of rim ca. 30 cm. Mirabello Fabric (reddish yellow, 7.5YR 6/6, with the core pinkish gray, 5YR 6/2). Thickened rim. From level 2, locus A601.5, object 121, residue 1005. *Date*: EM III.

222 (PAR 117, HN 14716; Fig. 43). Hole-mouthed jar, rim sherd. D. of rim ca. 36 cm. Mirabello Fabric (light reddish brown, 5YR 6/3). Thickened rim. From level 8, locus A604.13, object 131, residue 1176. *Date*: EM III.

223 (PAR 196, HN 14810; Fig. 43). Hole-mouthed jar, rim sherd. D. of rim ca. 28 cm. Mirabello Fabric (light reddish brown, 2.5YR 6/4). Thickened rim, dark slip on rim. From level 12, locus A604.29, object 410, residue 1067. Petrography sample 09/32. *Date*: EM III.

224 (PAR 209, HN 14689; Fig. 43). Hole-mouthed jar, complete profile. D. of rim elliptical (warped), ca. 22 x 26 cm. Mirabello Fabric (reddish yellow, 5YR 7/6). Thickened rim; one horizontal handle preserved. From level 12, locus A604.27, object 350, residue 1360. *Date*: EM III.

Cooking Vessels

Tripod Cooking Pots

Most Minoan tripod cooking pots are deep vessels with three legs, vertical or horizontal handles, and a small open spout on the rim. They are the most common cooking vessels in Crete (for general discussions, see Betancourt 1980; Martlew 1988). They first appear in EM I along with other shapes that suggest Anatolian predecessors (Shank 2005, 105). By EM II, the period immediately preceding the deposit in the rock shelter, the shape was very similar to the EM III form (Warren 1972, 176, fig. 60:P 327). The same shape continues into MM IA (MacGillivray et al. 1992, fig. 9:8 [Palaikastro]). Minoan tripods are almost always made of red-firing clay using recipes that include substantial amounts of aplastic rock fragments to create a coarse or semi-coarse fabric.

225 (PAR 377, HN 14804; Fig. 44; Pl. 43). Tripod cooking pot, almost complete. D. of rim ca. 14 cm. Mirabello Fabric (red to light red, 2.5YR 5/8 to 6/6). Thickened rim, one vertical handle with circular section. From levels 12–14, 19, 20; loci A604.40, 46, 53, 59, 64, 65; objects 416, 419; residues 1289, 1363. *Date*: EM III.

226 (PAR 438, HN 14796; Fig. 44). Tripod cooking pot, fragmentary. D. of rim ca. 28.3–28.9, d. of base ca. 20 cm. Mirabello Fabric (red, 2.5YR 5/6). Thickened, inturned rim, one vertical handle with circular section. From levels 5, 8–10; loci A604.6, 7, 13, 17–19; objects 72, 86, 91, 137, 206, 216, 223, 224, 244, 258; residues 1045, 1057, 1096, 1132, 1156, 1194, 1222, 1230, 1278. *Date*: EM III.

227 (PAR 397, HN 14797; Fig. 44; Pl. 43). Tripod cooking pot, lower part. D. of base 15 cm. Mirabello Fabric (red, 2.5YR 5/6, with the core dark gray, 5YR 4/1). Legs with thin oval sections. From levels 11, 17; loci A604.22, 59; objects 316, 497; residues 1302, 1333. *Date*: EM III.

228 (PAR 445, HN 14654; Fig. 44; Pl. 43). Tripod cooking pot, almost complete. Rest. h. ca. 30, d. of rim ca. 24, d. of base ca. 20–22 cm. Mirabello Fabric (reddish brown, 2.5YR 5/4). Legs with thin oval sections. From levels 9, 10, 11, 15, 20; loci A604.18, 20, 22, 28, 42, 46, 55, 64, 65; objects 222, 352, 417, 422, 435; residues 1141, 1213, 1232, 1310, 1359. *Date*: EM III.

229 (PAR 106, HN 14653; Fig. 44; Pls. 10A, 11B, 13B, 43). Tripod cooking pot, almost complete. Rest. h. ca. 25, d. of rim 19.8 x 20.3, d. of base 14.5 cm. Mirabello Fabric (red, 2.5YR 5/6). Thickened rim, rounded profile, one vertical handle with circular section opposite an open spout. From levels 3–8, locus A604.2, object 226, residue 1267. *Date*: EM III.

230 (PAR 109, HN 14680; Fig. 44; Pls. 14A, 18A, 43). Tripod cooking pot, almost complete. Rest. h. ca. 27–30, d. of rim 19.8 x 22.8, d. of base 15 cm. Mirabello Fabric (red, 2.5YR 5/6, to black on the interior and from reddish brown, 2.5YR 5/4, to black on the exterior). Thickened rim, rounded profile, one vertical handle with circular section neither opposite a leg nor opposite the space between two legs. From levels 5–12, locus A604.14, object 382, residue 1266. *Date*: EM III.

231 (PAR 114, HN 14660; Fig. 44; Pls. 10A, 11A, 14A, 43). Tripod cooking pot, almost complete. Rest. h. ca. 22–25, d. of rim 17.8–18.4, d. of base 14.0–14.7 cm. Mirabello Fabric (reddish brown, 2.5YR 5/4, to black). Thickened inturned rim, rounded profile, one vertical handle with circular section neither opposite a leg nor opposite the space between two legs; legs with thin elliptical sections. From levels 3–7, locus A604.4, residue 1285. *Date*: EM III.

232 (PAR 152, HN 14745; Fig. 45; Pl. 43). Tripod cooking pot, almost complete. Rest. h. ca. 27, d. of rim 21.5–22, d. of base 16 cm. Mirabello Fabric (red, 2.5YR 5/6, to reddish brown, 5YR 4/3). Thickened inturned rim, rounded profile, legs with thin oval sections. From levels 11, 12, 14, 19, 21; loci A604.21, 29, 40, 46, 53, 59, 61, 64, 65, 67; objects 153, 242, 404, 426; residues 1082, 1140, 1304, 1346. *Date*: EM III.

233 (PAR 332, HN 14688; Fig. 45). Tripod cooking pot, almost complete. Pres. h. (without legs) 25.5, d. of rim ca. 23.0–23.7, d. of base 15.0–15.5 cm. Mirabello Fabric (reddish brown, 5YR 5/6). Thickened inturned rim, rounded profile, legs with thin oval sections. From levels 1, 9, 11, 12; loci A604.17, 22, 40, 46, 53, 59, 64, 65; objects 219, 317, 322, 328, 329, 331, 400, 416, 419; residues 1187, 1202, 1208, 1234, 1235, 1289, 1343, 1355, 1363. Burned. *Date*: EM III.

234 (PAR 440, no HN number; Fig. 45). Tripod cooking pot, rim, body, and handle sherds. D. of rim ca. 16 cm. Mirabello Fabric (red, 2.5YR 5/6). Thickened rim, vertical handle with circular sections. From levels 12–14, 19, 21; loci A604.28, 40, 47, 53, 56, 59, 64, 65, 67; objects 503, 521; residues 1151, 1237. *Date*: EM III.

235 (PAR 173, HN 14646; Fig. 45). Tripod cooking pot, almost complete. H. ca. 37.2 cm. Mirabello Fabric (red, 2.5YR 5/6). Leg with thin oval section. From level 13, locus A604.46, object 465, residue 1326. *Date*: EM III.

236 (PAR 444, HN 14687; Fig. 45). Tripod cooking pot, rim and body sherds with handle. D. of rim ca. 22.8 cm. Mirabello Fabric (reddish yellow, 5YR 6/6, with the core gray, 5YR 5/1). Thickened inturned rim, vertical handle with circular section. From levels 8, 11–13, 15–18; loci A604.13, 22, 28, 29, 40, 42, 47, 55–57, 59, 61, 62; objects 490, 517; residues 1219, 1318. *Date*: EM III.

237 (PAR 149, HN 14701; Fig. 45). Tripod cooking pot (or jar?), rim sherd. D. of rim 22 cm. Mirabello Fabric (red, 2.5YR 5/6). Thickened rim. From level 10, locus A604.19, object 258, residue 1084. *Date*: EM III.

238 (PAR 221, HN 14455; Fig. 45). Tripod cooking pot, rim sherd with handle. D. of rim ca. 19.8 cm. Mirabello Fabric (red, 2.5YR 4/8). Thickened inturned rim. From level 14, locus A604.53, object 474, residue 1123. *Date*: EM III.

239 (PAR 158, HN 14694; Fig. 45). Tripod cooking pot (or jar?), rim sherd. D. of rim 22 cm. Mirabello Fabric (red, 2.5YR 5/6). Thickened rim. From level 11, locus A604.21, object 241, residue 1034. *Date*: EM III.

240 (PAR 168, HN 14825; Fig. 45). Tripod cooking pot (or jar?), rim sherd. D. of rim 20 cm. Mirabello Fabric (red, 2.5YR 4/6). Thickened rim. From level 9, locus A604.17, object 215, residue 1160. *Date*: EM III.

241 (PAR 5, HN 14786; Fig. 45). Tripod cooking pot (or jar?), rim sherd. D. of rim 24 cm. Mirabello Fabric (dark reddish gray, 5YR 4/2). From level 5, locus A604.6, object 80, residue 1180. *Date*: EM III.

242 (PAR 230, HN 14449; Fig. 46). Tripod cooking pot (or jar?), rim sherd. D. of rim 20–22 cm. Mirabello Fabric (red, 2.5YR 4/6). Thickened inturned rim. From level 12, locus A604.40, object 370, residue 1100. *Date*: EM III.

243 (PAR 511, HN 14537; Fig. 46). Cooking pot or jar, rim sherd. D. of rim 15.6 cm. A fine fabric (red, 2.5YR 5/6). Thickened outturned rim. From level 19, locus A604.64. No object or residue number. *Date*: EM III.

244 (PAR 145, HN 14703; Fig. 46). Tripod cooking pot (or jar?), rim sherd. D. of rim ca. 20 cm. A pale fabric (pink, 5YR 7/4). Thickened rim. From level 11, locus A604.21, object 243, residue 1217. *Date*: EM III.

245 (PAR 142, HN 14794; Fig. 46). Tripod cooking pot (or jar?), rim sherd. D. of rim ca. 24 cm. Mirabello Fabric (reddish yellow, 5YR 6/6). Thickened rim, rounded profile. From level 10, locus A604.19, object 270, residue 998. *Date*: EM III.

246 (PAR 161, HN 14818; Fig. 46). Tripod cooking pot (or jar?), rim sherd. D. of rim ca. 26–28 cm. Mirabello Fabric (red, 2.5YR, between 4/6 and 5/6). Thickened rim. From level 11, locus A604.21, object 244, residue 989. Petrography sample PAR 09/28. *Date*: EM III.

247 (PAR 74, HN 14588; Fig. 46). Tripod cooking pot (or jar?), rim sherd. D. of rim 54 cm. Mirabello Fabric (reddish yellow, 5YR 7/6). Thickened rim. From level 3, locus A601.7, object 29, residue 960. *Date*: EM III.

248 (PAR 138, HN 14708; Fig. 46). Tripod cooking pot (or jar?), rim sherd. D. of rim ca. 26–28 cm. Mirabello Fabric (reddish brown, 2.5YR 4/4). Thickened rim. From level 10, locus A604.9, object 273, residue 1285. *Date*: EM III.

249 (PAR 228, HN 14452; Fig. 46). Tripod cooking pot (or jar?), rim sherd. D. of rim ca. 24 cm. Mirabello Fabric (between light yellowish brown, 10YR 6/4, and pale brown, 10YR 6/3, with the surface reddish yellow, 7.5YR 6/6). Thickened rim. From level 10, locus A604.19, object 271, residue 996. *Date*: EM III.

250 (PAR 227, HN 14451; Fig. 46). Tripod cooking pot (or jar?), spout sherd. D. of rim ca. 22 cm. Mirabello Fabric (weak red, 2.5YR 4/2). Open spout. From level 16, locus A604.56, object 446, residue 1066. *Date*: EM III.

251 (PAR 241, HN 14463; Fig. 46). Tripod cooking pot (or jar?), spout sherd. Max. dim. 6.6 cm. Mirabello Fabric (red, 2.5YR 4/8, with the core reddish brown, 5YR 5/3). Open spout. From level 18, locus A604.62, object 511, residue 1090. *Date*: EM III.

252 (PAR 222, HN 14682; Fig. 46). Cooking pot, rim, body, base, and handle sherds. Pres. h. (without base) 19.1, d. of rim (warped) 18.2 x 21.5 cm. Mirabello Fabric (red, 2.5YR 5/6). Thickened inturned rim, vertical handle with circular sections. From levels 3, 9, 10–20, 21; loci A604.7, 15, 16, 17, 19, 20, 21, 28, 40, 46, 55, 56, 57, 59, 61, 62, 64, 65, 67; objects 239, 339, 371, 374; residues 1015, 1071, 1114, 1142. *Date*: EM III.

253 (PAR 85, HN 14723; Fig. 46). Tripod cooking pot, leg sherd. Pres. length of leg 3.7, max. w. 5.7 cm. Mirabello Fabric (between reddish brown, 2.5YR 4/4, and red, 2.5YR 4/6). Thin oval section. From level 3, locus A601.7, object 19, residue 952. *Date*: EM III.

254 (PAR 385, HN 14801; Fig. 46). Tripod cooking pot, base sherd with leg. Length of leg 11.1, max. w. of leg 6.5 cm. Mirabello Fabric (red, 2.5YR 4/6, with a dark core, reddish brown, 5YR 5/4). Thin oval section. From level 7, locus A604.11, object 179, residue 1357. *Date*: EM III.

255 (PAR 242, HN 14743; Fig. 47). Tripod cooking pot, leg sherd. Pres. length of leg 4.9, max. w. of leg 6.8 cm. Mirabello Fabric (from reddish yellow, 7.5YR 6/6, to black, 7.5YR N2/). Thin oval section. From level 21, locus A604.67, object 425, residue 1109. *Date*: EM III.

256 (PAR 122, HN 14713; Fig. 47). Tripod cooking pot, leg sherd. Pres. length of leg 6.8, max. w. of leg 6.1 cm. Mirabello Fabric (from dark reddish brown, 5YR 3/3, to brown, 7.5YR 5/4). Thin oval section. From levels 5–7, locus A604.16, object 198, residue 1032. *Date*: EM III.

257 (PAR 87, HN 14733; Fig. 47). Tripod cooking pot, leg sherd. Pres. length of leg 4.9, max. w. 6.1 cm. Mirabello Fabric (light reddish brown, 7.5YR 6/3, with the next layer very dark brown, 7.5YR 3/3, and the core reddish brown, 5YR 5/4). Thin oval section. From level 3, locus A601.7, object 18, residue 949. *Date*: EM III.

258 (PAR 340, HN 14770; Fig. 47). Tripod cooking pot, base sherd with edge of leg. D. of base ca. 16–8 cm. Mirabello Fabric (reddish brown, 5YR 4/3). From level 20, locus A604.66, object 508, residue 1212. *Date*: EM III.

259 (PAR 33, HN 14613; Fig. 47). Tripod cooking pot, leg sherd. Pres. length of leg 5.6, max. w. 5.7 cm. Mirabello Fabric (dark brown, 7.5YR 4/4). Thin oval section. From level 6, locus A604.9, object 102, residue 1007. *Date*: EM III.

260 (PAR 38, HN 14622; Fig. 47). Tripod cooking pot, leg sherd. Pres. length of leg 5.3, max. w. 6.7 cm. Mirabello Fabric (reddish yellow, 5YR 6/6, with the core light reddish brown, 5YR 6/4). Thin oval section. From level 4, locus A604.1, object 13, residue 979. *Date*: EM III.

261 (PAR 183, HN 14565; Fig. 47). Tripod cooking pot, leg sherd. Pres. length of leg 2.3, max. w. of leg 5.8 cm. Mirabello Fabric (red, 2.5YR 4/6, with the core dark red, 2.5YR 3/6). Thin oval section. From level 12, locus A604.29, object 412, residue 1063. *Date*: EM III.

262 (PAR 171, HN 14822; Fig. 47). Tripod cooking pot, leg sherd. Length of leg 13.2, max. w. of leg 9.5 cm. Mirabello Fabric (reddish brown, 2.5YR 5/4). Thin oval section. From level 12, locus A604.40, object 365, residue 1349. Petrography sample PAR 09/26. *Date*: EM III.

263 (PAR 80, HN 14731; Fig. 47). Tripod cooking pot, leg sherd. Pres. length of leg 5.1, max. w. 5.1 cm. Mirabello Fabric (red, 2.5YR 5/6, with the core darker, very dark gray, 2.5YR 4/2). Thin oval section. From level 3, locus A601.7, object 20, residue 953. *Date*: EM III.

264 (PAR 191, HN 14573; Fig. 47). Tripod cooking pot, leg sherd. Pres. length of leg 4.6 cm. Mirabello Fabric (yellowish red, 5YR 5/6). From level 11, locus A604.22, object 321, residue 1117. *Date*: EM III.

265 (PAR 37, HN 14617; Fig. 47). Tripod cooking pot, leg sherd. Pres. length of leg 8.6, max. w. 6.8 cm. Mirabello Fabric (red, 2.5YR 5/6). Thin oval section. From level 5, locus A604.6, object 78, residue 1307. *Date*: EM III.

266 (PAR 229, HN 14448; Fig. 47). Tripod cooking pot, leg sherd. Pres. length of leg 7.4, max. w. of leg 7.2 cm. Mirabello Fabric (reddish brown, 2.5YR 5/4). Thin oval section. From level 21, locus A604.67, object 423, residue 1216. *Date*: EM III.

267 (PAR 49, HN 14604; Fig. 48). Tripod cooking pot, leg sherd. Pres. length of leg 6.4, max. w. 7.0 cm. Mirabello Fabric (exterior surface gray, 10YR 5/1, with the core gray, 10YR 6/1). Thin oval section. From level 8, locus A604.13, object 133, residue 1239. *Date*: EM III.

268 (PAR 64, HN 14593; Fig. 48). Tripod cooking pot, leg sherd. Pres. length of leg 5.2, max. w. 7.0 cm. Mirabello Fabric (exterior surface yellowish red, 5YR 5/6, with the core darker, reddish brown, 2.5YR 5/4). Thin oval section. From level 10, locus A604.20, object 340, residue 1199. Burned. *Date*: EM III.

269 (PAR 51, HN 14598; Fig. 48). Tripod cooking pot, leg sherd. Pres. length of leg 6.9, max. w. 4.0 cm. Mirabello Fabric (exterior surface red, 2.5YR 5/6, with the core between light reddish brown, 2.5YR 6/4, and reddish brown, 5YR 5/4). Thin oval section. From level 8, locus A604.13, object 134, residue 1018. *Date*: EM III.

270 (PAR 129, HN 14711; Fig. 48). Tripod cooking pot, leg sherd. Pres. length of leg 10.5, max. w. of leg 6.6 cm. Mirabello Fabric (from gray, 7.5YR N5/0, to light red, 2.5YR 6/8, with the core pale red, 2.5YR 6/2). Thin oval section. From level 11, locus A604.21, object 232, residue 1281. *Date*: EM III.

271 (PAR 154, HN 14696; Fig. 48). Tripod cooking pot, leg sherd. Pres. length of leg 6.7, max. w. of leg 5.8 cm. Mirabello Fabric (yellowish red, 5YR between 4/6 and 5/6). Thin oval section. From levels 13–14, locus A604.52, object 516, residue 1164. *Date*: EM III.

272 (PAR 141, HN 14700; Fig. 48). Tripod cooking pot, leg sherd. Pres. length of leg 6.3, max. w. of leg 5.6 cm. Mirabello Fabric (red, 2.5YR 4/6). Thin oval section. From level 11, locus A604.21, object 251, residue 1027. *Date*: EM III.

273 (PAR 186, HN 14576; Fig. 48). Tripod cooking pot, leg sherd. Pres. length of leg 4.5, max. w. of leg 5.7 cm. Mirabello Fabric (red, 2.5YR 5/6, with the core light reddish yellow, 5YR 6/4). Thin oval section. From level 12, locus A604.29, object 411, residue 1106. *Date*: EM III.

274 (PAR 193, HN 14460; Fig. 48). Tripod cooking pot, leg sherd. Pres. length of leg 6.7, max. w. of leg 8.1 cm. Mirabello Fabric (yellowish red, 5YR 5/6, with the core light reddish brown, 5YR 6/3). Thin oval section. From level 12, locus A604.40, object 372, residue 1188. *Date*: EM III.

275 (PAR 190, HN 14574; Fig. 48). Tripod cooking pot, leg sherd. Pres. length of leg 4.1, max. w. of leg 7.5 cm. Mirabello Fabric (reddish brown, 5YR 4/4). Thin oval section, added vertical ridge on outside of leg. From level 12, locus A604.27, object 354, residue 1300. *Date*: EM III.

276 (PAR 175, HN 14832; Fig. 48). Tripod cooking pot, leg sherd. Pres. length of leg 6.5, max. w. of leg 7.4 cm. Mirabello Fabric (dark gray, 2.5YR N4/). Thin oval section. From level 21, locus A604.67, object 424, residue 1236. Petrography sample PAR 09/25. *Date*: EM III.

277 (PAR 217, HN 14459; Fig. 48). Tripod cooking pot, base sherd with leg. D. of base ca. 18–20, pres. length of leg 7.9, max. w. of leg 7.2 cm. Mirabello Fabric (red, 2.5YR 5/6, with the core weak red, 2.5YR 4/2). Thin oval section. From level 17, locus A604.59, object 496, residue 1329. *Date*: EM III.

278 (PAR 220, HN 14458; Fig. 48). Tripod cooking pot, leg sherd. Pres. length of leg 7.7, max. w. of leg 6.3 cm. Mirabello Fabric (reddish brown, 2.5YR 5/4). Thin oval section. From level 14, locus A604.53, object 478, residue 1292. *Date*: EM III.

279 (PAR 211, HN 14440; Fig. 48). Tripod cooking pot, leg sherd. Pres. length of leg 4, max. w. of leg 6.3 cm. Mirabello Fabric (reddish brown, 2.5YR 5/4, with the core reddish brown, 2.5YR 5/4). Thin oval section. From level 17, locus A604.61, object 532, residue 1189. Burned. *Date*: EM III.

280 (PAR 69, HN 14592; Fig. 49). Tripod cooking pot, leg sherd. Length of leg 11 cm. Mirabello Fabric (red, 2.5YR 4/6; with dark core, reddish gray, 5YR 5/2). Thin oval section. Object 116. Burned. *Date*: EM III.

281 (PAR 7, HN 14639; Fig. 49). Tripod cooking pot, base sherd with leg. Length of leg 9.5, max. w. of leg 5.7 cm. Mirabello Fabric (reddish brown, 5YR 5/3). Thin oval section. From level 7, locus A604.11, object 180, residue 1238. *Date*: EM III.

282 (PAR 15, HN 14638; Fig. 49). Tripod cooking pot, leg sherd. Pres. length of leg 4, max. w. of leg 7.3 cm. Mirabello Fabric (dark brown, 7.5YR 4/2). Thin oval section. From level 7, locus A604.11, object 182, residue 1088. *Date*: EM III.

283 (PAR 60, HN 14596; Fig. 49). Tripod cooking pot, leg sherd. Pres. length of leg 5.3, max. w. 7.4 cm. Mirabello Fabric (exterior surface reddish yellow, 5YR 6/6, with the core mottled gray, between gray, 5YR 5/1, and dark gray, 5YR 4/1). Thin oval section. From level 8, locus A604.13, object 135, residue 1170. *Date*: EM III.

284 (PAR 96, HN 14725; Fig. 49). Tripod cooking pot, leg sherd. Pres. length of leg 8.6, max. w. of leg 5.8 cm. Mirabello Fabric (mostly red, 2.5YR 4/6). Thin oval section. From level 14, locus A604.49, object 457, residue 1087. *Date*: EM III.

Cooking Dishes

Large shallow vessels with thick rims and often very thin walls, called cooking dishes, are very common in the assemblage. This class of vessel was used from EM I until the end of the Minoan period. Many of the dishes have been found with burn marks from cooking, confirming their use in preparing food. Like the cooking pots, they are regularly made of red-firing clay (for discussion, see Betancourt 1980). Cooking dishes come from virtually every Minoan domestic site. Parallels (among many others) include the following:

Chrysokamino: Betancourt 2006, 74, fig. 5.3:22, 23
Myrtos: Warren 1972, 161–163, figs. 45–47
Pseira: Betancourt and Davaras, eds., 1995, 81, no. AM 15, fig. 44; 1998a, 18, 21, 24, nos. AC 44, AC 45, AC 68, AC 97–AC 101, pls. 9, 11, 12; 1998b, 36, 55, nos. 59, 154, figs. 4, 9

285 (PAR 247, HN 14766; Fig. 49). Cooking dish, rim sherd. D. of rim ca. 39 cm. Mirabello Fabric (red, 2.5YR 5/6). Thickened rim. From level 17, locus A604.59, object 489, residue 1201. *Date*: EM III.

286 (PAR 23, HN 14631; Fig. 49). Cooking dish, rim sherd. D. of rim 44 cm. Mirabello Fabric (red, 2.5YR 4/6). Thickened rim. From level 5, locus A604.7, object 96, residue 1095. *Date*: EM III.

287 (PAR 25, HN 14632; Fig. 49). Cooking dish, rim sherd. D. of rim not measurable, max. dim. 6.6 cm. Mirabello Fabric (uneven color, ranging from dark red, 2.5YR 3/6, to very dark gray, 7.5YR N3/). Rim turned out to make a handle. From level 6, locus A604.9, object 111, residue 1022. *Date*: EM III.

288 (PAR 27, HN 14629; Fig. 49). Cooking dish, rim sherd. D. of rim ca. 40 cm. Mirabello Fabric (red, 2.5YR 5/8). Thickened below the rim. From level 5, locus A604.7, object 98, residue 1157. *Date*: EM III.

289 (PAR 54, HN 14595; Fig. 49). Cooking dish, sherd from the handle. D. of rim not measurable. Mirabello Fabric (red, 2.5YR 5/6). Flat handle added at the rim. From level 4, locus A604.1, object 15, residue 975. *Date*: EM III.

290 (PAR 231, HN 14453; Fig. 49). Cooking dish, rim sherd. D. of rim ca. 40 cm. Mirabello Fabric (reddish brown, 5YR 5/4). Thickened rim. From level 16, locus A604.58, object 523, residue 1149. *Date*: EM III.

291 (PAR 233, HN 14447; Fig. 49). Cooking dish, rim sherd. D. of rim ca. 50 cm. Mirabello Fabric (reddish brown, 5YR 5/3). Thickened rim. From level 10, locus A604.20, object 344, residue 1309. Burned. *Date*: EM III.

292 (PAR 235, HN 14444; Fig. 49). Cooking dish, rim sherd. D. of rim ca. 32 cm. Mirabello Fabric (dark reddish brown, 5YR 3/4). Thickened rim. From level 5, locus A604.6, object 82, residue 1154. *Date*: EM III.

293 (PAR 237, HN 14462; Fig. 49). Cooking dish, rim sherd. D. of rim ca. 52–54 cm. Mirabello Fabric (reddish brown, 5YR 5/4). Thickened rim. From level 13, locus A604.47, object 436, residue 1205. *Date*: EM III.

294 (PAR 238, HN 14465; Fig. 49). Cooking dish, rim sherd. D. of rim ca. 54 cm. Mirabello Fabric (reddish brown, 2.5YR 5/4, with the core weak red, 2.5YR 4/2). Thickened rim. From level 13, locus A604.47, object 434, residue 1233. *Date*: EM III.

295 (PAR 239, HN 14812; Fig. 49). Cooking dish, rim sherd. D. of rim ca. 54 cm. Mirabello Fabric (reddish brown, 5YR 5/4). Thickened rim. From level 12, locus A604.40, object 366, residue 1314. Petrography sample PAR 09/22. *Date*: EM III.

296 (PAR 240, HN 14464; Fig. 49). Cooking dish, rim sherd. D. of rim ca. 44 cm. Mirabello Fabric (red, 2.5YR 5/6, and partly reddish brown, 2.5YR 4/4). Thickened rim. From level 5, locus A604.7, object 94, residue 1058. *Date*: EM III.

297 (PAR 243, HN 14824; Fig. 49). Cooking dish, rim sherd. D. of rim not measurable, max. dim. 12 cm. Mirabello Fabric (reddish brown, 5YR 4/4). Thickened rim. From level 15, locus A604.55, object 487, residue 1134. Petrography sample PAR 09/23. *Date*: EM III.

298 (PAR 244, HN 14467; Fig. 49). Cooking dish, rim sherd. D. of rim ca. 54 cm. Mirabello Fabric (red, 2.5YR 4/6). Thickened rim. From level 17, locus A604.59, object 492, residue 1126. *Date*: EM III.

299 (PAR 246, HN 14468; Fig. 49). Cooking dish, rim sherd. D. of rim ca. 54 cm. Mirabello Fabric (red, 2.5YR 4/6). Thickened rim. From level 12, locus A604.40, object 368, residues 1180, 1190. *Date*: EM III.

300 (PAR 248, HN 14469; Fig. 50). Cooking dish, rim sherd. D. of rim ca. 54 cm. Mirabello Fabric (red, 2.5YR 4/6). Thickened rim. From level 8, locus A604.13, object 136, residue 1072. *Date*: EM III.

301 (PAR 249, HN 14471; Fig. 50). Cooking dish, rim sherd. D. of rim ca. 54 cm. Mirabello Fabric (weak red, 2.5YR 4/2). Thickened rim. From level 6, locus A604.9, object 101, residue 1184. *Date*: EM III.

302 (PAR 250, HN 14470; Fig. 50). Cooking dish, rim sherd. D. of rim ca. 54 cm. Mirabello Fabric (dark gray, 10YR 4/1, with the core reddish brown, 2.5YR 5/4). Thickened rim. From level 12, locus A604.28, object 519, residue 1169. *Date*: EM III.

303 (PAR 251, HN 14472; Fig. 50). Cooking dish, rim sherd. D. of rim ca. 54 cm. Mirabello Fabric (red, 2.5YR 4/6, with the core weak red, 2.5YR 5/2). Thickened rim. From level 12, locus A604.40, object 403, residue 1119. *Date*: EM III.

304 (PAR 252, HN 14473; Fig. 50). Cooking dish, rim sherd. D. of rim ca. 54 cm. Mirabello Fabric (red, 2.5YR 4/4). Thickened rim. From level 10, locus A604.19, object 257, residue 1167. *Date*: EM III.

305 (PAR 253, HN 14767; Fig. 50). Cooking dish, rim sherd. D. of rim ca. 48 cm. Mirabello Fabric (reddish brown, 2.5YR 5/4). Thickened rim. From level 12, locus A604.27, object 357, residue 1124. *Date*: EM III.

306 (PAR 254, HN 14474; Fig. 50). Cooking dish, rim sherd. D. of rim ca. 54 cm. Mirabello Fabric (yellowish red, 5YR 5/6). Thickened rim. From level 4, locus A604.1, object 16, residue 978. *Date*: EM III.

307 (PAR 255, HN 14475; Fig. 50). Cooking dish, rim sherd. D. of rim ca. 52 cm. Mirabello Fabric (reddish brown, 2.5YR 5/4). Thickened rim. From level 11, locus A604.21, object 249, residue 992. *Date*: EM III.

308 (PAR 256, HN 14476; Fig. 50). Cooking dish, rim sherd. D. of rim ca. 56 cm. Mirabello Fabric (weak red, 2.5YR 4/2). Thickened rim. From level 7, locus A604.11, object 191, residue 1179. *Date*: EM III.

309 (PAR 257, HN 14477; Fig. 50). Cooking dish, rim sherd. D. of rim ca. 50–52 cm. Mirabello Fabric (red, 2.5YR 4/6, with the core weak red, 2.5YR 4/2). Thickened rim. From level 14, locus A604.49, object 455, residue 1078. *Date*: EM III.

310 (PAR 258, HN 14478; Fig. 50). Cooking dish, rim sherd. D. of rim ca. 54 cm. Mirabello Fabric (red, 2.5YR 5/8, with the core brown, 7.5YR 5/2). Thickened rim. From level 16, locus A604.58, object 526, residue 1097. *Date*: EM III.

311 (PAR 259, HN 14837; Fig. 50). Cooking dish, rim sherd. D. of rim ca. 40–42 cm. Mirabello Fabric (dark reddish brown, 5YR 3/2). Thickened rim. From level 16, locus A604.57, object 515, residue 1165. Petrography sample PAR 09/24. *Date*: EM III.

312 (PAR 260, HN 14488; Fig. 50). Cooking dish, rim sherd. D. of rim ca. 52 cm. Mirabello Fabric (reddish brown, 2.5YR 5/4). Thickened rim. From level 10, locus A604.19, object 263, residue 1012. *Date*: EM III.

313 (PAR 262, HN 14486; Fig. 50). Cooking dish, rim sherd. D. of rim ca. 50 cm. Mirabello Fabric (light reddish brown, 5YR 6/4). Thickened rim. From level 13, locus A604.46, object 263, residue 1152. *Date*: EM III.

314 (PAR 263, HN 14484; Fig. 50). Cooking dish, rim sherd. D. of rim ca. 52 cm. Mirabello Fabric (yellowish red, 5YR 5/6). Thickened rim. From level 5, locus A604.6, object 85, residue 1121. *Date*: EM III.

315 (PAR 264, HN 14485; Fig. 50). Cooking dish, rim sherd. D. of rim ca. 54 cm. Mirabello Fabric (red, 2.5YR 5/6). Thickened rim. From level 10, locus A604.19, object 269, residue 1000. *Date*: EM III.

316 (PAR 265, HN 14483; Fig. 50; Pl. 22A). Cooking dish, rim sherd. D. of rim ca. 52 cm. Mirabello Fabric (red, 2.5YR 5/6). Thickened rim. From level 5, locus A604.16, object 201, residue 995. *Date*: EM III.

317 (PAR 266, HN 14482; Fig. 50). Cooking dish, rim sherd. D. of rim ca. 54 cm. Mirabello Fabric (weak red, 2.5YR 4/2). Thickened rim. From level 12, locus A604.40, object 329, residue 1092. *Date*: EM III.

318 (PAR 267, HN 14480; Fig. 50). Cooking dish, rim sherd. D. of rim ca. 54 cm. Mirabello Fabric (red, 2.5YR 5/6, with the core weak red, 2.5YR 5/2). Thickened rim. From level 9, locus A604.17, object 225, residue 1158. *Date*: EM III.

319 (PAR 268, HN 14481; Fig. 50). Cooking dish, rim sherd. D. of rim ca. 54 cm. Mirabello Fabric (reddish brown, 5YR 5/4). Thickened rim. From level 19, locus A604.64, object 499, residue 1133. *Date*: EM III.

320 (PAR 269, HN 14479; Fig. 50). Cooking dish, rim sherd. D. of rim ca. 52 cm. Mirabello Fabric (red, 2.5YR 4/6). Thickened rim. From level 9, locus A604.17, object 214, residue 1159. *Date*: EM III.

321 (PAR 270, HN 14431; Fig. 50). Cooking dish, rim sherd. D. of rim ca. 40 cm. Mirabello Fabric (yellowish red, 5YR 5/6). Thickened rim. From level 12, locus A604.29, object 413, residue 1023. *Date*: EM III.

322 (PAR 271, HN 14430; Fig. 50). Cooking dish, rim sherd. D. of rim ca. 46 cm. Mirabello Fabric (dark reddish gray, 5YR 4/2). Thickened rim. From level 10, locus A604.19, object 268, residue 997. *Date*: EM III.

323 (PAR 272, HN 14823; Fig. 50). Cooking dish, rim sherd. D. of rim ca. 50 cm. Mirabello Fabric (reddish yellow, 5YR 6/6). Thickened rim. From level 3, locus A601.7, object 21, residue 963. Petrography sample PAR 09/21. *Date*: EM III.

324 (PAR 273, HN 14775; Fig. 50). Cooking dish, rim sherd. D. of rim ca. 44 cm. Mirabello Fabric (reddish yellow, 5YR 6/6). Thickened rim. From level 21, locus A604.67, object 427, residue 1200. *Date*: EM III.

325 (PAR 274, HN 14429; Fig. 50). Cooking dish, rim sherd. D. of rim ca. 50 cm. Mirabello Fabric (red,

2.5YR 5/6). Thickened rim. From level 14, locus A604.53, object 472, residue 1144. *Date*: EM III.

326 (PAR 331, HN 14425; Fig. 50). Cooking dish, rim sherd. D. of rim ca. 56 cm. Mirabello Fabric (brown, 7.5YR 5/2). Thickened rim. From level 17, locus A604.61, object 534, residue 1192. *Date*: EM III.

327 (PAR 459, HN 14737; Fig. 50). Cooking dish, rim sherd with handle. D. of rim 46–48 cm. Mirabello Fabric (red, 2.5YR 5/8). Thickened outturned rim with a handle at the rim. From level 6, locus A604.10. No object or residue number. *Date*: EM III.

328 (PAR 492, HN 14533; Fig. 50). Cooking dish, rim sherd. D. of rim ca. 46 cm. Mirabello Fabric (red, 2.5YR 5/8). Thickened rim. From level 19, locus A604.64. No object or residue number. Burned on the bottom. *Date*: EM III.

329 (PAR 245, HN 14466; Fig. 50). Cooking dish, rim sherd. D. of rim ca. 42 cm. Mirabello Fabric (red, 2.5YR 5/6). Thickened rim. From level 12, locus A604.40, object 323, residue 1099. *Date*: EM III.

LARGE VATS

The vat, an industrial shape, is a large basin with a drain near the base. The shape from EM II is already mature, with a form that is very similar to the EM III vat from the rock shelter at Alatzomouri (Warren 1972, 190, fig. 74:P 529). Because the shape was used until LM I, parallels exist from many sites:

Epano Zakros: Platon 1965, 218–219, pl. 247A
Gournia: Hawes et al. 1908, pl. 1:14
Kato Zakros: Platon 1961, pl. 174B; 1963, 164, pl. 142:A
Kommos: Betancourt 1990, 183, no. 1860
Malia: Chapouthier and Demargne 1942, fig. 24
Myrtos: Warren 1972, 190, fig. 74:P 529
Prophetes Elias Tourtoulon: Platon 1960, pl. 239
Vathypetro: Platon 1974, pl. on p. 195; Marinatos and Hirmer 1976, pl. 62

330 (PAR 140, HN 14421; Fig. 51; Pl. 18B). Vat, complete profile. H. 33.1, d. of rim ca. 58–60, d. of base ca. 38–40 cm. Mirabello Fabric (light gray, 5YR 7/1). Slightly thickened, straight rim; conical shape; large drain above base. Dark band on rim, inside and out. From level 12, locus A604.43, object 350, residue 1369. *Date*: EM III.

331 (PAR 510, HN 14539; Fig. 51). Vat, spout sherd or part of a drain. Length of sherd 10.4 cm. Mirabello Fabric (pale brown, 10YR 6/3). Fragment from the large drain of a vat like **330**. Flat base with raised side. From level 16, locus A604.56. No object or residue number. *Date*: EM III.

332 (PAR 176, HN 14561; Fig. 51). Vat, spout sherd or part of a drain (see Ch. 9). Length of fragment 12.1 cm. Mirabello Fabric (weak red, 2.5YR 4/2, with the core reddish brown, 2.5YR 5/4). Section of the trough-like spout or drain. From level 12, locus A604.40, object 399, residue 1113. *Date*: EM III.

FLAT LIDS

Several flat disks that were probably lids were present in the deposit. None of them were complete. A minimum of two and a maximum of five lids were present. Flat lids are common from all Minoan periods. They usually have a single centrally placed pawn-shaped handle on the top, but examples without any handle and lids with loop handles are also known.

333 (PAR 506, HN 14541; Fig. 51). Lid, rim sherd. D. of rim 20 cm. Mirabello Fabric (yellowish red, 5YR 5/6, with the core pinkish gray, 5YR 6/2). Flat disk. From level 10, locus A604.20. No object or residue number. *Date*: EM III.

334 (PAR 507, HN 13538; Fig. 51). Lid, rim sherd. D. of rim 32 cm. Mirabello Fabric (light reddish brown, 5YR 6/3, with the core light gray, 5YR 7/1). Flat disk. From level 3, locus A604.5. No object or residue number. *Date*: EM III.

335 (PAR 466, HN 14555; Fig. 51). Lid, rim sherd. D. of rim ca. 28 cm. Mirabello Fabric (reddish yellow, 5YR 7/6). Flat disk. From level 13, locus A604.46. No object or residue number. *Date*: EM III.

336 (PAR 475, HN 13556; Fig. 51). Lid, rim sherd. D. of rim ca. 20 cm. Mirabello Fabric (reddish yellow, 5YR 7/6). Flat disk. Traces of slip on the edge and upper surface. From level 14, locus A604.53. No object or residue number. *Date*: EM III.

337 (PAR 451, HN 14557; Fig. 51). Lid, rim sherd. D. of rim ca. 24 cm. Mirabello Fabric (reddish yellow, 5YR 7/6). Flat disk. From level 13, locus A604.46. No object or residue number. *Date*: EM III.

338 (PAR 470, HN 14558; Fig. 51). Lid, rim sherd. D. of rim ca. 22 cm. Mirabello Fabric (reddish yellow, 5YR 7/6). Flat disk. From level 13, locus A604.47. No object or residue number. *Date*: EM III.

UNKNOWN SHAPES

339 (PAR 516, HN 14547; Fig. 51). Unknown vessel, possibly a bowl, base sherd. D. of base not measurable, max. dim. 7.2 cm. A fine fabric (light brown, 7.5YR 6/4). Basket impression on bottom of base. Dark slip on interior. From level 16, locus A604.56. No object or residue number. *Date*: EM III.

340 (PAR 431, HN 14813; Fig. 51). Unknown vessel, rim sherd. D. of rim ca. 54 cm. A gray coarse fabric (dark yellowish brown, 10YR 4/4). Slip (reddish yellow, 7.5YR 6/6) on exterior. From level 13, locus A604.46, object 140. No residue number. *Date*: EM III.

Closed Vessels

Many fragments are too small to attribute to a specific shape. Examples were cataloged for several reasons including as potential subjects for analysis.

341 (PAR 147, HN 14785; Fig. 51). Closed vessel, base sherd. D. of base ca. 12–14 cm. Mirabello Fabric (red, 2.5YR 4/6). From level 10, locus A604.19, object 276, residue 991. *Date*: EM III.

342 (PAR 58, HN 14601; Fig. 51). Closed vessel, base sherd. D. of base 21 cm. Mirabello Fabric (reddish brown, 2.5YR 5/4, with the core weak red, 2.5YR 4/2). From level 9, locus A604.17, object 217, residue 1107. *Date*: EM III.

343 (PAR 73, HN 14589; Fig. 51). Closed vessel, base sherd. D. of base 24 cm. Mirabello Fabric (light red, 2.5YR 6/6, with the core between light reddish brown, 5YR 6/3, and pink, 5YR 7/3). From levels 2–7, locus A601.5, object 125, residue 1013. *Date*: EM III.

344 (PAR 14, HN 14780; Fig. 51). Closed vessel, base sherd. D. of base 16 cm. Mirabello Fabric, with layers of color between interior and exterior: from interior, reddish brown (5YR 4/3), black, and reddish brown (5YR 4/3). From level 7, locus A604.11, object 185, residue 1182. *Date*: EM III.

345 (PAR 223, HN 14456; Fig. 51). Closed vessel, base sherd. D. of base ca. 16 cm. Mirabello Fabric (between light reddish brown, 5YR 6/4, and reddish yellow, 5YR 6/8). From level 10, locus A604.19, object 265, residue 1003. *Date*: EM III.

346 (PAR 383, HN 14817; Fig. 52). Closed vessel, base sherd. D. of base ca. 20.4 cm. Mirabello Fabric (brown, 7.5YR 5/4 to black). From level 10, locus A604.20, object 346, residue 1226. Petrography sample PAR 09/30. *Date*: EM III.

347 (PAR 29, HN 14744; Fig. 52; Pl. 13A). Closed vessel, base sherd. D. of base 13.8 cm. Mirabello Fabric (light reddish brown, 5YR 6/4). From levels 2–7, locus A601.5, object 123, residue 1006. *Date*: EM III.

348 (PAR 76, HN 14587; Fig. 52). Closed vessel, base sherd. D. of base 14 cm. Mirabello Fabric (red, 2.5YR 5/6, with the core reddish gray, 5YR 5/3). From level 3, locus A601.7, object 29, residue 964. *Date*: EM III.

349 (PAR 3, HN 14758; Fig. 52). Closed vessel, base sherd. D. of base 28 cm. A medium-coarse fabric (reddish yellow, 5YR 6/6). From level 6, locus A604.9, object 108, residue 1040. *Date*: EM III.

350 (PAR 8, HN 14783; Fig. 52). Closed vessel, base sherd. D. of base 19–20 cm. A pale fabric (reddish yellow, 5YR 7/6 to 6/6). From level 7, locus A604.11, object 188, residue 1183. *Date*: EM III.

351 (PAR 10, HN 14776; Fig. 52). Closed vessel, base sherd. D. of base 15–16 cm. Mirabello Fabric (pink, 5YR 7/3). From level 7, locus A604.11, object 187, residue 1026. *Date*: EM III.

352 (PAR 12, HN 14789; Fig. 52). Closed vessel, base sherd. D. of base 15 cm. Mirabello Fabric (between pink, 5YR 7/3, and light reddish brown, 5YR 6/3). From level 7, locus A604.11, object 189, residue 1038. *Date*: EM III.

353 (PAR 22, HN 14633; Fig. 52). Closed vessel, base sherd. D. of base 42 cm. Mirabello Fabric (light reddish brown, 5YR 6/4). Pronounced base. Dark band at base (or dark slip on exterior). From level 5, locus A604.6, object 82, residue 1035. *Date*: EM III.

354 (PAR 55, HN 14606; Fig. 52). Closed vessel, base sherd. D. of base ca. 32 cm. Mirabello Fabric (pink, 5YR 7/3). From level 15, locus A604.55, object 483, residue 1330. *Date*: EM III.

355 (PAR 32, HN 13612; Fig. 52). Closed vessel, base sherd. D. of base 17.4 cm. Mirabello Fabric (light brown, 7.5YR 6/4, with a dark core gray, 6/). Dark slip on exterior. Added white not preserved. From level 6, locus A604.9, object 103, residue 1174. *Date*: EM III.

356 (PAR 28, HN 14628; Fig. 52). Closed vessel, base sherd. D. of base 16 cm. Mirabello Fabric (light reddish brown, 5YR 6/4). Dark slip on exterior. Added white not preserved. From level 6, locus A604.9, object 106, residue 1016. *Date*: EM III.

357 (PAR 57, HN 14602; Fig. 52). Closed vessel, base sherd. D. of base 28 cm. Mirabello Fabric (pink, 7.5YR 7/4, with the core gray, N/0). From level 9, locus A604.17, object 210, residue 1313. *Date*: EM III.

358 (PAR 61, HN 14599; Fig. 52). Closed vessel, base sherd. D. of base 30–32 cm. Mirabello Fabric (between pink, 5YR 7/3, and reddish yellow, 5YR 7/8). From level 7, locus A604.11, object 184, residue 1077. *Date*: EM III.

359 (PAR 62, HN 14594; Fig. 52). Closed vessel, base sherd. D. of base 14 cm. Mirabello Fabric (pink, 5YR 7/4, with the core light reddish brown, 5YR 6/4). From level 4, locus A604.1, object 12, residue 971. *Date*: EM III.

360 (PAR 67, HN 14581; Fig. 52). Closed vessel, base sherd. D. of base 18–20 cm. Mirabello Fabric (pink, 7.5YR 7/4, with the core pinkish gray, 7.5YR 6/2). From levels 3–7, locus A604.3, object 229, residue 990. *Date*: EM III.

361 (PAR 68, HN 14580; Fig. 52). Closed vessel, base sherd. D. of base 18 cm. Mirabello Fabric (light reddish brown, 2.5YR 6/4). From levels 2–7, locus A601.5, object 124, residue 1004. *Date*: EM III.

362 (PAR 65, HN 14584; Fig. 52). Closed vessel, base sherd. D. of base 34 cm. Mirabello Fabric (pink, 5YR 7/3). From level 4, locus A604.1, object 20, residue 985. *Date*: EM III.

363 (PAR 70, HN 14591; Fig. 52). Closed vessel, base sherd. D. of base ca. 25–26 cm. Mirabello Fabric (light brown, 5YR 7/4). From level 3, locus A601.7, object 23, residue 955. *Date*: EM III.

364 (PAR 128, HN 14582; Fig. 52). Closed vessel, base sherd. D. of base ca. 22–23 cm. Mirabello Fabric (light brown, 7.5YR 6/4, with the core pinkish gray, 7.5YR 7/2). From level 8, locus A604.13, object 138, residue 1011. *Date*: EM III.

365 (PAR 84, HN 14727; Fig. 53). Closed vessel, base sherd. D. of base 13–14 cm. Mirabello Fabric (pink, 5YR 7/3, with the core light reddish brown, 5YR 6/3). From levels 2–7, locus A601.5, object 120, residue 1225. *Date*: EM III.

366 (PAR 111, HN 14773; Fig. 53). Closed vessel, base sherd. D. of base ca. 16 cm. Mirabello Fabric (reddish yellow, 5YR 7/6, with the second layer reddish yellow, 5YR 6/6, and the core pinkish gray, 5YR 6/2). From level 14, locus A604.49, object 456, residue 1102. *Date*: EM III.

367 (PAR 146, HN 14702; Fig. 53). Closed vessel, base sherd. D. of base 13–14 cm. Mirabello Fabric (reddish yellow, 7.5YR 6/6, with the core pale brown, 10YR 6/3). From level 11, locus A604.21, object 147, residue 1178. *Date*: EM III.

368 (PAR 155, HN 14697; Fig. 53). Closed vessel, base sherd. D. of base ca. 26 cm. Mirabello Fabric (reddish yellow, 5YR 6/6). From level 17, locus A604.58, object 525, residue 1128. *Date*: EM III.

369 (PAR 164, HN 14554; Fig. 53). Closed vessel, base sherd. D. of base ca. 20 cm. Mirabello Fabric (reddish yellow, 5YR 6/6). From level 11, locus A604.21, object 354, residue 1014. *Date*: EM III.

370 (PAR 118, HN 14718; Fig. 53). Closed vessel, base sherd. D. of base ca. 14–15 cm. Mirabello Fabric (pink, 5YR 7/4). From level 12, locus A604.29, object 407, residue 1168. *Date*: EM III.

371 (PAR 184, HN 14578; Fig. 53). Closed vessel, base sherd. D. of base ca. 22–23 cm. Mirabello Fabric (light reddish brown, 7.5YR 6/4). Dark vertical drip. From level 12, locus A604.27, object 360, residue 1177. *Date*: EM III.

372 (PAR 185, HN 14579; Fig. 53). Closed vessel, base sherd. D. of base ca. 13–14 cm. Mirabello Fabric (light reddish brown, 2.5YR 6/4). Slip on exterior, pinkish white, 7.5YR 8/2. From level 12, locus A604.27, object 361, residue 1044. *Date*: EM III.

373 (PAR 194, HN 14571; Fig. 53). Closed vessel, base sherd. D. of base ca. 25–26 cm. Mirabello Fabric (brown, 7.5YR 5/4). From level 12, locus A604.29, object 408, residue 1209. *Date*: EM III.

374 (PAR 205, HN 14442; Fig. 53). Closed vessel, base sherd. D. of base ca. 14 cm. Mirabello Fabric (pink, 5YR 7/4). From level 10, locus A604.20, object 342, residue 1163. *Date*: EM III.

375 (PAR 207, HN 14443; Fig. 53). Closed vessel, base sherd. D. of base ca. 6–7 cm. Mirabello Fabric (reddish yellow, 5YR 7/6). From level 16, locus A604.56, object 444, residue 1069. *Date*: EM III.

376 (PAR 213, HN 14778; Fig. 53). Closed vessel, base sherd. D. of base ca. 14–16 cm. Mirabello Fabric (between pink, 7.5YR 7/4, and light brown, 7.5YR 6/4, with the core light gray, 7.5YR N7/). Paint stroke near base. From level 10, locus A604.20, object 343, residue 1137. *Date*: EM III.

377 (PAR 225, HN 14454; Fig. 53). Closed vessel, base sherd. D. of base ca. 13–14 cm. Mirabello Fabric (light reddish brown, 5YR 6/3). From level 17, locus A604.59, object 495, residue 1089. *Date*: EM III.

378 (PAR 48, HN 14615; Fig. 53). Closed vessel, base sherd. D. of base 9–10 cm. Mirabello Fabric (pink, 5YR 7/3. From level 9, locus A604.17, object 220, residue 1080. *Date*: EM III.

379 (PAR 386, no HN number; Fig. 53). Closed vessel, base sherd. D. of base ca. 14 cm. Mirabello Fabric (reddish yellow, 5YR 7/6). From levels 2–7, locus A601.5, object 115, residue 1341. *Date*: EM III.

380 (PAR 437, HN 14756; Fig. 53; Pl. 38B). Closed vessel, base and body sherds. D. of base ca. 18.5–19 cm. Mirabello Fabric (red, 2.5YR 4/6, with the core dark gray, 2.5YR N4/). From levels 12, 18, 20, 21; loci A604.38, 62, 69; objects 389, 513, 530; residues 1110, 1253, 1265. *Date*: EM III.

381 (PAR 380, HN 14658; Fig. 53). Closed vessel, base sherd. D. of base ca. 14 cm. Mirabello Fabric (brown, 7.5YR 4/2 to light brown, 7.5YR 6/4). From levels 8, 10, 12, 13–15, 17, 18, 20; loci A604.12, 19, 40, 46, 53, 55, 59, 62, 65; objects 195, 382, 420, 481; residues 1139, 1362, 1367. *Date*: EM III.

382 (PAR 381, HN 14679; Fig. 54). Closed vessel, base sherd. D. of base ca. 14 cm. Mirabello Fabric (reddish yellow, 5YR 6/6). From levels 13, 15, 17, 21; loci A604.46, 47, 55, 59, 60, 67; objects 482, 528; residues 1347, 1352. *Date*: EM III.

383 (PAR 339, HN 14764; Fig. 54). Closed vessel, base sherd. D. of base ca. 16–18 cm. Mirabello Fabric (between reddish yellow, 5YR 6/6, and light reddish brown, 5YR 6/4). From levels 14, 21; loci A604.49, 67; objects 432, 461; residues 1118, 1210. *Date*: EM III.

384 (PAR 388, HN 14436; Fig. 54). Closed vessel, base sherd. D. of base ca. 15 cm. Mirabello Fabric (yellowish red, 5YR 5/6). From levels 11–12, locus A604.30, object 388, residue 1272. *Date*: EM III.

385 (PAR 77, HN 14586; Fig. 54). Closed vessel, body sherds. Max. dim. 10.2 cm. Mirabello Fabric (between light reddish brown, 2.5YR 6/4, and light red, 2.5YR 6/6). From level 5, locus A604.8, object 99, residue 1147. *Date*: EM III.

386 (PAR 78, HN 14585; Fig. 54). Closed vessel, body sherd. Max. dim. 10.2 cm. Mirabello Fabric (between light reddish brown, 2.5YR 6/4, and light red, 2.5YR 6/6). From levels 2–7, locus A601.5, object 122, residue 1288. *Date*: EM III.

387 (PAR 124, HN 14719; Fig. 54). Closed vessel, base sherd. D. of base 8 cm. Mirabello Fabric (reddish yellow, 5YR 5/6, with a dark core, dark gray, 5YR 4/1). From levels 5–7, locus A604.16, object 203, residue 1037. *Date*: EM III.

388 (PAR 127, HN 14714; Fig. 54; Pl. 22A). Closed vessel, base sherd. D. of base ca. 19–20 cm. Mirabello Fabric (red, 2.5YR 5/6, with a dark core, reddish brown, 2.5YR 5/4). From levels 5–7, locus A604.16, object 202, residue 994. *Date*: EM III.

389 (PAR 130, HN 14705; Fig. 54). Closed vessel, base sherd. D. of base ca. 25–26 cm. Mirabello Fabric (strong brown, 7.5YR 5/6, with the interior surface brown, 7.5YR 5/4, and the core pinkish gray, 7.5YR 5/2). From level 10, locus A604.19, object 266, residue 999. *Date*: EM III.

390 (PAR 162, HN 14553; Fig. 54). Closed vessel, base sherd. D. of base ca. 28–30 cm. Mirabello Fabric (reddish yellow, 5YR 6/6). Groove at base. From level 10, locus A604.19, object 225, residue 993. *Date*: EM III.

391 (PAR 403, HN 14659; Fig. 54). Closed vessel, base and lower part. D. of base 14.6 cm. Mirabello Fabric (red, 2.5YR 5/8). From levels 10, 13–16; loci A604.20, 46, 47, 53, 55, 56; objects 462, 466, 475; residues 1224, 1298, 1348. *Date*: EM III.

5

Chipped and Ground Stone Tools

by
Heidi M.C. Dierckx

Introduction

Unlike many other Minoan sites where the context of stone implements is uncertain and/or undatable except for "Minoan" in general, the rock shelter at Alatzomouri near Pacheia Ammos can be securely dated to one period, EM III. This situation provides a rare datable context for the stone implements. The stone tool assemblage from the rock shelter totals 109 tools, including 17 made of chipped stone, 75 ground stone tools, and 17 possible tools. From the cataloged implements, 21 ground stone tools were selected and sampled for phytoliths by Andrew Koh at the INSTAP Study Center as a part of his program of analysis for organic remains (Koh 2008). Unfortunately, the results were negative for any residue (A. Koh, pers. comm.), indicating that the ground stone implements from this site are not suitable for this kind of research analysis.

A summary of the tools found in the rock shelter is shown in Table 5.1. The distribution of tools shows a concentration in the lower(?) levels of the shelter (39 ground stone and 5 chipped stone from Strata 1–2 and 53 ground stone and 12 chipped stone from Stratum 3). Of note are the following loci (marked by * in the table):

A604.21: two pounders, one abrader, one grinder, and one working(?) slab

A604.46: one pounder-abrader, three querns (fragments), and one obsidian flake

A604.49: two pounder-abraders, one pestle, one chopper, two abraders, one polisher, and one quern (probably fragment of the same one from A604.46)

A604.56: two choppers, two grinders, one abrader, one mortar, one scraper, three possible tools, and one obsidian blade

A604.59: one pounder, three choppers, one abrader, one grinder, one quern, one weight, and one obsidian ridge blade

A604.67: two pounder-abraders, one abrader, one polisher, one obsidian core flake, and three initial flakes

Context	Ground Stone	Chipped Stone
A600.2	2	—
A601.1	1 + 1 PT	—
A604.1	0 + 1 PT	—
A604.2	1	—
A604.6	1 + 1 PT	—
A604.9	2	2
A604.10	1 + 1 PT	—
A604.11	2 + 2 PT	1
A604.12	1	—
A604.13	2	1
A604.16	1	—
A604.19	1	—
A604.20	2 + 1 PT	—
*A604.21	5	—
A604.22	1 + 1 PT	—
A604.28	3	—
A604.29	2	1
A604.40	2 + 1 PT	—
*A604.46	4	1
A604.47	1	—
*A604.49	8	—
A604.53	3	—
A604.55	2	—
*A604.56	7 + 3 PT	1
A604.57	1	—
A604.58	0	1
*A604.59	8	1
A604.61	3	—
A604.62	1 + 1 PT	1
A604.63	1	1
A604.65	1 + 2 PT	—
A604.66	0	2
*A604.67	4	4
A700.1	0 + 1 PT	—
A800.1.2	1 + 1 PT	—
Total	75 + 17 PT	17

Table 5.1. Number of stone implements by context from the rock shelter. PT = possible tool; * = context of note (see p. 69).

The catalog entries follow the systems used in the rest of the volume. Color follows the Munsell system.

The Chipped Stone

The chipped stone examined from the Alatzomouri Rock Shelter consisted of 17 obsidian pieces: six parallel-sided prismatic blades and 11 pieces of debitage, which included one retouched ridge blade, two retouched flakes, one chunk/flake, two initial flakes, four tertiary flakes, and one core flake. All belong to a typical Bronze Age chipped stone assemblage. The occurrence of both debitage, albeit a small amount, and finished products indicates that material was gathered from an area where knapping had taken place and was carried to the cave. Knapping in the cave is disproved by the complete excavation (including water sieving) that did not result in the complete record of knapping residue. The retouched debitage pieces suggest full use of the raw material, which was (most likely) imported from Melos.

At Melos, two quarries for the exploitation of obsidian have been located, Demenegaki and Sta Nychia (Carter and Kilikogolou 2007, 115; Carter 2008, 225–226). Macroscopically, the obsidian from each of the two quarries appear to be different in appearance. While the more translucent and lustrous obsidian comes from Demenegaki, the dark gray to black dense obsidian is found at Sta Nychia (Carter and Kilikoglou 2007, 119–122). Macroscopic analysis of the obsidian from the rock shelter appears to indicate that all finished products and retouched debitage pieces were made from obsidian coming from Sta Nychia. Only four tertiary flakes appear to have the translucent appearance of the Demenegaki source. Of course, X-ray flourescence analysis would be necessary to be certain. Carter does not mention a possible change of use of the quarries sometime during the Early Bronze I period from Demenegaki to Sta Nychia (V. Kilikoglou, pers. comm., in Carter 2008, 225). The obsidian from the rock shelter would support this idea. The majority of the pieces came from the lower levels of the cave: A604.56, 58, 59, 62, 63, 66, and 67.

Finished Products: Prismatic Blades

392 (PAR 281; Fig. 55). Blade, distal end with tip broken. Pres. length 2.61, max. w. 1.02, max. th. 0.26 cm. Obsidian (black, N1). Two ridges; marginal retouch on both edges of dorsal surface, including a notch on the right edge and concave curve on left edge. From level 6, locus A604.9, object 28. Parallel-sided prismatic blade.

393 (PAR 280; Fig. 55). Blade, medial and distal section with tip of distal end broken. Pres. length 1.16, max. w. 0.83, max. th. 0.23 cm. Obsidian (black, N1). Two ridges; retouched notch at lower left edge of ventral surface; chipped and worn edges from use. From level 7, locus A604.11, object 39. Parallel-sided prismatic blade.

394 (PAR 285; Fig. 55). Blade, distal end. Pres. length 2.95, max. w. 0.61, max. th. 0.15 cm. Obsidian (black, N1). Three ridges; retouched notch at upper right edge of ventral surface; serrated retouch and large notch on lower left edge of ventral surface; truncated at distal end creating a point. From level 8, locus A604.13, object 46. Parallel-sided prismatic blade.

395 (PAR 276; Fig. 55). Blade, medial section. Pres. length 1.87, max. w. 0.67, max. th. 0.17 cm. Obsidian (black, N1). Two ridges; chipped and worn edges from use. From level 12, locus A604.29, object 67. Parallel-sided prismatic blade.

396 (PAR 278; Fig. 55). Blade, proximal end. Pres. length 2.46, max. w. 0.85, max. th. 0.18 cm. Obsidian (black, N1). Two ridges; marginal retouch on left edge of ventral surface and upper left edge of dorsal surface. From level 16, locus A604.56, object 169. Parallel-sided prismatic blade.

397 (PAR 283; Fig. 55). Blade, medial section. Pres. length 2.5, max. w. 0.83, max. th. 0.2 cm. Obsidian (black, N1). Two ridges; chipped and worn edges from use. From level 16, locus A604.58, object 177. Parallel-sided prismatic blade.

Worked Debitage: Flakes and Blades

398 (PAR 282; Fig. 55). Flake, complete. Max. dim. 2.83 cm. Obsidian (black, N1). Marginal retouch on three edges on ventral surface creating rectangular shape. From level 6, locus A604.9, object 29. Scraper.

399 (PAR 287; Fig. 55). Blade, complete. Length 1.91, w. 0.98, th. 0.39 cm. Obsidian (black, N1). One ridge; cortex preserved on one side of ridge; retouch at both ends: the proximal end retouched on ventral surface to a point; retouched notch on upper right edge of dorsal surface. From level 17, locus A604.59, object 285. Worked ridge blade.

400 (PAR 277; Fig. 55). Flake, complete(?). Max. dim. 1.87 cm. Obsidian (black, N1). Some cortex preserved on dorsal surface; retouch at distal end creating a haft; serrated retouch on both edges of dorsal surface; broken(?) on proximal end. From level 21, locus A604.67, object 309. Worked initial flake.

Debitage

401 (PAR 284; Fig. 55). Flake, complete. Max. dim. 1.86 cm. Obsidian (grayish black and translucent, N2). Chipped edges. From level 13, locus A604.46, object 146.

402 (PAR 286; Fig. 55). Primary chunk/flake, complete. Max. dim. 2.46 cm. Obsidian (black, N1). Cortex preserved on one surface; small flake scars on platform at proximal end, which appears to be part of the platform preparation process of a core. From level 18, locus A604.62, object 292.

403 (PAR 288; Fig. 55). Flake, complete. Max. dim. 1.76 cm. Obsidian (black, N1). Some cortex preserved. From level 19, locus A604.63, object 293.

404 (PAR 290; Fig. 55). Flake, complete. Max. dim. 1.13 cm. Obsidian (grayish black and translucent, N2). From level 20, locus A604.66, object 303.

405 (PAR 291; Fig. 55). Flake, complete. Max. dim. 2.84 cm. Obsidian (grayish black and translucent, N2). From level 20, locus A604.66, object 303.

406 (PAR 275; Fig. 55). Flake, complete. Max. dim. 2.24 cm. Obsidian (black, N2). Some cortex preserved on dorsal surface. From level 21, locus A604.67, object 310. Initial flake.

407 (PAR 279; Fig. 55). Flake, complete. Max. dim. 2.19 cm. Obsidian (black, N1). Some cortex on dorsal surface. From level 21, locus A604.67, object 308. Initial flake.

408 (PAR 289; Fig. 55). Flake, complete. Max. dim. 2.49 cm. Obsidian (grayish black and translucent, N2). Some cortex preserved; distal end fragment of a core. From level 21, locus A604.67, object 306. Core flake.

The Ground Stone

The ground stone tool assemblage consists of the typical Bronze Age tools found elsewhere at Minoan sites, indicating domestic activity. The 75 implements entailed a variety of tool types, the majority being pounders, pounder-abraders, grinders, and querns. The raw materials used were primarily crystalline limestone cobbles with some examples of quartzite, metacarbonate (calcareous limestone with layered structure), sandstone, conglomerate, and pumice. One example each of meta-andesite (**456**) and andesite (**463**), which are common to northeastern Crete in the Siteia area, were also recovered. Unlike the other raw materials used that were available to the inhabitants nearby as cobbles on the beaches near Pacheia Ammos, metacarbonate and meta-andesite can be found in the Myrsini area, about 15–20 km to the east, and the green andesite came from a farther distance in the Siteia area (pers. obs.). The latter materials were likely brought to the rock shelter either because of aesthetics, in the case of the green color of the andesite, or for the abrasive properties of the rocks themselves, in the case of metacarbonate and meta-andesite.

The typology of the ground stone implements is largely based, with few variations, on Blitzer's ground stone typology from the site of Kommos in South-Central Crete (Blitzer 1995, 403–487). For easier comparison and consistency, this typology is also used in defining the typology of the nearby Minoan settlement of Vronda (Dierckx 2016). The two central attributes, which underlie the typology of the ground stone implements, are the wear marks (e.g., pecking, abrading, flaking, grooves) and/or the shape and size of the tools. These characteristics lead to a possible functional interpretation of the tool. Raw material may become a secondary variable creating a different type (in the case of Type 7, pumice abraders).

The tools from the rock shelter can be divided into 14 main groups or types. Each type is described below, including accession number and raw materials used. The following list of major ground stone tool types from the rock shelter includes the interpretative functions of the tools, some of which are used in the catalog description for identification, as well as the equivalent tool types found at Kommos (Blitzer 1995, 425–487) and Vronda (Dierckx 2016). Because Kommos provides the best up-to-date complete account for ground stone implements, this site is used as the primary reference for parallels for the Alatzomouri ground stone. For other recent bibliographic references to ground stone, see Evely 1984 (Knossos Unexplored Mansion), Evely 2003 (Knossos, the South House), Dierckx 1992 (Pseira), and Carter 2004 (Mochlos). In the catalogs below, the term "hand tool" is used as identification for Types 1, 2, 4–8, and 11.

Type 1. Implements with pecked or battered ends or circumference: pounders or hammer stones (Blitzer 1995, type 1; Dierckx 2016, type 1)

Type 2. Implements with pecked ends or circumference and abraded faces: pounder-abraders (Blitzer 1995, types 2, 7; Dierckx 2016, type 3)

Type 3. Implements with flaked end: choppers or hammers (Blitzer 1995, type 10; Dierckx 2016, type 4)

Type 4. Implements with pecked and abraded facets: facetted (Blitzer 1995, type 6; Dierckx 2016, type 5)

Type 5. Implements with abraded faces: abraders and/or grinders (Blitzer 1995, type 7; Dierckx 2016, type 6)

Type 6. Implements with one or two abraded faces: whetstones (Blitzer 1995, type 5; Dierckx 2016, type 7)

Type 7. Implements from pumice stone: abraders or polishers (Blitzer 1995, 509–510; Dierckx 2016, type 8)

Type 8. Implements with polished faces: polishers (Blitzer 1995, type 16C; Dierckx 2016, type 9)

Type 9. Pestles (Dierckx 2016, type 10)

Type 10. Weights (Blitzer 1995, type 12; Dierckx 2016, type 14)

Type 11. Other

Type 12. Querns (Blitzer 1995, type 17; Dierckx 2016, type 15)

Type 13. Mortars (Blitzer 1995, type 18; Dierckx 2016, type 16)

Type 14. Other large scale implements

Type 1: Pounders or Hammer Stones

Most commonly found at many Minoan sites, this type consists of 13 cobbles with pecked or battered ends and margins and/or center of faces and margins. Flakes and/or chipped secondary breakage are common. One quartzite cobble was used, and the remainder was of crystalline limestone.

409 (PAR 313, HN 14897; Fig. 56). Hand tool, intact. Length 6.86, w. 5.11, th. 4.4 cm; pres. wt. 198 g. Crystalline limestone (medium gray, N5). Small waterworn cobble, triangular-rounded shape; pecked on two corners; broken on remaining corner. From level 1, locus A600.2, object 1.

410 (PAR 312, HN 14872; Fig. 56). Hand tool, fragment. Max. dim. 8.14 cm; pres. wt. 109 g. Crystalline limestone (medium gray, N5). Fragment of waterworn cobble; pecked on center of preserved face. From level 3, locus A601.1, object 2.

411 (PAR 556, HN 18137; Fig. 56). Hand tool, complete. Length 6.73, w. 8.04, th. 5.99 cm; wt. 433 g. Crystalline limestone (medium light gray, N6). Triangular-rounded cobble; battered on one end; pecked on other end. From level 6, locus A604.9, object 30.

412 (PAR 302, HN 14846; Fig. 56). Hand tool, complete. Length 9.3, w. 8.28, th. 3.66 cm; wt. 433 g. Crystalline limestone (medium dark gray, N4). Waterworn cobble, rectangular-rounded and flat shape; flaked and battered on two adjacent margins from use. From level 6, locus A604.9, object 31.

413 (PAR 553, HN 18138; Fig. 56). Hand tool, intact. Pres. length 10.36, w. 8.33, th. 5.06 cm; pres. wt. 605 g. Crystalline limestone (medium dark gray, N4). Waterworn cobble, ovoid shape; pecked on both ends; broken on one end. From level 6, locus A604.10, object 33.

414 (PAR 328, HN 14860; Fig. 56). Hand tool, fragment. Pres. length 7.36, pres. w. 4.82, pres. th. 3.91 cm; pres. wt. 206 g. Crystalline limestone (medium dark gray, N4). Waterworn cobble, ovoid(?) shape; pecked central depression on one face. From level 7, locus A604.11, object 40.

415 (PAR 552, HN 18139; Fig. 56). Hand tool, half preserved. Pres. length 8.17, pres. w. 7.31, pres. th. 5.25 cm; pres. wt. 433 g. Crystalline limestone (very light gray, N8). Waterworn cobble, oblong shape; pecked heavily on preserved end; broken on other end. From level 10, locus A604.19, object 48.

416 (PAR 324, HN 14855; Fig. 56). Hand tool, complete. Length 7.54, w. 4.73, th. 3.67 cm; wt. 214 g. Crystalline limestone (medium gray, N5). Oblong cobble; pecked on small end. From level 10, locus A604.20, object 50.

417 (PAR 300, HN 14839; Fig. 56). Hand tool, intact. Pres. length 9.76, pres. w. 8.22, pres. th. 4.37 cm; pres. wt. 376 g. Quartzite (brownish gray, 5YR 4/1). Triangular-rounded cobble; pecked on one corner; pecked-flaked on one corner. From level 11, locus A604.21, object 52.

418 (PAR 547, HN 18140; Fig. 57). Hand tool, intact. Pres. length 10.4, w. 4.99, th. 3.39 cm; pres. wt. 304 g. Crystalline limestone (light gray, N7). Waterworn cobble, oblong shape; pecked on one end; broken on other end from use. From level 11, locus A604.21, object 62.

419 (PAR 337, HN 18141; Fig. 57). Hand tool, intact. Pres. length 7.19, w. 5.47, th. 3.35 cm; pres. wt. 204 g. Crystalline limestone (medium gray, N5). Irregularly shaped cobble; two pecked areas on one face; pecked at center of one margin; pecked-ground on one end; other end chipped. From locus A604.59, object 279.

420 (PAR 310, HN 14852; Fig. 57). Hand tool, intact. Length 10.64, pres. w. 8.49, pres. th. 4.64 cm; pres. wt. 626 g. Crystalline limestone (medium gray, N5). Waterworn cobble, ovoid shape; pecked on most of circumference and center of both faces; chipped along one margin. From level 17, locus A604.61, object 287.

421 (PAR 297, HN 14845; Fig. 57). Hand tool, complete. Length 9.27, w. 7.6, th. 3.35 cm; wt. 378 g. Crystalline limestone (medium gray, N5). Waterworn cobble, triangular-rounded shape; slightly pecked on small end, one margin, center of other margin, and center of both faces; unifacial flaking on large end with worn and chipped edge from use. From level 18, locus A604.62, object 291.

Type 2: Pounder-Abraders

This category consists of tools that combine pecked ends, corners, or circumferences and one or two abraded faces. Secondary flaking and pecking in the center of faces or margins can also be evident on some examples. Some faces are abraded smooth as if used as a polisher. These cobbles are equivalent to Blitzer's Type 2 and Type 7 "handstones." A total of 17 examples belong to this group. Although some of these tools were used in a percussive action, like Type 1 tools above, many were generally used as abraders, either for grinding (primarily) or polishing purposes based on size, shape, and degree of wear. One tool with a thin layer of light gray clay on one face may have been used as a burnisher (**435**). All but one cobble were of crystalline limestone. One quartzite cobble was used.

422 (PAR 326, HN 14869; Fig. 57). Hand tool, complete. Length 7.16, w. 5.18, th. 2.27 cm; wt. 135 g. Crystalline limestone (medium gray, N5). Small waterworn cobble, ovoid and flat shape; pecked on one end; one face abraded flat and smooth. From level 1, locus A600.2, object 3. Pounder-polisher; Blitzer 1995, type 7.

423 (PAR 548, HN 18142; Fig. 57). Hand tool, intact. Pres. length 9.89, w. 7.74, th. 5.56 cm; pres. wt. 697 g. Crystalline limestone (medium gray, N5). Waterworn cobble, oblong shape; pecked on one end; two faces abraded smooth; broken on one end. From level 12, locus A604.29, object 65. Blitzer 1995, type 7.

424 (PAR 296, HN 14870; Fig. 57). Hand tool, complete. Length 6.67, w. 5.43, th. 2.38 cm; wt. 137 g. Crystalline limestone (medium gray, N5). Small waterworn cobble, ovoid shape; pecked on both ends and one margin; one face abraded smooth; scratches visible on abraded face. From level 13, locus A604.46, object 143. Blitzer 1995, type 7.

425 (PAR 549, HN 18143; Fig. 57). Hand tool, complete. Length 10.15, w. 8.25, th. 4.24 cm; wt. 507 g. Crystalline limestone (light gray, N7). Waterworn cobble, ovoid shape; pecked slightly intermittently on circumference; one face abraded smooth and flat. From level 14, A604.49, object 143. Blitzer 1995, type 7.

426 (PAR 298, HN 14844; Fig. 57). Hand tool, half preserved. Pres. length 7.07, pres. w. 6.84, pres. th. 3.9 cm; pres. wt. 285 g. Quartzite (brownish gray, 5YR 4/1). Triangular-rounded cobble; pecked on two corners/ends; one face abraded flat. From locus A604.49, object 157. Blitzer 1995, type 7.

427 (PAR 554, HN 18144; Fig. 57). Hand tool, complete. Length 12.29, w. 9.94, th. 5.59 cm; wt. 889 g. Crystalline limestone (medium gray, N5). Waterworn cobble, ovoid shape; bifacial flaking on one end with worn edge from use; pecked on other end; one face abraded smooth; scratches visible on abraded face. From level 15, locus A604.55, object 163. Blitzer 1995, type 7

428 (PAR 322, HN 14850; Fig. 57). Hand tool, complete. Length 7.21, w. 6.07, th. 3.18 cm; wt. 186 g. Crystalline limestone (medium gray, N5). Small waterworn cobble, triangular-rounded shape; pecked on circumference and center of one face; other face abraded flat and smooth; unifacial flake on small end with worn edge from use. From level 16, locus A604.57, object 176. Blitzer 1995, type 7.

429 (PAR 551, HN 18145; Fig. 58). Hand tool, intact. Pres. length 10.16, w. 6.8, th. 3.48 cm; pres. wt. 345 g. Crystalline limestone (medium gray, N5). Waterworn cobble, oblong shape; pecked on small end; broken on large end from use; one face abraded; scratches visible on abraded face. From level 20, locus A604.65, object 300. Blitzer 1995, type 7.

430 (PAR 560, HN 18146; Fig. 58). Hand tool, intact. Pres. length 9.34, w. 9.08, th. 5.31 cm; pres. wt. 669 g. Crystalline limestone (medium gray, N5). Waterworn cobble, ovoid shape; pecked on preserved circumference; two faces abraded smooth; broken on one end. From level 17, locus A604.61, object 288. Blitzer 1995, type 7.

431 (PAR 314, HN 14876; Fig. 58). Hand tool, intact. Pres. length 6.57, w. 6.39, th. 3.46 cm; pres. wt. 231 g. Crystalline limestone (medium gray, N5). Small waterworn cobble, trapezoidal-rounded shape; pecked on small end; pecked and broken on large end; one face abraded smooth (concave). From level 21, locus A604.67, object 315. Pounder-polisher; Blitzer 1995, type 7.

432 (PAR 550, HN 18147; Fig. 58). Hand tool, intact. Pres. length 10.05, w. 10.2, th. 4.02 cm; pres. wt. 654 g. Crystalline limestone (medium gray, N5). Waterworn cobble, ovoid shape; pecked on one end; broken on other end; two faces abraded smooth and flat; scratches visible on abraded faces from use. From level 21, locus A604.67, object 305. Blitzer 1995, type 7.

433 (PAR 558, HN 18148; Fig. 58). Hand tool, half preserved. Pres. length 7, w. 3.7, pres. th. 2.05 cm; pres. wt. 82 g. Crystalline limestone (medium dark gray, N4). Waterworn cobble, rounded flat shape; pecked on preserved circumference; one face abraded smooth. From level 11, locus A604.21, object 63. Blitzer 1995, type 2.

434 (PAR 321, HN 14863; Fig. 58). Hand tool, fragment. Pres. length 5.87, pres. w. 5.61, pres. th. 3.92 cm; pres. wt. 204 g. Crystalline limestone (light gray, N7). Waterworn cobble; pecked on preserved circumference and center of one face; other face abraded smooth with scratches. From level 12, locus A604.40, object 75. Blitzer 1995, type 2.

435 (PAR 569, HN 18149; Fig. 58). Hand tool, intact. Pres. length 10.36, w. 7.27, th. 3 cm; pres. wt. 449 g. Crystalline limestone (medium gray, N5). Oblong waterworn cobble; pecked on circumference; two faces abraded, one face with traces of clay (light gray, N7). From level 14, ocus A604.49, object 151. Blitzer 1995, type 2.

436 (PAR 320, HN 14883; Fig. 58). Hand tool, half preserved. Pres. length 11.79, w. 6.43, th. 4.68 cm; pres. wt. 572 g. Crystalline limestone (medium gray, N5). Waterworn cobble, ovoid shape; pecked on preserved circumference; one face abraded smooth. From level 14, locus A604.49, object 156. Blitzer 1995, type 2.

437 (PAR 559, HN 18150; Fig. 58). Hand tool, intact. Pres. length 9.64, w. 9.39, th. 4.7 cm; pres. wt. 656 g. Crystalline limestone (medium gray, N5). Waterworn cobble, triangular-rounded shape; pecked on preserved circumference; one face abraded smooth; broken on one end. From level 17, locus A604.59, object 282. Blitzer 1995, type 2.

438 (PAR 561, HN 18151; Fig. 58). Hand tool, complete. Length 9.32, w. 9.28, th. 3.79 cm; wt. 474 g. Crystalline limestone (medium gray, N5). Waterworn cobble, triangular-rounded shape; pecked on preserved circumference; one face abraded; pecked on three corners. From level 19, locus A604.63, object 295. Blitzer 1995, type 2.

Type 3: Choppers or Hammers

The primary working edges of the six implements from the rock shelter, identified as choppers or hammers, consisted of flaked ends, both unifacial, bifacial, or multifacial flaking. Secondary wear marks consist of pecking on part of the circumference. The tools consisted mainly of crystalline limestone, but one example was metacarbonate. They were of various shapes and all cobble-sized.

439 (PAR 352, HN 14849; Fig. 59). Chopper, complete. Length 9.97, w. 10.53, th. 6.84 cm; wt. 962 g. Crystalline limestone (light olive gray, 5Y 6/1). Waterworn cobble, triangular shape; multi-facial flaking on small end with edge worn from use; flaked platform at opposite end. From level 14, locus A604.49, object 152.

440 (PAR 323, HN 14873; Fig. 59). Chopper, fragment. Max. dim. 9.57, pres. th. 2.42 cm; pres. wt. 280 g. Calcareous schist (greenish gray, 5GY 6/1 and 5G 6/1). Rounded and flat cobble; pecked center of preserved face and circumference; bifacial flaking at one end/margin with worn edge from use. From level 16, locus A604.56, object 165.

441 (PAR 301, HN 14864; Fig. 59). Chopper, half preserved. Pres. length 7.72, w. 8.39, th. 3.83 cm; pres. wt. 417 g. Crystalline limestone (medium gray, N5). Waterworn cobble, triangular-rounded(?) shape; unifacial flaking on preserved end with worn edge from use. From level 16, locus A604.56, object 172.

442 (PAR 308, HN 14859; Fig. 59). Chopper, fragment. Pres. length 5.77, pres. w. 8.71, pres. th. 3.76 cm; pres. wt. 274 g. Crystalline limestone (medium gray, N5). Waterworn cobble, ovoid shape; bifacial flaking on one end with worn edge from use. From level 17, locus A604.59, object 280.

443 (PAR 295, HN 14887; Fig. 59). Chopper, intact. Pres. length 8.53, w. 4.95, th. 3.49 cm; pres. wt. 215 g. Crystalline limestone (medium gray, N5). Waterworn cobble, oblong shape; bifacial flaking on one end with worn edge from use; pecked on one edge of broken end. From level 17, locus A604.59, object 281.

444 (PAR 330, HN 14900; Fig. 59). Chopper, complete. Length 6.3, w. 4.39, th. 3.7 cm; wt. 140 g. Crystalline limestone (medium gray, N5). Ovoid-irregular pebble; multifacial flaking on one end creating a small edge worn from use. From level 17, locus A604.59, object 284.

Type 4: Faceted Tool

One crystalline limestone facetted tool of irregular shape with battered and abraded facets was recovered from the rock shelter.

445 (PAR 555, HN 18152; Fig. 59). Facetted hand tool, complete. Length 6.78, w. 7.11, th. 6.89 cm; wt. 442 g. Crystalline limestone (dark medium to medium gray, N4/N5). Waterworn cobble, irregular shape; two opposite facets/ends abraded flat; battered on most of remaining surface. From level 5, locus A604.6, object 24.

Type 5: Abraders and/or Grinders

This category of tools contains eight implements with their main working surface consisting of one or two abraded faces. Of these, three tools, two crystalline limestone and one sandstone example, can generally be considered abraders. The remainder are recognized as grinders based on raw material, size, shape, and degree of wear. One each of quartzite, quartz sandstone, and metacarbonate and two made of shelly limestone were represented.

446 (PAR 317, HN 14840; Fig. 59). Hand tool, fragment. Pres. length 8.79, pres. w. 4.83, pres. th. 2.66 cm; pres. wt. 195 g. Crystalline limestone (medium gray, N5). Waterworn cobble, rectangular shape; one face abraded flat and smooth. From level 12, locus A604.28, object 70. Abrader.

447 (PAR 311, HN 14843; Fig. 59). Hand tool, fragment. Pres. length 10.11, pres. w. 4.3, pres. th. 3.61 cm; pres. wt. 231 g. Crystalline limestone (medium gray, N5). Waterworn cobble, ovoid shape; one face abraded smooth. From level 16, locus A604.56, object 120. Abrader.

448 (PAR 299, HN 14842; Fig. 60). Hand tool, complete. Length 7.41, w. 4.83, th. 2.8 cm; wt. 163 g. Sandstone (very pale orange, 10YR 8/2, medium-grained). Rectangular-rounded cobble; one face abraded smooth. From level 21, locus A604.67, object 312. Abrader.

449 (PAR 354, HN 14878; Fig. 60). Hand tool, intact. Length 18.97, w. 13.63, pres. th. 5.3 cm; pres. wt. 1761 g. Calcareous schist (light olive gray, 5Y 6/1). Large waterworn cobble, ovoid flat shape; pecked on preserved circumference; possibly one face abraded flat; broken along same face. From level 5, locus A604.16, object 44. Grinder.

450 (PAR 353, HN 14880; Fig. 60). Hand tool, complete. Length 8.72, w. 11.23, th. 10.11 cm; wt. 1414 g. Shelly limestone (yellowish gray, 5YR 8/1). Triangular-rounded cobble; three faces abraded with pecking on the edges/margins of the abraded faces. From level 11, locus A604.21, object 59. Grinder.

451 (PAR 563, HN 18153; Fig. 60) Hand tool, intact. Pres. length 13, w. 7.91, th. 5.27 cm; pres. wt. 681 g. Quartzite (brownish gray, 5YR 4/1). Oblong cobble; slightly pecked on one end; one face abraded; chipped on one corner. From level 16, locus A604.56, object 166. Grinder.

452 (PAR 327, HN 14862; Fig. 60). Hand tool, half preserved. Pres. length 12.5, w. 5.78, th. 4.32 cm; pres. wt. 344 g. Shelly limestone (yellowish gray, 5Y 8/1). Rounded(?) cobble; pecked on preserved circumference; one face abraded. From level 16, locus A604.56, object 167. Grinder.

453 (PAR 335, HN 14884; Fig. 60). Hand tool, complete. Length 10.32, w. 9.6, th. 5.52 cm; wt. 781 g. Quartz sandstone (medium grained and dense; medium dark gray, N4, to olive gray, 5Y 4/1). Rounded cobble; one face abraded. From level 17, locus A604.59, object 278. Grinder.

Type 6: Whetstones

Similar to Type 5 (above), this type of tool also has one or two abraded faces, but it can be distinguished by additional wear marks of grooves as well as the shape, size, and raw material used. The four cobbles are primarily oblong or rounded with a flat section. Raw materials used were metamorphic rocks such as quartzite, metacarbonate, and meta-andesite. Cobbles of these raw materials were likely found to the east of Pacheia Ammos—as part of the Phyllite-Quartzite geological nappe in the Myrsini area—and were chosen because of their abrasive properties.

454 (PAR 307, HN 14847; Fig. 60). Hand tool, complete. Length 6.93, w. 6.47, th. 1.1 cm; wt. 95 g. Quartzite (brownish gray, 5YR 4/1). Small cobble, rounded flat shape; one face abraded smooth (slightly concave); grooves on abraded face; chipped on edge. From level 7, locus A604.11, object 41.

455 (PAR 319, HN 14901; Fig. 60). Hand tool, fragment. Pres. length 6.68, pres. w. 4.65, pres. th. 2.06 cm; pres. wt. 98 gr. Calcareous schist (light bluish gray, 5B 7/1). Oblong cobble; pecked on preserved end; abraded flat one face. From level 8, locus A604.12, object 37.

456 (PAR 309, HN 14871; Fig. 61). Hand tool, fragment. Pres. length 6.94, w. 6.78, th. 2.45 cm; pres. wt. 134 g. Meta-andesite (olive gray, 5Y 4/1 [interior], and grayish red, 10R 4/1 [exterior]). Oblong(?) cobble; one face abraded smooth. From level 12, locus A604.29, object 68.

457 (PAR 316, HN 14853; Fig. 61). Hand tool, complete. Length 6.07, w. 2.54, th. 0.88 cm; wt. 23 g. Calcareous schist (olive gray, 5Y 4/1 [interior], and dark greenish gray, 5GY 4/1). Oblong flat pebble; one face abraded smooth (slightly concave surface). From locus A800.2, object 3.

Type 7: Pumice Abrader/Polishers

In addition to two relatively large pieces of pumice that were used as abraders, which can be evidenced by their abraded flat facets, several pieces of unworked pumice also came from the rock shelter.

458 (PAR 416, HN 18154; Fig. 61). Hand tool, complete. Length 7.32, w. 7.02, th. 4.14 cm; wt. 37 g. Pumice (light gray, N7 [core], and very pale orange, 10YR 8/2 [exterior]). Irregular cobble; one face abraded flat. From locus A604.13, object 42.

459 (PAR 417, HN 18155; Fig. 61). Hand tool, complete. Length 12.85, w. 9.24, th. 8.24 cm; wt. 173 g. Pumice (light gray, N7 [core], and very pale orange, 10YR 8/2 [exterior]). Triangular cobble; one face abraded flat; large V-shaped groove on one face (natural?). From level 17, locus A604.61, object 286.

Type 8: Polishers

This category consists of four tools with abraded smooth faces, two of which had scratches visible on the surface. They are generally flattish in section, but they have a variety of shapes. Two crystalline limestone and one calcite cobble were used. In addition, and of note, is one green andesite pebble with two polished smooth facets. This tool type can be compared to Blitzer 1995, type 16C.

460 (PAR 315, HN 14861; Fig. 61) Hand tool, complete. Length 5.67, w. 3.73, th. 1.09 cm; wt. 35 g. Crystalline limestone (medium dark gray, N4). Oblong waterworn cobble; one face abraded smooth (slightly concave) with scratches visible. From level 8, locus A604.13, object 45.

461 (PAR 566, HN 18156; Fig. 61). Hand tool, complete. Length 9.27, w. 8.55, th. 4.95 cm; wt. 503 g. Calcite (moderate to dark yellowish brown, 10YR 4/2–5/4). Waterworn cobble, rounded shape; one face polished (glossy) with scratches visible from use. From level 14, locus A604.49, object 153.

462 (PAR 325, HN 14866; Fig. 61). Hand tool, complete. Length 6.9, w. 5.73, th. 2.78 cm; wt. 160 g. Crystalline limestone (medium gray, N5). Small waterworn cobble, ovoid flat shape; pecked slightly on circumference; one face abraded flat and smooth. From level 14, locus A604.53, object 161.

463 (PAR 338, HN 14856; Fig. 61). Hand tool, intact. Pres. length 4.4, w. 3.65, th. 3.38 cm; pres. wt. 91 g. Andesite (dark greenish gray, 5GY 4/1). Square-rounded pebble; two facets abraded smooth and flat. From level 21, locus A604.67, object 313.

Type 9: Pestles

The two tools, one of crystalline limestone and one of limestone conglomerate, which belong to this group, are distinguished by their shape and placement of wear. Their oblong shape and pecking on the small end suggest their function as pestles.

464 (PAR 318, HN 14858; Fig. 61). Hand tool, fragment. Pres. length 5.86, pres. w. 4.82, pres. th. 3.63 cm; pres. wt. 156 g. Crystalline limestone (medium gray, N5). Waterworn cobble, oblong shape; pecked and flaked on one margin; preserved end pecked flat. From level 14, locus A604.49, object 155.

465 (PAR 333, HN 14865; Fig. 61). Hand tool, complete. Length 9.35, w. 4.43, th. 4.05 cm; wt. 183 g. Limestone conglomerate (very light gray, N8). Waterworn cobble, oblong shape with triangular section; pecked on small end. From level 14, locus A604.53, object 159.

Type 10: Other Tools

Two unusual tools, a piercer of metacarbonate and a quartzite scraper, form part of the assemblage of ground stone implements.

466 (PAR 303, HN 14851; Fig. 61). Hand tool, complete. Length 10.24, w. 7.72, th. 2.5 cm; wt. 301 g. Quartzite (brownish gray, 5YR 4/1). Trapezoidal cobble; bifacial flaking on two adjacent margins and one corner with worn edges from use. From level 16, locus A604.56, object 171. Scraper.

467 (PAR 334, HN 14874; Fig. 61). Piercer, complete. Length 9.19, w. 6.48, th. 1.29 cm; wt. 129 g. Calcareous schist (greenish gray, 5GY 6/1). Triangular-rounded flat cobble; bifacial flaking on one corner creating a point. From level 14, locus A604.53, object 158.

Type 11: Weights

Three cobbles, two of crystalline limestone and one of sandstone, with a drilled or pecked hole in the upper portion of the stone can be considered weights.

468 (PAR 292, HN 14854; Fig. 61). Weight, half preserved. Pres. length 6.9, pres. w. 4.65, pres. th. 2.25 cm; pres. wt. 84 g. Crystalline limestone (medium gray, N5). Small waterworn cobble, ovoid shape; drilled hole partially preserved near edge in upper portion (d. of hole 1.14 cm). From level 3, locus A604.2, object 4.

469 (PAR 293, HN 14867; Fig. 61). Weight, complete. Length 7.29, w. 6.68, th. 2.45 cm; wt. 149 g. Crystalline limestone (very light gray, N8). Waterworn cobble, trapezoidal-rounded flattish shape; pecked hole near edge in upper portion. From level 12, locus A604.40, object 72.

470 (PAR 294, HN 14848; Fig. 62). Weight, complete. Length 8.96, w. 6.44, th. 3.51 cm; wt. 227 g. Sandstone (light brown, 5YR 5/6), fine grained. Irregularly shaped cobble; drilled hole near edge in upper portion (d. of hole 1.14 cm). From level 17, locus A604.59, object 277.

Type 12: Querns

A dozen quern fragments, mostly preserved as small pieces, were recovered. One of these was a fragment of a sandstone saddle quern. Five fragments of conglomerate probably are from two separate querns. Limestone, quartzite, conglomerate, and sandstone were represented.

471 (PAR 345, HN 14868; Fig. 62). Quern, fragment. Max. dim. 10.12, pres. th. 3.64 cm; pres. wt. 238 g. Conglomerate granular in white matrix (subrounded and poorly sorted). Slab; fragment of the saddle and some of working surface. From level 10, locus A604.20, object 49. Probably came from the same quern as **475** and **479**.

472 (PAR 562, HN 18157; Fig. 62). Quern, fragment. Pres. length 10.66, pres. w. 11.22, pres. th. 4.17 cm; pres. wt. 776 g. Quartzite (medium dark gray, N4, coarse-grained). Slab, wedge-shaped fragment; upper working surface abraded smooth; pecked-flaked to shape. From level 11, locus A604.22, object 58.

473 (PAR 351, HN 14879; Fig. 62). Quern, fragment. Pres. length 18.5, pres. w. 8.8, pres. th. 7.15 cm; pres. wt. 1,940 g. Calcareous schist (medium light gray, N6). Slab, oblong(?) shape; working surface abraded flat. From level 12, locus A604.28, object 64.

474 (PAR 349, HN 14882). Quern, fragment. Max. dim. 11.56, pres. th. 5.95 cm; pres. wt. 1,028 g. Conglomerate, granular-pebble with white matrix (subrounded and poorly sorted). Slab; flat working surface. From level 12, locus A604.28, object 71. Probably came from the same quern as **476**.

475 (PAR 347, HN 14886; Fig. 62). Quern, fragment. Max. dim. 5.79, pres. th. 2.59 cm; pres. wt. 73 g. Conglomerate, granular in white matrix (subrounded and poorly sorted). Slab; working surface abraded. From level 13, locus A604.46, object 141. Probably came from same quern as **471** and **479**.

476 (PAR 348, HN 14885). Quern, fragment. Max. dim. 12.26, pres. th. 5.15 cm; pres. wt. 488 g. Conglomerate, granular-pebble with white matrix (subrounded and poorly sorted). Slab; flat working surface. From level 13, locus A604.46, object 141. Probably came from same quern as **474**.

477 (PAR 567, HN 18158; Fig. 62). Saddle quern, half preserved. Pres. length 17, w. 23, th. 7.4 cm; pres. wt. 4,675 g. Sandstone (yellowish gray, 5Y 7/2, coarse-grained, poorly sorted). Slab, ovoid shape; one saddle-ridge preserved; upper working surface abraded (concave); pecked-flaked to shape. From level 13, locus A604.47, object 148.

478 (PAR 329, HN 14881; Fig. 62). Quern, fragment. Pres. length 10.06, pres. w. 12.63, pres. th. 3.44 cm; pres. wt. 772 g. Crystalline limestone (medium gray, N5). Large cobble, rectangular(?) shape; concave pecked-abraded working surface. From level 13, locus A604.46, object 142.

479 (PAR 344, HN 14857; Fig. 62). Quern, fragment. Max. dim. 9.23, pres. th. 3.44 cm; pres. wt. 285 g. Conglomerate, granular in white matrix (subrounded and poorly sorted). Slab; working surface abraded. From level 14, locus A604.49, object 149. Probably came from same quern as **471** and **475**.

480 (PAR 565, HN 18159). Quern, fragment. Pres. length 11.27, pres. w. 8.97, pres. th. 7.95 cm; pres. wt. 1,331 g. Crystalline limestone (medium light gray, N6). Slab, wedge-shaped fragment; flat working surface; pecked edge. From level 15, locus A604.55, object 165.

481 (PAR 350, HN 14838; Fig. 62). Quern, fragment. Pres. length 4.87, w. 19.38, th. 10.58 cm; pres. wt. 1,910 g. Quartzite (brownish gray, 5YR 4/1). Slab; preserved working surface abraded flat. From level 17, locus A604.59, object 283.

Type 13: Mortar

One mortar fragment of conglomerate was found.

482 (PAR 564, HN 18160; Fig. 63). Mortar, fragment. Pres. length 12.64, pres. w. 9.88, pres. th. 6.4 cm; pres. wt. 935 g. Conglomerate (granular-pebble with white matrix, subrounded and poorly sorted). Slab; wedge-shaped fragment, rounded(?) shape with shallow round depression in center (d. ca. 8 cm); rim abraded flat (ca. 4 cm wide). From level 16, locus A604.56, object 54.

Type 14: Other Large-Scale Implements

483 (PAR 568, HN 18161; Fig. 63). Working(?) slab, complete. Length 20, w. 17, th. 6.5 cm; wt. 2,963 g. Sandstone (dusky yellow, 5Y 6/4; medium coarse, poorly sorted). Large cobble, triangular-rounded shape; pecked on one surface. From level 11, locus A604.21, object 56.

Possible Tools

These 17 cobbles consist of complete and fragmentary waterworn stones with no signs of use or wear. The fact that many were broken could be a sign of having been used at some point in time.

484 (PAR 369, HN 14894). Possible hand tool, fragment. Pres. length 6.57, pres. w. 4.2, pres. th. 2.75 cm; pres. wt. 123 g. Crystalline limestone (medium gray, N5). Waterworn and weathered cobble, ovoid shape; no visible signs of use wear. From locus A700.1, object 1.

485 (PAR 373, HN 14875). Possible hand tool, fragment. Pres. length 7.09, pres. w. 6.16, pres. th. 1.54 cm; pres. wt. 126 g. Calcareous schist (light olive gray, 5Y 6/1). Oblong cobble; no visible signs of use wear. From locus A800.1, object 2.

486 (PAR 371, HN 14904). Possible hand tool, fragment. Max. dim. 6.65 cm; pres. wt. 76 g. Crystalline limestone (medium gray, N5). Fragment of large flake of waterworn cobble; no visible signs of use wear. From level 2, locus A601.1, object 1.

487 (PAR 363, HN 14896). Possible hand tool, fragment. Max. dim. 6.02 cm; pres. wt. 47 g. Crystalline limestone (medium gray, N5). Fragment of large flake of waterworn cobble; no visible signs of use wear. From level 14, locus A604.1, object 3.

488 (PAR 372, HN 14877). Possible hand tool, fragment. Max. dim. 9.46 cm; pres. wt. 171 g. Crystalline limestone (medium gray, N5). Fragment of large flake of waterworn cobble; no visible signs of use wear. From level 5, locus A604.6, object 27.

489 (PAR 362B, HN 14888). Possible hand tool, fragment. Max. dim. 10.53 cm; pres. wt. 372 g. Crystalline limestone (medium gray, N5). Fragment of large flake of waterworn cobble; no visible signs of use wear. From level 6, locus A604.10, object 34.

490 (PAR 374, HN 14892) Possible hand tool, fragment. Max. dim. 10.69 cm; pres. wt. 209 g. Crystalline limestone (medium gray, N5). Waterworn cobble, ovoid or oblong shape; no visible signs of use wear. From level 7, locus A604.11, object 36.

491 (PAR 376, HN 14893). Possible hand tool, fragment. Pres. length 5.59, pres. w. 5.76, pres. th. 3.21 cm; pres. wt. 142 g. Crystalline limestone (dark medium gray, N4). Waterworn cobble; no visible signs of use wear. From level 7, locus A604.11, object 38.

492 (PAR 375, HN 14889). Possible hand tool, complete. Length 6.02, w. 5.68, th. 3.43 cm; wt. 170 g. Crystalline limestone (medium gray, N5). Waterworn and weathered pebble, triangular-rounded shape; possible abraded flat large end; possible pecked small end. From level 10, locus A604.20, object 51.

493 (PAR 355, HN 14841). Possible hand tool, fragment. Pres. length 11.4, w. 12.58, th. 5.05 cm; pres. wt. 891 g. Crystalline limestone (medium gray, N5). Large waterworn cobble; ovoid flattish shape; possible one face abraded flat; broken at ends and along possible abraded face. From level 11, locus A604.22, object 55. Possible grinder.

494 (PAR 370, HN 14890). Possible hand tool, complete; length 6.37, w. 4.63, th. 2.93 cm; wt. 130 g. Crystalline limestone (medium gray, N5). Waterworn pebble, ovoid shape; no visible signs of use wear. From level 12, locus A604.40, object 74.

495 (PAR 364, HN 14898). Possible hand tool, fragment. Max. dim. 10.2 cm; pres. wt. 195 g. Crystalline limestone (medium gray, N5). Waterworn cobble; no visible signs of use wear. From level 16, locus A604.56, object 175.

496 (PAR 367, HN 14902). Possible hand tool, complete. Length 5.67, w. 5.43, th. 4.25 cm; wt. 188 g. Crystalline limestone (medium light gray, N6). Waterworn pebble, triangular-rounded shape; no visible signs of use wear. From level 16, locus A604.56, object 173.

497 (PAR 368, HN 14895). Possible hand tool, fragment. Pres. length 6.78, w. 4.48, th. 3.02 cm; pres. wt. 123 g. Crystalline limestone (medium light gray, N6). Waterworn cobble; no visible signs of use wear. From level 16, locus A604.56, object 174.

498 (PAR 365, HN 14903). Possible hand tool, fragment. Pres. length 10.04, w. 3.1, th. 3.62 cm; pres. wt. 151 g. Crystalline limestone (medium gray, N5). Waterworn cobble, ovoid shape; possible one face abraded smooth. From level 16, locus A604.62, object 289.

499 (PAR 557, HN 18162). Possible hand tool, half preserved. Pres. length 6.64, w. 6.45, th. 2.17 cm; pres. wt. 151 g. Crystalline limestone (medium dark gray, N4). Waterworn and weathered cobble, trapezoidal flat shape; broken on one end and two corners; no visible signs of use wear. From level 20, locus A604.65, object 298.

500 (PAR 366, HN 14899). Possible hand tool, complete. Length 6.62, w. 5.63, th. 3.27 cm; wt. 183 g. Crystalline limestone (medium light gray, N6). Small waterworn cobble, triangular-rounded shape; no visible signs of use wear. From level 20, locus A604.65, object 299.

Conclusions

As mentioned above, the types of ground stone implements from the rock shelter are typical tools found in domestic contexts at other Minoan settlement sites. Of these tools, 55% were fragmentary, either broken on the ends or the margins. Even if one examined the tools from the lowest levels (Stratum 3), only 50% of those tools were fragmentary. The breaks, however, appear not to have been of deliberate nature. In fact, in many cases it appears the tools broke from being used. Breakage of ground stone tools is common at many Minoan settlement sites, so it is problematic to use the evidence of the ground stone implements in interpreting the function of the rock shelter.

6

Textile Tools

by

Thomas M. Brogan and Joanne E. Cutler

Eleven tools for textile production were recovered from different parts of the rock shelter. They include eight discoid loomweights and one cuboid loomweight for use on the vertical warp-weighted loom, a spindle whorl, and a clay vessel that has been identified as a spinning bowl. Together they form an important addition to the small corpus of Cretan textile tools from the Early Bronze Age.

The loomweights were analyzed using the methodology established by the Center for Textile Research (CTR) in Copenhagen (Andersson Strand and Nosch, eds., 2015), which estimates the most suitable thread types for use with the various forms and sizes of weights used on a warp-weighted loom. Loomweights apply tension to the warp threads, which hang from the upper beam of the loom. The weight and the thickness of a loomweight both have an effect on the type of cloth that is made with it. The weight governs how many warp threads needing a particular tension can be attached to it; weavers generally consider that 10 and 30 threads are the practical lower and upper limits for the number of threads that can be fastened to a single loomweight. The loomweight's thickness affects how closely the warp threads will be spaced in the finished fabric: the maximum number of warp threads per centimeter using thread of a given tension is estimated by multiplying the number of threads per loomweight by the number of rows of loomweights used; the product is divided by the thickness of the loomweight (in cm).

The eight discoid weights weigh between 69 g and 195 g, with thicknesses ranging from 1.2 cm to 2.4 cm (Fig. 64). The cuboid loomweight is both heavier and thicker than the discoid weights.

The discoid loomweights would be best suited for use with very thin thread needing ca. 5–10 g of tension (Table 6.1). All except the heaviest weight, which is 195 g, would work well with thread needing 5 g of tension; however, in a tabby weave (the most basic weave, in which the horizontal weft thread passes alternatively over one warp thread and under the next) using two rows of loomweights, the resulting thread count would be ca. 17–32 threads per cm, which is a very large range

Type	Weight (g)	Thickness (cm)	Thread Count per cm with Various Tensions				
			5 g	7.5 g	10 g	15 g	20 g
Discoid	195	24	—	22*	17**	11	8
	119	18	27	18*	13**	—	—
	90	18	20*	13**	—	—	—
	101	20	20*	13**	10	—	—
	135	17	32	21*	16**	—	—
	112	15	29	20*	15**	—	—
	80	19	17*	12**	—	—	—
	69	12	23*	15**	—	—	—
Cuboid	224	36	—	17	12	8	6

Table 6.1. Early Minoan III discoid and cuboid loomweights from Alatzomouri Rock Shelter: weight and thickness and the resulting thread count per cm when used with threads requiring different tensions in a tabby weave. *Warp thread count per cm for thread needing tension of 5–7.5 g. **Warp thread count per cm for thread needing tension of 7.5–10 g (original weights estimated for the incomplete weights, based on the percentage preserved).

and would not be optimal. Alternatively, all of the discoid loomweights could be used with thread needing ca. 7.5 g of tension, but the resulting warp thread count range of ca. 12–22 threads per cm is also large. Thread needing slightly different amounts of tension can be used in the same loom setup (Andersson Strand and Nosch, eds., 2015), and this group of loomweights would thus work best with thread needing tensions of ca. 5–7.5 g or 7.5–10 g. With a tension range of 5–7.5 g, the discoid weights could be used to weave a textile with ca. 17–23 warp threads per cm (Table 6.1); with a tension range of 7.5–10 g, the thread count range would be ca. 12–17 threads per cm (Table 6.1). The fabrics made with the loomweights could be relatively balanced (i.e., with approximately the same number and thickness of weft threads and warp threads per cm^2). In a tabby setup, the eight discoid loomweights would only be sufficient to weave a very narrow fabric, ca. 7 cm wide, which makes it likely that the loomweights were originally part of a much larger group.

With thread needing ca. 7.5 g of tension, the single cuboid weight found with the group could produce a fabric with ca. 17 warp threads per cm (Table 6.1). In spite of the difference in type, it therefore would be possible to use the cuboid weight in the same loom setup with the discoid weights with thread needing ca. 5–7.5 g or 7.5–10 g of tension (although this would require 30 threads tied to it, which is the upper limit of what could be considered practical). The difference in shape would, however, suggest that the cuboid weight was not originally intended for use with the other weights even if it was added to the set later. It thus appears likely that the loomweights recovered from the rock shelter were taken from a larger collection of weights used elsewhere.

One spindle whorl was recovered from the rock shelter, and it appears similar in shape and size to the group of 28 spindle whorls recovered in the EM IIB levels at Myrtos Phournou Koriphi (Warren 1972, 228–229). Generally, spindle whorls, which are used on a spindle to add momentum when spinning thread, are far less common than loomweights; only rare examples have been reported from the Middle and Late Minoan periods, and during this time some whorls may have been made of perishable materials such as wood, or an alternative method of spinning may have been used. Spindle whorl **510** weighs 43 g, and on the basis of experiments by CTR would have been optimal for spinning thicker threads needing ca. 40 g of tension. This point is interesting because threads of 40 g of tension would not appear to be a good fit for use with loomweights from the rock shelter (with warp threads suitable for 5–10 g of tension), but they could have been used to spin thicker weft threads (or for warp threads meant for use with other loomweights or possibly even for use on another type of loom).

The final tool from the rock shelter is a spinning bowl (or fiber-wetting bowl; **121**; Fig. 33; for the catalog entry, see Ch. 4), an object that is commonly depicted in Egyptian wall paintings and which is associated with an alternative method of producing linen thread (Dothan 1963, 105; Barber 1991, 45, 48, figs. 2.5, 2.10; Cutler 2011). This tool typically consists of a shallow bowl with one or more interior handles. A number of Middle Kingdom tomb paintings and models, as well as a New Kingdom scene, show women twisting together lengths of overlapping flax fibers to form a continuous thread—a technique known as "splicing." The spliced thread is then wound into balls or coils, and single threads are either given extra twist to strengthen them or are plied together to make a thicker thread, using a top-whorl suspended spindle that is set in motion by rolling it along the thigh (see Barber 1991, 44–48; Kemp and Vogelsang-Eastwood 2001, 68–77). The balls or coils of spliced thread are drawn through a bowl with an internal loop handle; these "spinning bowls" likely contained water because bast fibers are easier to work and hold a twist better when wet (Barber 1991, 70–73; Kemp and Vogelsang-Eastwood 2001, 74, 76). Similar bowls with an interior handle have been recovered from EM II contexts at Myrtos Phournou Koriphi (Warren 1972, 78, 153, 209) and from Middle Bronze Age contexts at Phaistos, Palaikastro, Kommos, and Archanes, as well as in a tomb at Drakones (Barber 1991, 74; the function of similar bowls at Malia has been debated; see Poursat 1984). A fragment of a spinning bowl of uncertain date was also recovered from the South House at Knossos (Evely 2003, 193–194). The presence of these bowls suggests that the technique of splicing flax (before either adding extra twist or plying threads together) was sometimes practiced on Crete during these periods.

The new assemblage from the EM III rock shelter at Alatzomouri forms an important addition to the small number of textile tools thus far reported from Early Minoan sites in East and Central Crete. Two discoid weights and five spindle whorls were found in the EM IIA level at Myrtos Phournou Koriphi (Warren 1972, 212), while eight discoid and 19 spherical and cylindrical weights, 28 spindle whorls, and spinning bowls were recovered in the EM IIB level (Warren 1972, 220–222, 228–229). An even larger number of tools, including discoid, spherical, and cuboid weights, came from the EM IIB destruction levels of the Red House and Southwest House at Vasiliki (Seager 1908, 49 [many weights]; Zois 1980, 332–333 [at least 100 weights in Room 27]; 2007, 181–182; Watrous and Schultz 2012a, 23). This concentration suggests that the residents of these large dwellings were specializing in textile production in addition to the manufacture of metal tools and pottery (Watrous and Schultz 2012a, 28). In contrast, the recent excavation of four EM II houses at Mochlos has uncovered impressive evidence for the production of stone vases, obsidian blades, and metal tools, but not a single textile tool (pers. comm., T.M. Brogan). The presence of a MM II site at Pefka specializing in dyeing fiber and cloth located 75 m southeast of the EM III rock shelter indicates that the inhabitants of the northern isthmus of Ierapetra had a long tradition in textile manufacture (Apostolakou, Betancourt, and Brogan 2007–2008). Finally, a group of cuboid weights recovered from an EM III house at Petras (Burke 2006, 280 n. 4) provides the best parallels for the cuboid weight from the rock shelter.

Loomweights

501 (PAR 418, HN 14751; Fig. 65). Discoid weight, complete. H. 8.7, w. 8.8, th. 2.4 cm; wt. 195 g. Mirabello Fabric (reddish yellow, 5YR 6/6). Rounded top, two holes. From level 6, locus A604.9, object 32. Slightly worn.

502 (PAR 419, HN 14494; Fig. 65). Discoid weight, three-quarters complete. H. 6.7, w. 7.4, th. 1.8 cm; wt. 89 g. Mirabello Fabric (reddish yellow, 5YR 6/6). Flat top, with slight groove, two holes. From level 19, locus A604.63, object 294.

503 (PAR 420, HN 14498; Fig. 65). Discoid weight, complete. H. 7.1, w. 6.6, th. 1.8 cm; wt. 90 g. Mirabello Fabric (red, 2.5YR 5/6). Rounded top, with groove, two holes. From level 14, locus A604.49, object 160.

504 (PAR 421, HN 14505). Discoid weight, complete. H. 7.1, w. 7.5, th. 2.0 cm; wt. 101 g. Mirabello Fabric (reddish yellow, 7.5YR 6/6). Flat top, with groove, two holes. From level 16, locus A604.58, object 178. Chipped on one edge.

505 (PAR 423, HN 14507; Fig. 65; Pl. 22B). Discoid weight, complete. H. 8.5, w. 8.7, th. 1.7 cm; wt. 135 g. Mirabello Fabric (reddish yellow, 7.5YR 6/6). Rounded top, two holes. From level 5, locus A604.7, object 25.

506 (PAR 425, HN 13500; Fig. 65). Discoid weight, half complete. Pres. h. 8, th. 1.5 cm; wt. 56 g. Mirabello

Fabric (red, 2.5YR 5/8). Bottom half preserved. From level 5, locus A604.7, object 26.

507 (PAR 426, HN 14506; Fig. 65). Discoid weight, complete. H. 7.4, w. 8.2, th. 1.9 cm; wt. 80 g. Mirabello Fabric (red, 2.5YR 5/6). Flat top, with groove, two holes. From level 19, locus A604.63, object 297.

508 (PAR 427, HN 14508; Fig. 65). Discoid weight, three-quarters complete. H. 5.9, th. 1.2 cm; pres. wt. 55 g. Mirabello Fabric (from reddish brown, 2.5YR 4/4, to reddish yellow, 5YR 6/6). Flat top, with groove, two holes. From level 19, locus A604.63, object 296.

Cuboid Loomweight

509 (PAR 424, HN 14749; Fig. 65; Pl. 19B). Cuboid weight. Pierced slab, three-quarters complete. Length 7.1, w. 3.6, h. 5.3 cm; wt. 168 g. Mirabello Fabric (light reddish brown, 2.5YR 6/4). Rectangular clay slab with slightly rounded corners, four holes (one in each corner). From level 12, locus A604.28, object 69.

Spindle Whorl

510 (PAR 422, HN 14509; Fig. 65). Spindle whorl, complete. H. 3.1, d. 4.2 cm; wt. 43 g. Mirabello Fabric (reddish yellow, 5YR 6/6). Flattened sphere with central hole. From level 11, locus A604.21, object 57.

7

Marine Animals

by
Dimitra Mylona

The sea shells and the single cephalopod from the Alatzomouri Rock Shelter were collected by hand with the exception of a small number that were found in the residue of soil samples after water flotation. The fragmentation of the assemblage is low; most specimens are either complete or preserved as large fragments. All the remains belong to the same EM III horizon. The majority were found in the fill below the floor of the rock shelter (Stratum 3) with the layer of sherds and well-preserved vessels or in the soil outside the wall blocking the entrance to the chamber, and no spatial pattern is evident. This distribution suggests that the assorted shells in the fill had been used and discarded elsewhere or were consumed and thrown into the fill while the rock shelter was being built.

Remains of marine animals are the only surviving faunal remains from the rock shelter (no animal bones were recovered, but their preservation may have been adversely affected by the marly soil of the rock shelter). The marine animals include 318 shells, which belong to 17 different species and/or genera (Table 7.1) and a single fragment of a cephalopod, a sepia bone (common cuttlefish). A few dozen sea urchin spines that were collected are not considered here because they could have been incorporated in the soil accidentally. The vast majority of the shells are limpets (*Patella* sp.), followed by the pisania shells (*Pisania maculasa*), the monodonts (*Monodonta* sp.) and one of the purple shellfish (*Bolinus brandaris*). Even though the standard methodology applied in this analysis leads to an estimation of the minimum number of purple shellfish individuals to be 15, most of the remains of this taxon are fragments, and it is possible that we are actually dealing with the remains of fewer individuals. The rest of the shell taxa that have been identified in the assemblage are represented by one or two shells each. The cuttlefish (*Sepia officinalis*) is also limited to a single bone fragment.

The preponderance of the limpets (Table 7.2) and, to a smaller degree, of monodonts in the assemblage is a common feature of Bronze Age shell assemblages on Crete (e.g., Pseira: Reese 1995a, 1998, 1999, 2009; Mochlos: Reese et al. 2004, 118–121, table 24; Kommos: Reese 1995b; Ruscillo

Taxon	MNI	%
Patella, indeterminate	64	20.1
Patella caerulea	133	41.8
Patella aspera	23	7.2
Patella sp. (total)	(220)	(69.18)
Monodonta sp.	20	6.3
Clanculus sp.	1	0.3
Astrea rugosa	1	0.3
Cassidaria sp.	1	0.3
Hexaplex trunculus	1	0.3
Bolinus brandaris	15	4.7
Columbella rustica	2	0.6
Pisania maculosa	31	9.7
Buccinulum corneum	15	4.7
Fasciolaria lignaria	2	0.6
Arca noae	2	0.6
Pinna nobilis	3	0.9
Pecten jacobaeus	1	0.3
Spondylus gaederopus	1	0.3
Mactra corallina	1	0.3
Sepia officinalis	1	0.3
TOTAL	318	100

Table 7.1. Taxonomic representation of marine invertebrates from the rock shelter. MNI = minimum number of individuals.

Taxon	"Nicking" MNI	"Nicking" %	Broken Lips MNI	Broken Lips %	Broken Tip MNI	Broken Tip %
Patella sp.	16	25.0	5	7.8	—	—
Patella caerulaea	25	18.8	20	15.1	1	0.7
Patella aspera	7	30.4	6	26.0	—	—
Total	48	21.8	31	14.1	1	0.4

Table 7.2. Modification of limpets from the rock shelter. Percentages are based on the total number of individuals for each taxon in Table 7.1. MNI = minimum number of individuals.

2006, esp. 800–802; Papadiokambos: Brogan et al. 2013; pers. obs.) and the southern Aegean in general (Karali 1999, 55–56, table 1).

Shellfish for Food and Other Uses

The edibility of the animals whose shells were found in the rock shelter has been determined on the basis of ethnographic observations in modern Greece (e.g., Syrides, forthcoming) and on size (Table 7.3). Small taxa, such as the *Pisania* sp. for example, are not considered edible even if their flesh is not toxic.

Some of the shells bear traces of deliberate collection or consumption. A large portion of the limpets preserve "nicking" on their lips, or part of their lip appears to have been broken in antiquity in a manner that is not consistent with the natural flaking of the limpet's lips due to trampling or erosion (Pl. 44A). A few of the limpets have a broken tip (Pl. 44B). Experimental collection of limpets by the author has shown that these marks are traces left on the limpet shell in the process of removing them from the rock, especially when the limpets are dry and fully exposed above the water line. In this condition they tend to attach themselves firmly on the rock, and collection requires extra force that often leads to breaking part of the shell. On the contrary, when limpets are underwater or regularly drenched with water, they can be easily dislodged either by hand or with the use of an empty shell. In this case no traces are left on the collected limpet.

In the rock shelter assemblage, 21.8% of the shells exhibit traces of "nicking" on their periphery, 14.1% have deliberately broken lips, and only one shell has a broken tip (Table 7.2). *Patella aspera* is the limpet that is modified slightly more often than the *Patella caerulea* variety.

Some of the shells may have had decorative value. The fan mussel (*Pinna nobilis*) is a good example. It is the largest bivalve shellfish found in the Mediterranean, and it has a pearly shell that has been used in the past to make decorative inlays or pendants (e.g., Theodoropoulou 2007, 318). This was probably the use of the pierced pen shell fragment found in the rock shelter (Pl. 44C). The pilgrim scallop (*Pecten jacobaeus*), with its distinctive purple color, although edible, has also been used in the Bronze Age Aegean as a decorative item (Karali 1999, 21). The fact that only a single shell, albeit unmodified, was recovered from the rock shelter suggests that this is probably

Taxon	Common Name	Edibility
Patella, indeterminate	Limpet	Edible
Patella caerulea	Limpet	Edible
Patella aspera	Limpet	Edible
Monodonta sp.	Monodont	Edible
Clanculus sp.	"Strawberry" top shell	Inedible?
Astrea rugosa	Rough star	Inedible?
Cassidaria sp.	Rugose bonnet	Edible
Hexaplex trunculus	Purple shellfish	Edible
Bolinus brandaris	Purple shellfish	Edible
Columbella rustica	Dove snail	Inedible
Pisania maculosa	Whelk	Inedible
Buccinulum corneum	Whelk	Edible
Fasciolaria lignaria	Spindle snail	Inedible?
Arca noae	Noah's ark	Edible
Pinna nobilis	Fan mussel	Edible
Pecten jacobaeus	Pilgrim scallop	Edible
Spondylus gaederopus	Thorny oyster	Edible
Mactra corallina	Duck clams	Edible
Sepia officinalis	Cuttlefish	Edible

Table 7.3. Edibility of the shellfish recorded on site.

its use here. If we accept color and brilliance as criteria for the choice of certain decorative shells, we might also suggest that the single spiny oyster (*Spondyus gaederopus*), which is also bright purple in color, was probably also used decoratively.

The smaller seashells in the assemblage, which are mostly inedible, have a more enigmatic origin. These shells are regularly found on the beach in small numbers, and they could have been brought to the site with sand and/or sea pebbles.

Fishing and Gathering

The shellfish found in the rock shelter essentially represent two different marine zones. One is the rocky zone on the wave line, at the border between water and dry land. Limpets and monodonts inhabit this zone. They can be gathered from the shore with very little effort by hand, or by using very simple tools (Leukadites 1941, 168–169). The gatherers are not required to enter the sea or master swimming and diving skills. Traditionally in southern Greece, these are shellfish that are collected not only by fishermen, but also by women and children. In the rock shelter assemblage these shellfish make up the majority of remains.

The second marine zone is the submersed littoral zone, of various depths, of various substrates, rocky, muddy, sandy, or covered by Poseidonia (a species of seagrass). The shellfish attested in the rock shelter are nowadays found at various depths, and some of them, in considerable depths. The fan mussel, for example, is usually considered a deep water shellfish, but recent research in the Aegean has indicated that this is probably the result of overfishing, and that in certain cases they are found in waters as shallow as 0.5 m (Koutsoubas et al. 2007, 81; Antoniadou et al. 2013). The case of the purple shells (both *Hexaplex trunculus* and *Bolinus brandaris*; formerly referred to as murex), pilgrim scallop, thorny oyster, whelk, and others is probably similar. We are not able to ascertain from what depth these shellfish had been fished, but it is certain that in order to collect the shellfish from the submerged zone, special skills and technology were required. Dredging the sea bottom, use of bait baskets, incidental capture in the fishing nets, and gathering by diving are some such methods (for an ethnographic description of these methods, see Leukaditis 1941). In this case entering the sea, swimming, and probably diving are required.

It should be noted, however, that many of the shells (apart from the limpets and monodonts) could well have been collected dead from the beach (not necessarily being worn), thus representing an activity unrelated to food procurement. This idea is further supported by the fact that a large number of small pebbles were found in the same contexts as the seashells.

The soil conditions in the rock shelter favored the preservation of seashells but not animal bones. Moreover, the shells probably represent a secondary deposition after they were consumed or used elsewhere. Most of the shells collected from the rock shelter are edible, but inedible species are also present. Some of them, especially the smaller ones, appear to have been brought to the site with other materials of marine origin (e.g., sand and/or pebbles). Others, such as the fan shell and the pilgrim scallop, may have had other, non-culinary uses as decorative objects, inlays, or jewelry.

Among the edible shellfish, the limpets and the monodonts predominate in a pattern that is typical for all Bronze Age sites in the southern Aegean. These animals lived on the rocks of the shore and were easily collected with minimal effort, skill, and equipment. The presence of other shellfish that live in various depths indicates that fishing was also involved in the procurement of shellfish.

8

Archaeobotanical Remains

by
Evi Margaritis

Plant remains at the site of Alatzomouri were retrieved by flotation. The processing of the samples and the residue sorting was undertaken at the INSTAP Study Center for East Crete. Very few seeds are preserved from the site, and they represent only fruit trees, olives, and grapes (Table 8.1). More specifically, olives derive from locus A604.22 (Pl. 44D), which is the lower level of the main deposit of objects inside the rock shelter, and only from the lower part of it. The cavity contained various stone tools and lots of pottery, dated to the EM III period. Olives also come from A604.46, which represents the locus exposed after removing the enclosure wall behind which all the well preserved objects were placed. The area was split into northern and southern halves, and A604.46 was the northern half inside the chamber of the rock shelter. This locus contained large numbers of sherds, organic remains including shells, a mortar fragment, other stone tools, and a complete rounded cup (**124**), which was probably one of the first vessels placed in the rock shelter. The olive pits were recovered in fragments, the result of a post depositional process (Margaritis and Jones 2008; Margaritis 2013b). Locus A604.53 comes from the same area. It contained not olives this time, but grape pips (Pl. 44D), together with pottery sherds, three stone tools, and some shells. The last two loci represent a fill which was deposited in the rock shelter as a single episode of action.

In order for the analysis of the archaeobotanical data to contribute to the interpretation of the site, it is imperative to examine the taphonomy of the plant remains, which consists of the routes they followed to enter the archaeological record. There could be different possibilities for the use and subsequent preservation of the olives and the grapes at the site. They are all carbonized. Given their very limited quantities, they could have entered the archaeological record as part of the residues of consumption, which ended up in the fires created at the rock shelter for lighting, heating, or cooking. It should be noted, however, that no visible burned areas or hearths were excavated and, moreover, the samples contained very few pieces of charcoal. If cooking had been a regular activity at the site, we

Locus Sample	Olive	Grape
A604.22	0/6	—
A604.46	0/11	—
A604.53	—	2/9

Table 8.1. Complete seeds/fragments from the rock shelter.

would expect to find more burned areas or hearths and consequently more charcoal, with the possibility of the preservation of more plant remains of staple foods such as cereals and legumes. The olives could represent food preserved to be eaten at any time and brought to the site; there is no indication that the grape pips might be part of raisins (Margaritis and Jones 2006), and so they could have been gathered from the vicinity of the site. The absence of animal bones, expected as part of the residue of eating, the presence of fruits, which can be consumed raw and do not require cooking, and the shells, which could also have been gathered easily in the vicinity of the site, could point toward the use of the site, suggesting an infrequent use of the area. These remains could represent residue of the last episode of use of the rock shelter, the last meal of the people who sealed the site with a wall. It can be suggested that the last use of the site could have been in the late summer or early autumn, when fresh grapes could be gathered and consumed.

Another explanation for the use of plants could also be posited in combination with the rest of the material culture of the site. The plant remains could be the remnants of food consumed not as part of daily domestic practice but rather as part of a ritual routine, deposited with the rest of the material deliberately in the rock shelter as part of a votive or a ceremonial deposit. According to the excavators, the character of the site could point to ritual activities, mainly due to the presence of human bones in the vicinity of the site and the lack of clear indications of daily use.

It has been suggested that plant remains were deposited charred in ritual deposits at prehistoric sites both in Crete and also in mainland Greece. The common element at these sites is the act of burning prior to deposition, with evidence preserved at Petras, Mochlos, Livari, Avgi, and Aspis (Margaritis 2014, 2017). At the site of Livari the recovered olive pits, if not deliberately deposited, could have been used as part of the fuel during cremation, remains of which have been found at the site. In the same study it was also suggested that the burning of the plant remains did not aim at their complete destruction but rather at their preservation through fire, in order to retain their form as part of a ritual deposit. The plant remains could have acted as mnemonic signifiers of food and related aspects of life; they could represent a portion of a larger food consumption episode, which would have taken place somewhere else, while the smaller portion was charred for deposition within the rock shelter with the rest of the material.

Whatever the role of plants at the site, as remnants of the last meal after the construction of the wall that sealed the site or as intentional deposition as part of a ritual action, the Alatzomouri assemblage adds to the evidence for the presence of the grape and the olive in third millennium Greece. It comes as an addition to data from other Early Bronze Age sites, suggesting that these fruit trees are recovered wherever suitable recovery techniques are employed and, moreover, they are a core part of the plant repertoire of the period (Margaritis 2013a). The material from Alatzomouri comes as an addition to the plant remains from a handful of published Early Minoan sites including Debla (Greig and Warren 1974), Vasiliki (Zois 1991), Myrtos Phournou Koriphi (Renfrew 1972), Chrysokamino (Jones and Schofield 2006), and Aphrodite's Kephali (Margaritis 2013a). More recent studies at Priniatikos Pyrgos (Molloy et al. 2014), Gournia (Watrous et al. 2015), Mochlos, Livari, and Petras (all under study by the author) will add to the reconstruction of the agricultural practices of the period, the information for which was, until recently, very scarce.

9

Miscellaneous Objects

by

Thomas M. Brogan and Philip P. Betancourt

Several miscellaneous objects were found in the rock shelter. They are of special interest because they contribute information on the composition and interpretation of this unusual assemblage of artifacts. All of the objects are broken and incomplete. The presence of a piece of a broken drain, an artifact that is an architectural element, shows a larger range of objects than household items, and the broken and incomplete condition of these objects indicates that there was never any intent to reuse the artifacts buried here.

Drain Fragment

One possible fragment of an open drain was found in the rock shelter (**332**). As is normal for this class of artifact, it was made of clay. Drains were common in Minoan towns, and they were useful both for drainage of rain and for the disposal of waste water. The best discussion of the drains themselves is by Shaw (1973, 201–204; 2009, 137–139). The various uses for drains in Minoan buildings are discussed by Graham (1987, 219–221).

The example of a drain found at Alatzomouri consisted of an open channel with a flat base and vertical sides with rounded upper edges. It was not decorated in any way. Such drains were of variable length, and they were used whenever water needed to be conducted somewhere, either to or away from a building. Drain sections were simply laid end to end. They could be either above ground or buried in the soil. Similar forms could be used as the drains for vats, and the fragment is listed there as well (Ch. 4).

The fragment is made of Mirabello Fabric, a clay recipe using fragments of igneous rocks that are only found in the region of the Gulf of Mirabello (for the fabric, see Myer, McIntosh, and Betancourt 1995, 144–145, with additional bibliography above, p. 38; for the geological formations that yielded the inclusions in this fabric, see Tsikouras, Dierckx, and Hatzipanagiotou 2008). The fabric is important because it identifies the drain as a local product. It also proves a relationship with the pottery and the other miscellaneous objects described in this chapter that were manufactured from a similar clay recipe (Ch. 10). Mirabello Fabric was almost certainly made at Gournia.

The presence of a broken drain is an interesting bit of evidence for the interpretation of this enigmatic context. It helps demonstrate that everything here is not necessarily a personal possession of a deceased individual or a part of a domestic kitchen or household workshop, but an assemblage of mostly fragmentary objects and artifacts that might be present in a destroyed Minoan building.

Potter's Batts or Lids

Four pieces of almost flat clay disks were present in the assemblage. They were manufactured as flat pieces of clay and painted with lines across them forming an X before being fired in a potter's kiln. Their clay is the normal Mirabello Fabric used for most of the pottery from the site as well as for the other objects described in this chapter. No traces of handles or other clay additions are preserved on the surviving pieces.

Similar artifacts were found at the EM II site of Myrtos Phournou Koriphi, and Peter Warren suggested a use as potters' turntables (Warren 1972, 245, fig. 98:10–15). He suggested that the dark cross on the upper surface could have been useful in centering the clay vessel as it was shaped. Modern experimental archaeology indicates that clay disks are useful in making handmade vessels because it is easier to rotate the vase during manufacture when it is attached to an easily turned flat object (Betancourt, Gosser, and Sapareto 1984, 128). The parallel with Myrtos (Fig. 67, lower image) is close because of the crossed lines on one side of the disk and the painted outer edge. As an alternative, the disks could be lids without handles because Minoan covers for vessels were sometimes flat circular pieces of fired clay (Betancourt 1990, 152, no. 1114).

511 (PAR 428, HN 14511; Fig. 66). Disk fragment, one-third complete. D. ca. 24–25 cm; pres. wt. 186 g. Mirabello Fabric (between pink, 7.5YR 7/4, and light brown, 7.5YR 6/4). Flat disk. Crossed lines of dark slip on upper surface. From level 12, locus A604.42, object 145.

512 (PAR 429, HN 14666; Fig. 66). Disk fragment, one-third complete. D. ca. 20–21 cm; pres. wt. 216 g. Mirabello Fabric (pink, 5YR 7/4). Flat disk. Crossed lines of dark slip on the upper surface. From Level 17, locus A604.59.

513 (PAR 467, HN 14518; Fig. 66). Disk fragment, rim sherd. D. ca. 24–26, th. 1.2 cm. Mirabello Fabric (light reddish brown, 5YR 6/4). Flat disk. From Level 5, locus A604.8.

514 (PAR 468, HN 14514; Fig. 67). Disk fragment, rim sherd. D. ca. 18–20 cm, th. 2.3 cm. Mirabello Fabric (pink, 5YR 7/4, and light reddish brown, 5YR 6/6). Flat disk. Dark line across upper surface. From Level 13, locus A604.46.

Clay Triangle

A broken piece of a triangular clay slab that had been fired like a vessel was present in the rock shelter. It is made of Mirabello Fabric, so it is a local product. No wear marks are present on the object, and its use is not known.

The best parallels for this type of artifact come fom Myrtos (Warren 1972, 217–218, 254, fig. 107: 46–55). The triangles from Myrtos are regarded as possibly counting devices for some type of game. The examples from Myrtos were all made from sherds.

515 (PAR 430, HN 14750; Fig. 67). Clay triangle, half complete. Max. dim. 8 cm. Mirabello Fabric (light red, 2.5YR 6/6). Triangular flat clay object. From locus A600/700.1, object 1.

10

Pottery Statistics

by

Philip P. Betancourt, Thomas M. Brogan, and Susan C. Ferrence

The statistics from the assemblage of pottery from the Alatzomouri Rock Shelter provide important evidence for the site's interpretation (Table 10.1). The authors compiled the statistics after the conservators had completed the work of joining sherds together to make whole vessels, partial vases, and sections. Criteria for sorting included fabrics, dates, styles, shapes, and degree of preservation. Pieces were counted and recorded in the table along with their position on a vessel (rim, handle, base, spout, leg, or body). The resulting numbers and percentages allowed informed decisions on several subjects including the minimum numbers of specific classes and shapes.

A total of 4,044 sherds and vases were in the deposit after mending. Within this total, 54 vessels were whole or largely restorable, and the rest were sherds or larger fragments of vases. By far the majority of items were isolated small pieces that did not mend into larger sections and vases.

In terms of fabric, a large majority of the sherds and vases were made of Mirabello Fabric (53 of 54 whole and restorable vases and 3,966 of 4,044 sherds). This fabric is recognized by the presence of igneous rock fragments in the diorite-to-granodiorite series that derive from one or more outcrops along the coast of the Gulf of Mirabello from Gournia to Priniatikos Pyrgos (Betancourt et al. 1979, 4–5, fig. 1, table I; Day 1991, 92–94; 1997, 225; Myer, McIntosh, and Betancourt 1995, 144–145; Whitelaw et al. 1997, 268; Vaughan 2002, 153–154; Barnard 2003, 7; Day, Joyner, and Relaki 2003, 17–18; Day et al. 2005, 183–187; Nodarou and Moody 2014). The fabric can be recognized without analysis because of the distinctive appearance of the inclusions (see Types 2 and 3 in Haggis and Mook 1993). The diorite from northern Crete is a granular igneous rock consisting principally of white plagioclase feldspar, a black member of the amphibole group of silicates, and dark mica, while granodiorite has a small percentage of K-feldspar as well as the plagioclase and more quartz than occurs in the diorite (Hurlbut and Klein 1977, 458, 460). Both rocks have been identified in the Minoan pottery. The dark mica was formerly called biotite (no longer considered a distinct species;

Fabric, Decoration, Date, and Vessel Type	Complete	Largely Restored	Rim Sherds	Handle Sherds	Base Sherds	Spout Sherds	Leg Sherds	Body Sherds	Total Number of Sherds	Percentage of Sherds	Minimum Number of Vessels (MNV)	Percentage of MNV
CALCITE-TEMPERED FABRIC												
Monochrome Style, EM I												
Cup, **1**					1				1	0.02%	1	0.35%
GRAY FABRIC												
Pyrgos Style, EM I												
Pyxis lid, **2**				1					1	0.02%	1	0.35%
GRAY COARSE FABRIC												
COARSE REDDISH-YELLOW FABRIC												
Monochrome Style, EM II–III												
Jug				1					1	0.02%	1	0.35%
RED FABRICS												
Monochrome Style, EM II–III												
Cup				1					1	0.02%	1	0.35%
Closed vessel					1			38	39	0.96%	1	0.35%
SOFT REDDISH-YELLOW FABRIC												
Monochrome Style, EM II–III												
Jug				1				29	30	0.74%	1	0.35%
MISCELLANEOUS FABRICS												
Monochrome Style, EM II–III												
Jug, **149**				1					1	0.02%	1	0.35%
Small jar			1						1	0.02%	1	0.35%
Open vessel, **107**				1					1	0.02%	1	0.35%
SOUTH COAST FABRICS												
Monochrome Style, EM IIB												
Jug **8**	1								1	0.02%	1	0.35%
MIRABELLO FABRIC												
Vasiliki Ware, EM IIB												
Shallow conical bowl, **3, 4**			4						4	0.05%	4	1.39%
Jug, **5, 6**				1	1			5	7	0.17%	3	1.04%
East Cretan White-on-Dark Ware, EM III												
Shallow conical bowl, **9–22**			19	5	15			8	47	1.16%	25–30	8.68%
Conical bowl, without spout, **23**	1								1	0.02%	1	0.35%
Conical bowl, **24–43**		3	15	4	10	2		14	48	1.19%	21–27	7.29%
Conical bowl with frying pan handle, **44, 45**				3					3	0.07%	3	1.04%
Spouted conical jar, **46–60**	1	3		2	5	2		27	40	0.99%	16–20	5.56%
Cup with handle, **61**			6					1	7	0.17%	5	1.74%

Table 10.1. Statistics of the pottery assemblage based on fabric, decoration, date, and vessel type. The percentage of minimum number of vessels (MNV) is based on a total number of 288 vessels, which is represented by the lower number of MNV when a range is given in that column. The sherds where MNV are not applicable (N/A) are primarily undiagnostic and thus assumed to have come from other plain and slightly decorated vessels in the table.

Fabric, Decoration, Date, and Vessel Type	Complete	Largely Restored	Rim Sherds	Handle Sherds	Base Sherds	Spout Sherds	Leg Sherds	Body Sherds	Total Number of Sherds	Percentage of Sherds	Minimum Number of Vessels (MNV)	Percentage of MNV
East Cretan White-on-Dark Ware, EM III, cont.												
Vessel, **62**			1						1	0.02%	1	0.35%
Jar, **63, 64**			1		1				2	0.05%	2	0.69%
Rounded cup, tiny handles, **65–67**			3						3	0.07%	3	1.04%
Rounded cup, **68–70**			4	6				19	29	0.72%	4–5	1.39%
Spouted two-handled cup, **71**	1								1	0.02%	1	0.35%
Bridge-spouted jar, **72–83**	1		9	12	4	3		408	437	10.81%	10–15	3.47%
Jar or jug, **84–86**		1	1	1	2			27	31	0.77%	4–5	1.39%
Jug, **87–92**		1		1		2		4	8	0.20%	5	1.74%
Collared jar, **93**								1	1	0.02%	1	0.35%
Vessel with elliptical base, **94**					1				1	0.02%	1	0.35%
Closed vessel, **95**					1				1	0.02%	1	0.35%
Teapot, **96**	1								1	0.02%	1	0.35%
Plain and Slightly Decorated Vessels, EM III												
Conical bowls and basins, **97–106, 108–114**	1	3	10	4	8			30	56	1.38%	13–15	4.51%
Basin with scoring inside, **115–120**			1	11				11	23	0.57%	3–6	1.04%
Spinning bowl, **121**		1							1	0.02%	1	0.35%
Undecorated cup, **122, 123**			1	1					2	0.05%	2	0.69%
Rounded cup, **124–134**	3		16	4	11			67	101	2.50%	14–18	4.86%
Cup or bowl			2						2	0.05%	2	0.69%
Jug, **135–149**	2			17		11			30	0.74%	20–24	6.94%
Wide-mouthed jug, **150–152**	3	1		1	1				6	0.15%	5	1.74%
Pithos/jar, **153–187**			138	7	7			19	171	4.23%	15–18	5.21%
Pithos with drain, **188–191**				4					4	0.10%	N/A	—
Spouted jar, **192–197**	1					7			8	0.20%	8	2.78%
Collared jar, **198–203**			6	1				1	8	0.20%	6	2.08%
Small jar with knobs, **204**					1				1	0.02%	1	0.35%
Bridge-spouted jar, **205–207**						3			3	0.07%	3	1.04%
Wide-mouthed jar, **208–224**	5	1	10	2	9			7	34	0.84%	10–15	3.47%
Tripod cooking pot, **225–284**	11	7	79	27	31	4	61	252	472	11.67%	40–45	13.89%
Closed vessel, **341–391**			1	110	181			1726	2018	49.90%	N/A	—
Cooking dish, **285–329**			96	12	5			228	341	8.43%	12–15	4.17%
Vat, **330–332**		1			4				5	0.12%	3	1.04%
Lid, **333–338**			6						6	0.15%	6	2.08%
Unknown vessel, **339, 340**			1	1					2	0.05%	2	0.69%

Table 10.1, cont. Statistics of the pottery assemblage based on fabric, decoration, date, and vessel type. The percentage of minimum number of vessels (MNV) is based on a total number of 288 vessels, which is represented by the lower number of MNV when a range is given in that column. The sherds where MNV are not applicable (N/A) are primarily undiagnostic and thus assumed to have come from other plain and slightly decorated vessels in the table.

see Back and Mandarino 2008, 25). It occurs as tiny plates that can provide bronzy sparkles when a sherd is examined in sunlight. The pure white plagioclase and the black amphibole and mica can occur either as rock fragments with two or all three constituents present or as isolated individual grains. Because the diorite and granodiorite outcrops are only present in Crete along the Gulf of Mirabello (Dierckx and Tsikouras 2007), the fabric demonstrates that most of the pottery from the assemblage was made in the region just west of the Alatzomouri Rock Shelter.

In terms of date, the pottery is very uniform (Table 10.1). Only 2% of the sherds and vases are from EM I to EM II–III. Except for one large jug, they are all small sherds. All the rest can be assigned to EM III. A total of 661 entries in the tables (16% of the total) are East Cretan White-on-Dark Ware, which is the most distinctive style for EM III (Betancourt 1984). The deposit was apparently placed in the cave in EM III, and it included only a few earlier pieces.

Percentages of vessel classes are shown at the extreme right in Table 10.1. They are computed using the number of sherds and vases. The second column from the right gives the minimum estimate of specific shapes. The percentages presented in these two columns provide different types of information. The minimum numbers are estimated using the numbers of specific features like handles, legs, and spouts as well as judgments based on the thickness of the wall or the size of the vessels.

A comparison between the second and third columns from the left and the next six columns demonstrates the general composition of the deposit. The comparison shows that it is bimodal, with a group of complete and partly complete vases and another group of highly fragmentary pieces including many containers represented only by a single sherd. Fifty-four vessels (ca. 1% of the total) are complete or partly complete. The other 3,990 entries (almost 99%) are fragments, and most of them are small. They were placed in the cave as sherds, and the excavation showed that many of them were put in the cave before the underground room was filled with the more complete pieces.

The complete vases have no signs that they are necessarily later than the EM III sherds, but they must have had a different history. Unlike the more complete examples, the sherds come from vessels whose broken pieces were scattered before they were placed underground. The complete and partly complete vessels include the following:

Plain or slightly decorated
   18 cooking pots
   six wide-mouthed jars
   four wide-mouthed jugs
   four bowls and basins
   three rounded cups
   two jugs
   one spouted jar
   one spinning bowl
   one vat
East Cretan White-on-Dark Ware
   four spouted conical jars
   four conical bowls
   one spouted kantharos
   one bridge-spouted jar
   one teapot
   one jug
   one jar or jug
South Coast Style
   one jug

This assemblage is unusual in comparison with other deposits from Minoan Crete. We now have a large number of Minoan deposits whose ceramics have been analyzed statistically (for funerary deposits, see Betancourt and Davaras, eds., 2002; for settlements, see Watrous 1992; Betancourt and Davaras, eds., 1995; 1998a; 1998b; 1999; 2009; Barnard and Brogan 2003). This deposit is especially interesting because of its closed nature. Nothing except soil, stones, and a single burial was found near it. The material inside the cave was self-contained and complete.

The complete and partly complete vessels suggest that the character has both some domestic features and a strong industrial and craftwork component. The White-on-Dark Ware vases would be appropriate either for serving or for display in a Minoan household, but too few drinking vessels are present to suggest feasting or even casual dining. The absence is especially striking in view of the large amount of food that could have been contained in the many storage vessels in this assemblage. The industrial and craftwork vessels include the vat, the spinning bowl, perhaps the medium-sized storage vessels, and the tripod cooking pots.

The large number of cooking vessels in comparison with drinking cups does not have a clear explanation. Among the whole and partial vases, surely 18 tripod cooking pots is too many to go with only three cups. Deep bowls set on tall legs are suitable for cooking (Betancourt 1980), but they can also be used for any other activity involving the heating of liquids. If they are used to prepare food, the meals are consumed along with a beverage, and drinking vessels are common in Minoan houses. More than 20% cups is not unusual for a household ceramic assemblage (Floyd 2009b, 173–175), though no statistics are available from an exclusively EM III context. The tripod vessels must have been used for something other than preparing normal meals in a household setting. They were either used to prepare large amounts of food for persons who were not part of the household, or they were an industrial component of the assemblage.

The same situation is present when statistics are computed for the full assemblage including the sherds. The tripod vessels are again a high percentage (15.4%), the storage shapes are high (56%), and additional vats are present (four more sherds, representing an additional two vats). Pieces from drinking cups are present, but the number is low for ordinary household debris (under 15%). The conclusion is that this assemblage is not just a selection from a purely domestic setting. Many of the vessels must have come from workshop environments.

11

Ceramic Petrography

by

Eleni Nodarou

The pottery fabric with granitic-dioritic inclusions is among the most easily recognizable Minoan fabric in hand specimens due to the mixture of black-and-white ("salt and pepper") grains in a fine matrix (Haggis and Mook 1993; Haggis 2005, 169; see also Ch. 10). This fabric has been described at many instances macroscopically and microscopically; it is probably among the few (if not the only) fabric whose origin is well known and has been named after the area where the outcrops for the main tempering agent are located—the central and western regions around the Gulf of Mirabello (hence the name "Mirabello Fabric" [Day 1991]), in particular the broader area of Gournia and the northern part of the Ierapetra Isthmus (for bibliographic overview, see Betancourt 2008, 30–31; Nodarou and Moody 2014, 91).

The macroscopic study shows that the pottery assemblage consists almost exclusively of vessels in granitic-dioritic fabrics. Exceptions include one EM IIB vessel manufactured in South Coast Fabric (**8**) and several EM sherds (see the pottery statistics in Ch. 10 and Table 10.1). This situation is explained by the location of the rock shelter less than one kilometer from Gournia and also near the sources of the raw materials. The petrographic analysis of selected pottery samples from Alatzomouri, therefore, was not dictated by the need to characterize the fabrics and associate them with local or distant geologies as is usually the case in petrographic projects. This material belongs almost entirely to a single chronological phase, and it provides a unique opportunity to investigate recipes of manufacture for the Mirabello Fabric.

The analysis of the material from Alatzomouri was incorporated into a broader analytical project investigating raw materials in the bay of the Mirabello. A multi-disciplinary approach combining petrology, thin section petrography, and scanning electron microscopy (SEM) was implemented to explore granitic and dioritic raw materials used for the manufacture of both stone tools (Dierckx and Tsikouras 2007) and pottery (Georgotas 2013; Nodarou 2013a; Nodarou and Moody 2014). Extensive raw material sampling was also carried out for clays and tempering materials, and an experimental project was performed (for preliminary results, see Georgotas 2013). The ceramic analysis was employed on an array of sherds deriving from the Vrokastro survey and extending chronologically from the Neolithic period to Venetian times. They were identified macroscopically as being manufactured in granitic-dioritic fabrics and then studied under the microscope. In this way a number of recipes have been identified within the granitic-dioritic pottery, and associations were made between vessel shapes and clay recipes as well as between clay recipes and the origin of the raw materials from specific areas within the Mirabello region (Nodarou and Moody 2014).

Within this framework the analysis of the pottery from Alatzomouri was expected to contribute information on the following issues:

1. The investigation of recipes for plain domestic versus decorated fine wares, especially the white-on-dark painted pottery.
2. The examination of whether the patterns of the clay recipes detected in the material from the rock shelter would confirm those seen in the pottery from the Vrokastro survey.
3. The comparison of the patterns of the clay recipes from Vrokastro against a chronologically well-defined and well-dated assemblage in order to confirm whether some recipes are date specific.

For this reason 33 pottery samples from the Alatzomouri Rock Shelter were submitted for petrographic analysis using a LEICA DMLP polarizing microscope (for a list of the samples analyzed, see Table 11.1). All the selected sherds were identified through macroscopic study as manufactured in a fabric with granitic-dioritic inclusions, and this was further confirmed through thin section petrography.

Summary of the Geology of the Mirabello Region

The geology of the central and western Mirabello area consists of limestones of Cretaceous age intruded by granitic to granodioritic rocks (see Ch. 2). There is also an onlap conglomerate of Miocene Age with basalt, dolerite, gabbro, chert, and various metamorphic rocks (I.G.S.R. 1959; Dierckx and Tsikouras 2007). These materials constitute the main nonplastic inclusions encountered in pottery fabrics, and they were sampled for the experimental part of the project. Moreover, various sources of red alluvial and white/gray marl clays were also sampled in the area, and experimental briquettes were manufactured with various combinations of clays and tempering materials (Georgotas 2013).

Results of the Analysis

As expected, the majority of the samples were grouped into the two main granitic-dioritic fabrics: the jar fabric and the cooking fabric. Both variants have been identified under the microscope and described by various scholars since 1979 on the occasion of assemblages varying chronologically from EM I to LM III (Day 1995, 159–161; Whitelaw et al. 1997, 270; Day et al. 2005, 183–185; 2006, 150; Nodarou 2007, 79, to name just a few). Briefly, the jar fabric is characterized by a very fine red to reddish-brown firing matrix that is optically inactive. The nonplastic inclusions consist of angular granitic and dioritic rock fragments sparsely distributed in the groundmass, and there are also characteristic rounded clay pellets. In the cooking fabric the matrix is optically active, and it ranges from yellowish brown to brown after firing. The nonplastic inclusions are mainly granitic (diorite is almost absent); they are more abundant and more densely and unevenly packed in the clay base. These two main recipes were further refined through the study of the Vrokastro survey material, and it is against these smaller groups that the pottery from Alatzomouri is compared.

Fabric Group 1: Jar Fabric

SAMPLES: PAR 09/03 (**55**), 04 (**38**), 07 (**90**), 08 (**73**), 09 (**99**), 10 (**119**), 11 (**116**), 13 (**171**), 14 (**183**), 15 (**195**), 16 (**175**), 19 (**95**), 27 (**218**), 29 (**207**), 30 (**346**), 31 (**219**), 32 (**223**), 33 (**220**)

This is the most common fabric in the Alatzomouri assemblage. It is characterized by a dark

Sample Number	Catalog Number	Description	Decoration	Petrographic Fabric Group
PAR 09/01	6	Jug	Vasiliki Ware	Loner low-fired fabric
PAR 09/02	11	Shallow bowl	White-on-Dark Ware	Loner calcareous fabric
PAR 09/03	55	Spouted conical jar	White-on-Dark Ware	F1 main jar fabric
PAR 09/04	38	Conical bowl or cup	White-on-Dark Ware	F1 main jar fabric
PAR 09/05	53	Spouted conical jar	White-on-Dark Ware	Loner dark fabric
PAR 09/06	72	Bridge-mouthed jar	White-on-Dark Ware	Loner dark fabric
PAR 09/07	90	Jug	White-on-Dark Ware	F1 main jar fabric
PAR 09/08	73	Bridge-spouted jar		F1 main jar fabric
PAR 09/09	99	Basin	Dark band on rim	F1 main jar fabric
PAR 09/10	119	Basin with scoring inside		F1 main jar fabric
PAR 09/11	116	Basin with scoring inside		F1 main jar fabric
PAR 09/13	171	Pithos with handle	Dark slip	F1 main jar fabric
PAR 09/14	183	Pithos/jar		F1 main jar fabric with organics
PAR 09/15	195	Spouted jar	Dark slip on rim and spout	F1 main jar fabric
PAR 09/16	175	Pithos/jar		F1 main jar fabric
PAR 09/19	95	Closed vessel	Traces of red slip on exterior	F1 main jar fabric
PAR 09/21	323	Cooking dish		F2 cooking fabric Sub C
PAR 09/23	297	Cooking dish		F2 cooking fabric Sub C
PAR 09/24	311	Cooking dish		F2 cooking fabric Sub B
PAR 09/25	276	Tripod cooking pot		F2 cooking fabric Sub C
PAR 09/26	262	Tripod cooking pot		F2 cooking fabric Sub B
PAR 09/27	218	Hole-mouthed jar		F1 main jar fabric
PAR 09/28	246	Tripod cooking pot		F2 cooking fabric Sub A
PAR 09/29	207	Bridge-spouted jar		F1 main jar fabric
PAR 09/30	346	Closed vessel		F1 main jar fabric
PAR 09/31	219	Hole-mouthed jar		F1 main jar fabric
PAR 09/32	223	Hole-mouthed jar		F1 main jar fabric
PAR 09/33	220	Hole-mouthed jar		F1 main jar fabric

Table 11.1. Concordance of petrographic samples with vessel shapes and petrographic fabric groups. Five samples were destroyed during thin section manufacture (PAR 09/12 [**145**], 09/17 [**200**], 09/18 [**201**], 09/20 [**190**], 09/22 [**295**]). F = fabric group, Sub = subgroup.

reddish-brown to brown firing clay matrix that is optically inactive (Pl. 45A). The size, angular shape, and distribution of the nonplastics indicate that they were added as temper. The inclusions consist mainly of diorite fragments, composed of plagioclase feldspar, and a dark mica, and occasionally amphibole and rare quartz. There are also a few fragments of plagioclase feldspar, but the component differentiating this fabric from the others is the frequent biotite fragments distributed in the clay matrix. The biotite lath-shaped fragments are the dark mica specks giving the vessel surface a shiny appearance. The majority of vessels represented in this group are jars (spouted conical, bridge spouted, hole mouthed), pithoi, and basins.

*Samples: PAR 09/08 (**73**), 09 (**99**)*

Two finer vessels are in this fabric group: a bowl and a jug with white-on-dark decoration, which are also characterized by small biotite fragments (Pl. 45B). In these two samples, however, the diorite and biotite fragments are more abundant than in the rest of the group, and the groundmass of the samples exhibits moderate optical activity.

*Sample: PAR 09/14 (**183**)*

This pithos/jar shares the same composition and texture as the rest of the material in this fabric group, but it has also been tempered with organics as indicated by the characteristic large voids oriented parallel to the vessel margins (Pl. 46A).

COMMENT

The amount of the material represented in this fabric group and the overall compositional and textural homogeneity leave no doubt that this is a variant of the jar recipe. Comparison with jars from the Vrokastro survey showed that this specific recipe, rich in biotite, first appears in later EM III to MM I but continues throughout LM I (Nodarou and Moody 2014, 95).

Fabric Group 2: Cooking Fabric

Three variants of cooking fabric were identified in the Alatzomouri assemblage. They are differentiated from each other in terms of firing temperature and the amount and distribution of the nonplastic inclusions.

SUBGROUP A: LOW FIRED

*Sample: PAR 09/28 (**246**)*

This subgroup is characterized by an orangish-brown firing matrix that is optically active (Pl. 46B). The nonplastic inclusions consist primarily of granite fragments composed of plagioclase feldspar, all unevenly distributed in the clay base. The varying size and the irregular distribution of the inclusions leads to the assumption that the nonplastics were not added purposefully as temper, but they were naturally occurring in the clay mix. The vessel represented is a tripod cooking pot.

SUBGROUP B: HIGH FIRED WITH BIOTITE

*Samples: PAR 09/24 (**311**), 26 (**262**)*

This subgroup is characterized by a dark reddish-brown firing matrix that is optically inactive (Pl. 47A). The nonplastic inclusions consist almost exclusively of large angular granite fragments evenly distributed in the clay matrix, indicating that they were added as temper. There are also a significant number of small fragments and laths of biotite. The vessels represented are a cooking dish and a tripod cooking pot.

SUBGROUP C: WITH WEATHERED GRANODIORITE

*Samples: PAR 09/21 (**323**), 23 (**297**), 25 (**276**)*

This subgroup is characterized by a brown firing matrix that is optically inactive (Pl. 47B). The non-plastic inclusions consist of granitic and dioritic fragments and a considerable amount of biotite, as is the case for Fabric Group 1 when used for jars. However, in the cooking variant the granitic and dioritic fragments are broken down into very small pieces and look weathered, which is indicative of a different origin of the raw material. The vessels represented are two cooking dishes and a tripod cooking pot.

COMMENT

The three subgroups identified in the Alatzomouri assemblage are all encountered in the Vrokastro survey material (Nodarou and Moody 2014, 93–94). Subgroup A is the most frequent and the one displaying longevity since it occurs from the MM to the LM IIIC period. Subgroup B cannot be associated directly with any of the cooking fabrics at Vrokastro because it is compositionally and texturally closer to the jar rather than to any of the cooking fabrics identified in the survey material. The closest parallel is the main MM fabric, which is characterized by similar nonplastics (i.e., granitic and dioritic rocks and plagioclase feldspar) and the frequent biotite fragments and biotite mica. However, in the Vrokastro cooking recipes the granitic-dioritic fragments are much smaller, and they are densely packed in the clay matrix. In Subgroup B the rock fragments are rather large, angular, and sparsely distributed in the groundmass. The presence of dark brown pellets is one more factor bringing this small group closer to the jar fabric. The number of samples is small, and not much can be said about this recipe and its connection with the jar fabric. Subgroup C is also well represented in the Vrokastro survey material, and it reflects a rather narrow chronological span, MM I–II. The material from Alatzomouri shows that the recipe with the weathered granite starts slightly earlier, in the EM III period.

Small Groups and Loners

In this category are included fabrics with granitic-dioritic inclusions that do not belong to the jar or the cooking classes presented above but reflect different recipes.

SAMPLES: PAR 09/05 (**53**), 06 (**72**)

These samples are characterized by a very fine dark (almost black) matrix and densely packed granitic-dioritic inclusions, frequent small fragments of plagioclase feldspar, and rare fragments of biotite and volcanic rocks (Pl. 48A). This small group differs from the jar fabric (Fabric Group 1) in that it contains less biotite in the clay matrix, and the nonplastic inclusions are more densely packed. Moreover, the dark color of the matrix indicates a reduced firing atmosphere, which is in accordance with the white-on-dark decoration of the surface. The vessels represented are a spouted conical jar and a bridge-spouted jar.

SAMPLE: PAR 09/01 (**6**)

This sample is characterized by an orangish-brown firing matrix that is optically active (Pl. 48B). The nonplastic inclusions are rather small in size and fairly irregularly spread in the clay matrix. They consist of granitic and dioritic rock fragments, some plagioclase feldspar, and rare biotite and pyroxene fragments. The vessel represented is a jug of Vasiliki Ware dated to the EM IIB period.

SAMPLE: PAR 09/02 (**11**)

This sample differs from the rest of the assemblage in that the raw material used is calcareous as indicated by the golden brown color of the clay matrix, which is also optically highly active (Pl. 49A). There are very few nonplastic inclusions consisting of fragments of weathered granite, diotite, plagioclase feldspar, biotite, and little amphibole. The vessel represented is a shallow bowl with white-on-dark decoration. A calcareous fabric is also identified in several samples of the Vrokastro survey material, primarily in jars and pithoi, covering a rather narrow chronological span, that is, MM I–II (Nodarou and Moody 2014, 95). The presence of this sample in the Alatzomouri assemblage indicates that this recipe started slightly earlier, in the EM III period.

Discussion

The pottery from Alatzomouri belongs to a single fabric group, considering that compositionally all the studied vessels are manufactured in fabrics with granitic-dioritic rock inclusions. The petrographic analysis demonstrated the presence of two main recipes, the jar and the cooking fabric, and their subgroups, which were compared to those identified in the Vrokastro survey pottery. The outcome of this research provided interesting insights into the ceramic technology of the EM III to MM I period.

The first observation concerns the recipes identified and their continuity in time. The predominant component of the Alatzomouri fabrics is the biotite-rich clay pastes, present in two variants, one used for jars (Fabric Group 1) and one for cooking wares (Fabric Group 2, Subgroup B). These recipes were also identified in the Vrokastro survey material as representing production of the MM I–II period, but the examples from Alatzomouri confirmed that their emergence can be dated slightly earlier, at least in the EM III period. Another interesting aspect of the biotite-rich recipe as revealed from the study of the Vrokastro material is that it seems to be confined to the northern and central part of the Mirabello region (Nodarou and Moody 2014, 93, fig. 2), which is also confirmed by their presence in the Alatzomouri material. Another MM I–II recipe occurring also in the Vrokastro material is the cooking fabric with weathered granodiorite (Fabric Group 2, Subgroup C), which is also proven to be starting in the EM III period. In terms of its distribution, its presence in the Alatzomouri assemblage confirms the observation that it occurs in the central and western part of the Mirabello region (Nodarou and Moody 2014, 94, fig. 3). This distribution of the granitic-dioritic fabrics on a micro-scale within the Mirabello region is hoped to be refined further through the analysis of other well-dated assemblages such as Gournia and Priniatikos Pyrgos. If there appear to be consistent patterns of distribution within the Mirabello region, they might be indicative of the micro-provenance of the pottery found not only in the broader area of the Mirabello but also farther afield.

The study of the EM III material from Alatzomouri also allowed exploration of technological aspects of the decorated pottery of the late EM and early MM periods. The main painted ware encountered at Alatzomouri is East Cretan White-on-Dark Ware (WOD), characterized by white decoration on a dark red- to brown- and black-firing ground. This ware was among the first classes of pottery studied with the application of a variety of archaeometric techniques (Betancourt 1984). In his pioneering petrographic study, G.H. Myer (1984) examined sherds bearing WOD decoration from five sites in East Crete and one site in Central Crete, all dated in the EM III period, with the samples from Gournia extending to MM I–III. Although the amount of the material included in that project was small, and it was difficult at the time to investigate pottery recipes, the analysis was indicative of their presence: it was demonstrated that there are compositional differences among the East Cretan sites of EM III as well as between EM III and the MM I–III phase at Gournia. This observation has now been confirmed with the analysis of the material from Alatzomouri, reflecting the recipes of the later EM III–MM I period in the north-northwestern part of the Mirabello region. Six samples of WOD were analyzed, and they belong to three different subgroups of the granodiorite fabric: three samples are included in the main jar fabric (Fabric Group 1), two form a small group with a fine dark matrix but are not included in any of the main fabrics, and one is a loner with calcareous matrix. This indicates that there is no standardized recipe for the WOD vessels, and probably that there is no centralized production for this fine ware either.

With regard to the technology of manufacture, the presence of the granitic-dioritic rock fragments is the common characteristic among the samples examined, but there are differences regarding the raw materials and their manipulation. The clay base for the majority of the vessels analyzed seems to be a red alluvial clay from the Mirabello area, and only for sample PAR 09/02 (**11**) was a calcareous clay used. The presence of rounded clay pellets indicates that the red clay has gone through levigation or sieving. For the main jar fabric (Fabric Group 1) and Subgroup B of the cooking fabric, the granitic-dioritic rock fragments were added as temper, whereas in the other two subgroups of the cooking fabric the nonplastics seem to be occurring naturally in the clay mix. There is a single sample of a jar that is also tempered with organic material, a practice known since the Neolithic period.

As to the origin of the raw materials, comparison with experimental briquettes (e.g., CS 13/26) manufactured with clays and tempering materials from across the area of the Mirabello region showed that the dark red-firing clay is compatible with alluvial clays from the area of Priniatikos

Pyrgos and granodiorite collected from the area of Phrouzi only a couple of kilometers northwest of Gournia (Pl. 49B).

Finally with regard to the firing technology, although SEM has not been performed yet, there are indications of lower or higher firing temperatures through the presence of optical activity. This seems to be the case for the cooking fabric Subgroup A, which displays an optically active groundmass, whereas its absence in the jar fabric and the cooking Subgroups B and C is indicative of a higher firing temperature. For the painted ware from Alatzomouri, an evaluation of the firing temperature is attempted with the aid of comparative data from the bibliography. The SEM analysis of EM III white-on-dark pottery (Matson 1984, 57) and of Protopalatial polychrome pottery (Faber et al. 2002, 132) indicated that the microstructure of the surface slip ranges from initial vitrification to total vitrification thus setting the firing temperature between 800°C and >1000°C. The White-on-Dark Ware from Alatzomouri belongs to fabric groups that are optically inactive reflecting a rather high firing temperature, probably around 800°C. Future SEM analysis is hoped to shed more light on EM III and MM I pyrotechnology.

The petrographic analysis of selected pottery samples from the assemblage of the rock shelter at Alatzomouri allowed for the investigation of the pottery recipes in granitic-dioritic fabrics in plain and painted wares, jars and cooking pots, and within a well-defined chronological frame, that of the EM III to MM I periods. Comparison of the fabric groups from Alatzomouri with the results from the Vrokastro survey allowed the refinement of certain issues, such as the emergence and continuity of certain recipes, and it is hoped that further data from other sites will contribute to a better understanding of pottery production and distribution in the Gulf of Mirabello throughout the Minoan period.

12

Evidence for Chronology

by
Philip P. Betancourt, Thomas M. Brogan, and Vili Apostolakou

The Alatzomouri Rock Shelter provides new information that can be used to assist in an understanding of Minoan chronology just before the period of the foundation of the Middle Minoan palaces. The clay vases come from a very restricted time period. Unlike many Minoan deposits, this context has very few objects from before its main period, and because the cave was closed after the deposition, the final period is also precise. The assemblage is very rich in the decorated vessels whose styles changed more quickly in the Minoan period than plainer containers. Also, the assemblage is a substantial one, with almost 100 well-decorated vessels that are sufficiently preserved to be certain of their style of ornament as well as their physical form.

Although the style of EM III and MM IA ceramics in East Crete has been known for over a century, and it survives in hundreds of examples, most of the pottery comes from mixed contexts. Enough uncertainty has existed on its chronology to prompt one scholar to title a major article, "Does the Early Minoan III Period Exist?" (Zois 1968a).

The main problems involve dividing EM III from MM IA and establishing a synchronism between the East Cretan style and the chronological development of pottery from elsewhere in Crete and in the Aegean. The deposit provides no new evidence for absolute calendar year synchronisms.

Defining the East Cretan Workshop Production

Before discussing chronological problems, it is necessary to carefully define the workshop production to which this deposit belongs. For this discussion, a pottery production is defined as the products of one or more workshops manufacturing clay vases using closely related raw materials, forming and decorating methodologies, and artistic styles. Although the use of Mirabello Fabric and the similarity in some of the shapes and in the techniques of production suggest that the workshops in question also made plainer pottery, only the fine decorated production has relevance for fine-tuned chronological synchronisms based on our present level of knowledge. In this deposit, the finest pottery belongs to the tradition called East Cretan White-on-Dark Ware. The potters who worked in this tradition made pottery by hand with the aid of turntables, covered the exterior and the inside of the rim with a coating of slip that would fire dark brown to black in a reducing atmosphere, and applied pale colored linear decoration over the undercoat. The finished products were mainly distributed over a broad region in Crete from Palaikastro at the eastern end of the island to Malia in East-Central Crete, with the largest concentration of finds at the eastern side of the Gulf of Mirabello region.

The standard vases that were produced consist of a limited number of forms, including both open and closed shapes:

> Open shapes suitable for eating and drinking
>> Shallow conical bowl
>> Conical bowl
>> Spouted conical bowl
>> Conical bowl with frying pan handle
>> Straight-sided cup with vertical handle
>> Rounded cup with vertical handle
>> Rounded cup with tiny lugs or horizontal handles low on the body
>> Open and partly open shapes suitable for holding liquids
>> Collared jar
>> Small amphora
> Shapes with spouts suitable for pouring liquids
>> Spouted conical jar
>> Bridge-spouted jar
>> Spouted kantharos
>> Jug
>> Teapot

Although a few rare shapes were also made, the forms that were standard for this pottery production leave little doubt that it was primarily produced as an elite table ware designed for serving and feasting. All of the decorated vases would be suitable for serving and consuming food and drink. The number of deep bowls is especially appropriate for this purpose because Minoans used many tripod cooking pots, indicating that soups and stews played a major role in the diet (Tzedakis and Martlew, eds., 1999, 79–92). Missing shapes are as important as what is present for an evaluation of the character of the tradition. No cooking vessels (cooking pots and cooking dishes), no large storage vessels (pithoi, large jars, and large amphoras), no ritual vessels (miniature tripods, kernoi, offering bowls, and cylindrical stands), and no specialized vessels for coals (lamps and fireboxes) are known from East Cretan White-on-Dark Ware. The potters manufactured an elite and specialized production with vessels that provided an attractive display in addition to a practical function for eating and drinking.

A few shapes are not present at the cave. Forms that are not present include the pyxis (found at Myrtos Phournou Koriphi: Warren 1972, pl. 62A), the cover with two handles at the side (found at Gournia: Hall 1904–1905, pl. 32:1), and the shallow spouted bowl (found at Vasiliki: Seager 1906–1907, 122, fig. 5c). None of these shapes is common enough to be regarded as a standard product. The presence of a few rare shapes does not change the character of the main tradition as a whole, which was clearly designed principally for a particular use. Considering that feasting has been suggested as an important aspect of Minoan palatial prestige (i.e., Hitchcock, Laffineur, and Crowley, eds., 2008), it is interesting to see that the pottery to support the practice was already present in this early period.

The ceramic forms that help define this local pottery production can be contrasted with the forms used for similar purposes elsewhere. In regard to drinking vessels, for example, the White-on-Dark Ware production manufactured many cups with vertical handles. At Knossos a similar purpose was met with tall conical containers without handles called tumblers and with small goblets with low pedestals (Momigliano 1991). In South-Central Crete, the most common drinking vessel was the conical cup without handles (Betancourt 1990, 63–64). Tumblers, goblets, and conical cups without handles are completely missing from the assemblage manufactured by the potters who made East Cretan White-on-Dark Ware.

The workshop tradition that produced White-on-Dark Ware can be identified as early as EM IIB by its use of the same specific assemblage of vessel shapes. Almost all of the shapes used in White-on-Dark Ware were already present in the Vasiliki

Ware of EM IIB (Betancourt 1979). In addition, Vasiliki Ware was made using a clay fabric containing the same granodiorite as White-on-Dark Ware (Betancourt 1979, 4, citing Myer), and the geographic distribution of the pottery was also similar. Some unusual and highly distinctive shapes, like the teapot with an exaggerated spout and the spouted conical bowl with a small rim-lug opposite the spout, are identical in both wares. The EM III adoption or inheritance of shapes from EM IIB, however, was not 100%. The small goblet, a standard shape in Vasilike Ware (Betancourt 1979), was not produced in East Cretan White-on-Dark Ware.

The painted vases from this workshop tradition were decorated using a mature philosophy of design that was inherited from EM IIA. This traditional Minoan philosophy of ceramic ornament can be illustrated by a teapot from Koumasa (Fig. 68). The vessel is decorated in the Koumasa Style of EM IIA by using an iron-rich slip that will fire to a dark brown color in a reducing atmosphere and will keep its color when the kiln is re-oxidized at the end of the firing cycle because it has become slightly vitrified during the reducing cycle (for a summary of the technological difference between the dark linear painted styles of the Hagios Onouphrios Style, with red lines, and the Koumasa Style, with dark lines, see Betancourt 2008, 47–48). In this philosophy of ornament, the vase and its physical parts are used as the basis for painted symmetry. Individual elements composed of painted lines are applied bilaterally as mirror images on both sides of physical parts like the spout and handle. The result, which can be clearly seen from above (Fig. 68), gives the vase a rational, balanced appearance as it is held in the hands and moved to view its ornament. This philosophy that used the form to establish the composition was an essential element of the EM III artistic style.

Several pieces of evidence suggest that the production of East Cretan White-on-Dark Ware was probably made by several closely related workshops that shared a common philosophy of design and similar methods of production (Ch. 11). Materials seem to have been local because analysis of the white paint used for the decoration has identified different compositions between the white paints used at Palaikastro and in the Mirabello region (Swann, Ferrence, and Betancourt 2000; Ferrence, Swann, and Betancourt 2001). Several settlements existed in the eastern region of the Gulf of Mirabello, which is where the granodiorite that occurs in Mirabello Fabric occurs geologically (Dierckx and Tsikouras 2007). At least two of these places, Gournia and Pera Alatzomouri, had pottery workshops in later periods (for potter's wheels from Gournia, see Hawes et al. 1908, pl. 8:33; for a potter's wheel from Pera Alatzomouri, see Watrous 2012, 110). Both of these sites were already settled by EM II. The large amount of White-on-Dark Ware excavated from the North Trench deposit at Gournia (Hall 1904–1905) leaves little doubt that Gournia was a major production site for the ware, but Pera Alatzomouri is still unexcavated, so we have no evidence one way or the other for a workshop this early for that site.

Chronological Development

Although he did not publish his evidence, Richard Seager wrote in 1905 that an early stage of White-on-Dark Ware used only rectilinear decoration, and that the complex corpus of curvilinear elements used on the pottery was added later (Seager 1904–1905, 218). He was excavating at Vasiliki at the time, and the evidence presumably came from the stratigraphy there. This is exactly the situation that has now been proved to be correct by the deposit from the rock shelter. One can now go further, however, and the chronological development can be understood much more fully and in more detail.

Stratigraphy forms the basis for the main outlines of East Cretan chronology from this period. The pottery decorated as White-on-Dark Ware appears first at Myrtos where its stratum is firmly dated to EM IIB (Warren 1972). The more developed phases, which we can recognize as EM III and MM IA, are stratigraphically above EM IIB at several sites (for Vasiliki, see Seager 1904–1905, 218–220; 1906–1907, 114, 118–119; for Palaikastro, see Dawkins 1903–1904, 199–200; 1904–1905, 273; Bosanquet and Dawkins 1923, 8; for Pseira, see Seager 1910, 17; for Mochlos, see Seager 1909, 278–279). Comparisons with Central Crete prove that the end of the development is contemporary with early MM I in Central Crete (Warren 1965, 25–27). That this stage in the development in East

Crete is contemporary with MM IA at Knossos is proved by the presence of a Central Cretan MM IA goblet along with White-on-Dark Ware in a deposit from Palaikastro (MacGillivray et al. 1992, 131, fig. 9). East Cretan White-on-Dark Ware also occurs at Knossos in a MM IA context (Momigliano 1991, 227, vase no. 26).

The internal chronological development within the EM IIB to MM IA stages can now be clarified with this new deposit. In summary, the white-painted designs from EM IIB were all very simple patterns (Fig. 69). More complex linear designs were added by the time of the deposit at the Alatzomouri Rock Shelter, and almost all of them were rectilinear (Fig. 70). These vases can be assigned to EM III. A more complex corpus of designs was added by the time of the North Trench deposit at Gournia (Fig. 71), which can now be placed in MM IA. The MM IA elements include a large number of new curvilinear designs including circle motifs (Fig. 71:m–p), semicircles formed by lines and dots (Fig. 71:k, l), and even some zoomorphic elements (Fig. 71:w, x). By MM IB, the style of vase painting had developed further in this workshop, and it included additional motifs and the painting of the interior of the vessel (Betancourt 1984, 27, fig. 3-6, Late Phase); the ornament used at some of the other workshops in East Crete was even more complex (Haggis 2007, 2012; Tsipopoulou 2012a, 2012b).

The division by decorative motifs shown in Figures 69 to 71 suggests that only two deposits known from this local ceramic production can be assigned to EM III: the present deposit and probably the tiny deposit from the copper smelting workshop at Chrysokamino (Betancourt 2006, 67–97). In both cases, the assemblages do not include the elaborate circle motifs and other complex designs found elsewhere.

The development from EM IIB to MM IA was not a change of motifs but an addition of more complicated designs to an existing tradition. Several examples illustrate the process. Chevrons, for example, were present in EM IIB (Fig. 69:a), and they were still being used in EM III (Fig. 70:d) and in MM IA (Fig. 71:a). By EM III the hatched triangle had been added to the repertoire (Fig. 70:a, b), and it, also, continued into MM IA in several variations (Fig. 71:b, d, e) alongside many newly added ornamental schemes. The additions in MM IA included many curvilinear designs, a few designs using zoomorphic images (Fig. 71:w, x), and even an occasional use of an area of color rather than a line (Fig. 71:t). As these examples demonstrate, the use of simple hatched triangles and bands, introduced early in the sequence, continued without a break alongside the more dynamic designs that were added later. Other aspects of the tradition, like the shapes of the vessels and the philosophy of the composition, did not change much. The tradition can be seen as a smooth development in the same workshop production. The elements shown in Figure 71 at the height of the mature tradition only represent a selection of the many new elements added to this rich assemblage of motifs (for a longer list, see Betancourt 1984, 21–32).

With these new insights we can better understand the development of the Minoan ceramic production of EM IIB to MM IA in the Gulf of Mirabello workshops. The tradition began as tentative experiments with contrasting pale colored slips on a darker undercoat of paint, and it gradually developed in a very dynamic and creative way. In this creative development, one cannot minimize the social aspects. It is important that this was an elite pottery designed from the beginning for elegant display in social events involving communal eating and drinking. The motivation for increasingly elegant ornament must have come at least partly from the patrons who used this pottery and appreciated the care that was being given to the vessels and the complexity of the ornaments that were being invented for their enjoyment.

13

Interpretation of the Rock Shelter

by

Vili Apostolakou, Thomas M. Brogan, Susan C. Ferrence, and Philip P. Betancourt

The rock shelter at Alatzomouri has several ambiguous characteristics that make its interpretation enigmatic and uncertain. In many ways, it is unique in Minoan archaeology. The exact function of the small deposit and the reasons for its burial will never be known for certain, but several possibilities can be suggested.

The rock shelter is located several hundred meters from the nearest settlements, the sites of Alatzomouri Site 17 and Pera Alatzomouri, both situated to the north on other parts of the same hill. These unexcavated habitation sites have been recorded by the survey of Vance Watrous and his colleagues (Watrous et al. 2012, 110–111). The rock shelter is, thus, a feature that was used to deliberately place its contents away from the community to which it belonged.

The outward form of the modest monument is very straightforward. It consists of a small underground room, which was carved into soft marl and filled with pottery and other objects. A wall was used to close the front of the space, and natural erosion eventually covered it from view. Almost all of the pottery was from Early Minoan III, and nothing was later than this date. The deposit had not been touched or disturbed after its closure until modern times.

A piece of a human skull and two clay vases found nearby might or might not be related to the rock shelter itself. The date of the two vases is EM III, which makes them contemporary with the date of the deposition of the objects in the rock shelter. This circumstance is strongly suggestive of a relationship between the two contexts, but it does not necessarily prove it.

The rock shelter was excavated with modern protocols for retrieval and study, with extensive processing of the soil with a water separation machine (water sieving). All artifacts were plotted by stratigraphic location, saved, and subsequently mended, conserved, cataloged, and studied. The pottery was strewn for mending, and all joins that were found were mended or recorded. The selection of objects to be published included all whole or restorable vessels, all sections of vessels, all tools and other categories of artifacts, and samples of all classes of pottery fragments. Many samples were taken and stored for later analysis.

Three general contexts could be recognized: a bed of sherds and small fragments that was mostly at the bottom of the shelter; more complete objects on top of this bed of sherds; and objects from outside the cave. Humble storage and industrial vessels, used cooking vases, and fine elite decorated pieces were mixed at random. Sometimes objects in individual classes (like loomweights) were near one another as if they had been in a bag. Many almost whole vessels were not deposited in one piece. In many cases, vessels were assembled from two or even from many pieces found at a distance from one another.

The objects from the rock shelter were almost all broken with one or more missing pieces. Many objects survived as single sherds. The complete record of the pottery is documented statistically (Table 10.1), and it is essential for any interpretation. The missing fragments were not buried inside the rest of the deposit, and this circumstance has to be taken into consideration for the meaning of the context. As found, the assemblage consisted only of imperishable goods made of clay or stone or shell, and the soil conditions may be responsible for the absence of animal bones, which are routinely present in Minoan household debris. Nothing was burned (aside from the cooking vessels, olive pits, and grape seeds), and very little carbonized wood was present.

Much of the pottery consisted of objects that would be used within a large Minoan household. Functions represented included household storage, food preparation and serving, and light industry, including weaving, tool making, pottery making(?), and the processing of grapes to make wine. The pottery included both fine serving vessels and undecorated storage and cooking vases, and even part of a large spouted vat. The assemblage of objects, however, was not complete for a normal Cretan household unit from the close of the Early Bronze Age. No metal objects or stone vessels were present, and both of them would have been normal for even a poor household from this period (for the Minoan household, see Watrous and Heimroth 2011). No personal jewelry items were in the deposit, and the cave contents did not include any pyxides (the Minoan clay box used for personal items). No ceremonial pieces, such as offering stands or effigy vases, were included.

The deposit is clearly a specialized one, and it is possible to suggest several possibilities for it.

Was It a Tomb?

A few characteristics are similar to a burial. The architecture is certainly a common one for burials in this part of Crete at the close of the Early Bronze Age. Rock shelters are used as tombs at many cemeteries in eastern Crete in the region of the eastern Gulf of Mirabello, including Sphoungaras (Hall 1912), Mochlos (Seager 1912), and Pseira (Betancourt and Davaras, eds., 2003). The shallow artificial cavities are carved into the soft local bedrock in the same way as the rock shelter here, and similar walls are used to close them off (although the walls for tombs seldom survive into modern times). In addition, the fragment of a human skull and two clay vessels found near the rock shelter indicate that even if they were unrelated, the deposit was in a funerary area.

The character of the deposit, however, argues against the idea that it was a burial. No bones were found, and even if the soil conditions destroyed them, one would expect teeth to survive. The objects inside the chamber, however, bear no relation to a Minoan burial assemblage. In Minoan tombs from this period, the preponderance of grave goods consists of personal possessions. Jewelry, tools, weapons, sealstones, and pottery that could have contained food or drink form the main part of every burial assemblage. Personal possessions like jewelry are precisely the objects that are missing here. Minoan burials never include large numbers of storage vessels or cooking pots or industrial vessels like those found in this context.

Another possibility is that the piece of skull and the two clay vases from outside the rock shelter could be regarded as a burial with the cave as an adjunct. In this theory, the cave would be a funerary votive associated with a jar burial. Placing the deceased inside jars was a tradition that was just beginning in EM III. The communal burial of earlier times would continue alongside individual interment in jars and clay chests (called larnakes) for some centuries, but jar burials would become more popular during the Middle Minoan period. For this part of Crete, the best known cemeteries with large numbers of jar burials are Sphoungaras (Hall 1912) and the Pacheia Ammos beach (Seager 1916). Similar practices were used for a few individuals in the cemeteries at Mochlos (Seager 1912, 87–89) and Pseira (Betancourt 2003, 128–129). In all cases except for Pacheia Ammos, the jar burials occurred alongside other classes of interment. The problem with this theory is that although over a hundred jar burials have been documented for the region, not a single one has an associated deposit of domestic objects like the example discussed here. In all other cases, the burial is a self-contained and singular unit.

Was It a Cenotaph?

A cenotaph is a burial monument without the body of the deceased. It can be built for various reasons, such as to commemorate a death that occurs at sea or somewhere else so that the body is not present for burial. The same arguments for and against a tomb can be used for a cenotaph. Such a theory explains the missing body, but it does not account for the fact that the assemblage does not fit the Minoan custom for burial assemblages. No cenotaphs have been positively identified for EM Crete.

Was It a Cache Deposited for Safekeeping?

Superficially, the idea of a buried cache might seem like a secure depository for items to be used later, but in this case the details deny the concept. Almost everything was broken and not useable. The items were not accidentally broken after burial by an earthquake because in many cases the fragments were spread apart. They were broken before burial, and pieces were retained outside the rock shelter. It could not have been a cache for future use.

Was It a Cleaning Operation after an Earthquake?

Certainly if pottery and other items are damaged in an earthquake or some other disaster like war, they need to be disposed. Earthquakes are common enough in Crete to require periodic architectural demolition and rebuilding. The problem with this theory is that archaeological evidence for earthquakes is common for Minoan Crete, and the damaged goods are never known to have been deposited this way. In fact, it is counter-productive to spend substantial time digging out an underground space, depositing the material, and building a wall to block the entrance. Broken pottery can just be leveled out and used as a base for the next architectural phase, and that is what was routinely accomplished.

Was It a Votive or Other Ceremonial Deposit?

Deliberately removing a group of objects from use could be done for many social reasons, and it could be an attempt to achieve either positive or negative results. For example, a positive reason might involve a votive gift to a deity or the traditional symbolic burial of one's past life before embarking on a voyage to a new land or before entering the service of a religious obligation. A negative motivation might include removing a household from the community in a desire to prevent a disease from spreading or removing the household of someone who was cursed or ostracized. A ceremonial explanation would explain both the breaking of the pottery and other objects and their removal from society, and this act of deliberate destruction could also be positive or negative. A positive attitude might lead to breaking objects to release their spirit or to consecrate them to a deity. For the negative alternative, the breakage might be a desire to kill the spirits of the objects or quell their negative energy. This theory does not explain why some categories, like metals and jewelry, are not present in the assemblage. With no written documents for guidance, social practices are difficult to demonstrate beyond a theoretical suggestion.

Discussion

From the presentation of alternative scenarios for the deposit in the rock shelter, it is clear that some of the questions about its meaning and purpose will always remain unanswered. Social motivations are difficult to address without written records. Several of the aspects of the assemblage, however, suggest a symbolic act. One of these characteristics is the pattern of broken and incomplete objects. The discovery of buried cult deposits with broken objects with some pieces missing has puzzled scholars before. The classic case is the Temple Repositories at Knossos (Panagiotaki 1999) where the presence of votive pits filled with broken faience figurines and other objects with pieces missing along with intact pottery vessels has still never been adequately explained. Special deposits at Keros from the Keros-Syros and Kastri phases of the Cyclades, the period immediately before EM III in Crete, have yielded hundreds of fragments of marble figurines and bowls that do not join with one another (Renfrew 2012, 35–36). The deposit suggests a pattern of ritual breakage and separation of certain fragments for ceremonial burial away from other parts of the original, including possible retention by the worshippers of some of the fragments. From a later Minoan period, Paul Rehak has suggested that stone vessels, including bull's head rhyta, were ceremonially broken and their fragments dispersed (Rehak 1995). The symbolic removal of objects from the community at Alatzomouri is by no means unique in the Aegean, and many other cases are surely masked by the normal processes of accidental or deliberate breakage in contexts that do not involve ritual. The example from the Alatzomouri Rock Shelter must be added to the list of enigmatic cases of this practice.

Among the possibilities discussed above, the most likely scenario is that the cave represents some type of deliberate symbolic act. Objects representing many (but not all) of the possessions of a large Minoan household were gathered together and buried. It is as if the carpets were gathered up with whatever fragments happened to be lying on them or the actual floor was scooped up and placed in the cave before more complete objects were added. After the act, the cave's mouth was closed with a wall. Although the exact motivation behind the event is unknown, the assemblage can still be examined as an example of an Early Minoan communal act. In addition, of course, the rock shelter is also important for many other reasons, including the contributions it makes to chronology, artistic style, domestic economy, and ceramic history.

References

Abbreviations follow the conventions used by the *American Journal of Archaeology*.

Ambraseys, N. 2009. *Earthquakes in the Mediterranean and Middle East: A Multidisciplinary Study of Seismicity up to 1900*, Cambridge.

Amouretti, M.-C. 1970. *Fouilles exécutées à Mallia: Le centre politique II. La crypte hypostyle (1957–1962)* (*ÉtCrét* 18), Paris.

Andersson Strand, E., and M.-L. Nosch, eds. 2015. *Tools, Textiles, and Contexts: Investigating Textile Production in the Aegean and Eastern Mediterranean Bronze Age* (*Ancient Textile Series 21*), Oxford.

Antoniadou, Ch., D. Vafidis, E. Voultsiadou, and Ch. Chindiroglou. 2013. "Εκτίμηση Πληθυσμιακών Παραμέτρων του Προστατευομένου Είδους *Pinnanobilis* με Μη-Καταστρεπτικές Τεχνικές στη Θαλάσσια Περιοχή της Δωδεκανήσου," Πρακτικά Πανελλήνιου Συνεδρίου Ιχθυολόγων 15, pp. 21–24.

Apostolakou, V., P.P. Betancourt, and T.M. Brogan. 2007–2008. "The Alatzomouri Rock Shelter: Defining EM III in Eastern Crete," *Aegean Archaeology* 9 [2010], pp. 35–48.

———. 2010. "Ανασκαφικές έρευνες στην Παχειά Άμμο και τη Χρυσή Ιεράπετρας," in Αρχαιολογικό Έργο Κρήτης 1, M. Andreanakis and I. Tzachili, eds., Rethymnon, pp. 143–154.

Armijo, R., H. Lyon-Caen, and D. Papanastassiou. 1992. "East–West Extension and Holocene Normal-Fault Scarps in the Hellenic Arc," *Geology* 20, no. 6, pp. 491–494. doi:10.1130/0091-7613(1992)020 <0491:EWEAHN>2.3.CO;2

Back, M.E., and J.A. Mandarino. 2008. *Fleischer's Glossary of Mineral Species*, Tucson.

Banou, E.S. 1995. "Building AA: The Pottery," "Building AM: The Pottery," and "Building AP: The Pottery," in Betancourt and Davaras, eds., 1995, pp. 19–20, 63–64, 78–80.

Barber, E.J.W. 1991. *Prehistoric Textiles: The Development of Cloth in the Neolithic and Bronze Ages with Special Reference to the Aegean*, Princeton.

Barnard, K.A. 2003. "A Macroscopic Analysis of the Neopalatial Fabrics," in Barnard and Brogan 2003, pp. 3–12.

Barnard, K.A., and T.M. Brogan. 2003. *Mochlos IB: Period III. Neopalatial Settlement on the Coast: The*

Artisans' Quarter and the Farmhouse at Chalinomouri. The Neopalatial Pottery (Prehistory Monographs 8), Philadelphia.

Betancourt, P.P. 1979. *Vasilike Ware: An Early Bronze Age Pottery Style in Crete* (SIMA 56), Göteborg.

———. 1980. *Cooking Vessels from Minoan Kommos: A Preliminary Report*, Los Angeles.

———. 1983. *The Cretan Collection in the University Museum, University of Pennsylvania I: Minoan Objects Excavated from Vasilike, Pseira, Sphoungaras, Priniatikos Pyrgos, and Other Sites* (University Museum Monograph 47), Philadelphia.

———. 1984. *East Cretan White-on-Dark Ware: Studies on a Handmade Pottery of the Early to Middle Minoan Periods* (University Museum Monograph 51), Philadelphia.

———. 1985. *The History of Minoan Pottery*, Princeton.

———. 1990. *Kommos II: The Final Neolithic through Middle Minoan III Pottery*, Princeton.

———. 1999. "Area BR: The Pottery," in Betancourt and Davaras, eds., 1999, pp. 141–154.

———. 2003. "Interpretation and Conclusions," in Betancourt and Davaras, eds., 2003, pp. 123–139.

———. 2006. *The Chrysokamino Metallurgy Workshop and Its Territory* (Hesperia Suppl. 36), Princeton.

———. 2008. *The Bronze Age Begins: The Ceramics Revolution of Early Minoan I and the New Forms of Wealth That Transformed Prehistoric Society*, Philadelphia.

———. 2013. *Aphrodite's Kephali: An Early Minoan Defensive Site in Eastern Crete* (Prehistory Monographs 41), Philadelphia.

Betancourt, P.P., and C. Davaras, eds. 1995. *Pseira I: Minoan Buildings on the West Side of Area A* (University Museum Monograph 90), Philadelphia.

———, eds. 1998a. *Pseira II: Building AC (the "Shrine") and Other Buildings in Area A* (University Museum Monograph 94), Philadelphia.

———, eds. 1998b. *Pseira III: The Plateia Building* (University Museum Monograph 102), Cheryl R. Floyd, Philadelphia.

———, eds. 1999. *Pseira IV: Minoan Buildings in Areas B, C, D, and F* (University Museum Monograph 105), Philadelphia.

———, eds. 2002. *Pseira VI: The Pseira Cemetery 1. The Surface Survey* (Prehistory Monographs 5), Philadelphia.

———, eds. 2003. *Pseira VII: The Pseira Cemetery 2. Excavation of the Tombs* (Prehistory Monographs 6), Philadelphia.

———, eds. 2009. *Pseira X: The Excavation of Block AF* (Prehistory Monographs 28), P.P. Betancourt, Philadelphia.

Betancourt, P.P., T.K. Gaisser, F.R. Matson, G.H. Myer, and C.P. Swann. 1979. "Definition of Vasilike Ware and Analytical Tests," in Betancourt 1979, pp. 3–11.

Betancourt, P.P., G. Gosser, and S. Sapareto. 1984. "Reconstruction of Potting Techniques and Pyrotechnology," in Betancourt 1984, pp. 126–129.

Betancourt, P.P., D.S. Reese, and W.F. Schoch. 2003. "Tomb 2: Catalog of Objects," in Betancourt and Davaras, eds., 2003, pp. 23–30.

Betancourt, P.P., and J.S. Silverman. 1991. *The Cretan Collection in the University Museum, University of Pennsylvania*, vol. II: *Pottery from Gournia* (University Museum Monograph 52), Philadelphia.

Blinkenberg, C., and K.F. Johansen. 1924. *Corpus vasorum antiquorum: Danemark. Copenhague, Musée National* 1 (CVA, Denmark 1), Paris.

Blitzer, H. 1995. "Minoan Implements and Industries," in *Kommos* I: *The Kommos Region and Houses of the Minoan Town*. Part 1: *The Kommos Region, Ecology, and Minoan Industries*, J.W. Shaw and M.C. Shaw, eds., Princeton, pp. 403–535.

Bosanquet, R.C., and R.M. Dawkins. 1923. *The Unpublished Objects from the Palaikastro Excavations 1902–1906* (BSA Suppl. 1), London.

Boschini, M. 1641. *Il regno tutto di Candia: delineato à parte, à parte, et intagliato da Marco Boschini venetiano*, Venice.

Brogan, T.M. 2013. "'Minding the Gap': Re-examining the Early Cycladic III Gap from the Perspective of Crete. A Regional Approach to Relative Chronology, Networks and Complexity in the Late Prepalatial Period," *AJA* 117, pp. 555–567.

Brogan, T.M., Ch. Sofianou, J.E. Morrison, E. Margaritis, and D. Mylona. 2013. "Living Off the Fruit of the Sea in House A.1 at Papadiokampos," in *Diet, Economy and Society in the Ancient Greek World: Towards a Better Integration of Archaeology and Science* (Pharos Suppl. 1), S. Votzaki and S.M. Valamoti, eds., Leuven, pp. 123–132.

Burke, B. 2006. "Textile Production at Petras: The Evidence from House 2," in Πεπραγμένα Θ' Διεθνούς Κρητολογικού Συνεδρίου Α' (1), Herakleion, pp. 279–295.

Cadogan, G. 1978. "Pyrgos, Crete 1970–1977," *AR* 24, pp. 70–84.

Caputo, R., S. Catalano, C. Monaco, G. Romagnoli, G. Tortorici, and L. Tortorici. 2010. "Active Faulting on the Island of Crete (Greece)," *Geophysical Journal International* 183, pp. 111–126. doi:10.1111/j.1365-246X.2010.04749.x

Carter, T. 2004. "The Stone Implements," in *Mochlos IC: Period III. Neopalatial Settlement on the Coast: The Artisans' Quarter and the Farmhouse at Chalinomouri. The Small Finds* (Prehistory Monographs 9), J.S. Soles, C. Davaras, J. Bending, T. Carter, D. Kondopoulou, D. Mylona, M. Ntinou, A.M. Nicgorski, D.S. Reese, A. Sarpaki, W.H. Schoch, M.E. Soles, V. Spatharas, Z.A. Stos-Gale, D.H. Tarling, and C. Witmore, Philadelphia, pp. 61–107.

―――. 2008. "The Consumption of Obsidian in the Early Bronze Age Cyclades," in *Horizons: A Colloquium on the Prehistory of the Cyclades* (McDonald Institute Monographs), N. Brodie, J. Doole, G. Gavalas, and C. Renfrew, eds., Cambridge, pp. 225–235.

Carter, T., and V. Kilikoglou. 2007. "From Reactor to Royalty? Aegean and Anatolian Obsidians from Quartier Mu, Malia (Crete)," *JMA* 20, pp. 115–143.

Chapouthier, F., and P. Demargne. 1942. *Fouilles exécutées à Mallia, troisième rapport: Exploration du palais, bordures orientale et septentrionale (1927, 1928, 1931, 1932)* (ÉtCrét 6), Paris.

Chapouthier, F., P. Demargne, and A. Dessenne. 1962. *Mallia, Quatrième Rapport (1929–1935, 1946–1960)* (ÉtCrét 12), Paris.

Christakis, K. 2005. *Cretan Bronze Age Pithoi: Traditions and Trends in the Production and Consumption of Storage Containers in Bronze Age Crete* (Prehistory Monographs 18), Philadelphia.

―――. 2008. *The Politics of Storage: Storage and Sociopolitical Complexity in Neopalatial Crete* (Prehistory Monographs 25), Philadelphia.

Creutzburg, N. 1977. General Geological Map of Greece: Crete. 1:200,000. Institute of Geology and Mining Research, Athens.

Cutler, J. 2011. *Crafting Minoanisation: Textiles, Crafts, Production and Social Dynamics in the Bronze Age Southern Aegean*, Ph.D. diss., University College London.

Dawkins, R.M. 1902–1903. "Excavations at Palaikastro II: 8. The Pottery," *BSA* 9, pp. 297–328.

―――. 1903–1904. "Excavations at Palaikastro III," *BSA* 10, pp. 192–226.

―――. 1904–1905. "Excavations at Palaikastro IV," *BSA* 11, pp. 258–292.

Day, P.M. 1991. *A Petrographic Approach to Pottery in Neopalatial East Crete*, Ph.D. diss., University of Cambridge.

―――. 1995. "Pottery Production and Consumption in the Siteia Bay Area during the New Palace Period," in *Achladia: Scavi e Ricerche della Missione Greco-Italiana in Creta Orientale (1991–1993)*, M. Tsipopoulou and L. Vagnetti, eds., Rome, pp. 149–173.

―――. 1997. "Ceramic Exchange between Towns and Outlying Settlements in Neopalatial East Crete," in *The Function of the "Minoan Villa." Proceedings of the Eighth International Symposium at the Swedish Institute at Athens, 6–8 June 1992* (ActaAth 4°, 46), R. Hägg, ed., Stockholm, pp. 219–228.

Day, P.M., L. Joyner, and M. Relaki. 2003. "A Petrographic Analysis of the Neopalatial Pottery," in Barnard and Brogan 2003, pp. 13–32.

Day, P.M., L. Joyner, E. Kiriatzi, and M. Relaki. 2005. "Petrographic Analysis of Some Final Neolithic–Early Minoan II Pottery from the Kavousi Area," in *Kavousi I: The Archaeological Survey of the Kavousi Region* (Prehistory Monographs 16), D.C. Haggis, Philadelphia, pp. 177–195.

Day, P.M., V. Kilikoglou, L. Joyner, and G.C. Gesell. 2006. "Goddesses, Snake Tubes, and Plaques: Analysis of Ceramic Ritual Objects from the LM IIIC Shrine at Kavousi," *Hesperia* 75, pp. 137–175.

Demargne, P. 1945. *Fouilles exécutées à Malia: Exploration des nécropoles (1921–1933)* (ÉtCrét 7), Paris.

Demargne, P., and H. Gallet de Santerre. 1953. *Fouilles exécutées à Mallia: Exploration des maisons et quartiers d'habitation (1921–1948)* (ÉtCrét 9), Paris.

Dierckx, H.M.C. 1992. *Aspects of Minoan Technology, Culture, and Economy: The Bronze Age Stone Industry of Crete*, Ph.D. diss., University of Pennsylvania

―――. 2016. "The Ground and Chipped Stone Implements from the Settlement," in *Kavousi IIC: The Late Minoan IIIC Settlement at Vronda. Specialist Reports and Analyses* (Prehistory Monographs 52), G.C. Gesell and L.P. Day, eds., Philadelphia, pp. 137–153.

Dierckx, H.M.C., and B. Tsikouras. 2007. "Petrographic Characterization of Rocks from the Mirabello Bay Region, Crete, and Its Application to Minoan Archaeology: The Provenance of Stone Implements from Minoan Sites," *Proceedings of the 11th International Congress, Athens, May, 2007* (Bulletin of the Geological Society of Greece 40), pp. 1768–1779.

Dothan, T. 1963. "Spinning Bowls," *Israel Exploration Journal* 13, pp. 97–112.

Evely, R.D.G. 1984. "The Other Finds of Stone, Clay, Ivory, Faience, Lead, Etc.," in *The Minoan Unexplored Mansion at Knossos* (*BSA Suppl.* 17), M.R. Popham, London, pp. 223–259.

———. 2003. "The Stone, Bone, Ivory, Bronze and Clay Finds," in *Knossos: The South House* (*BSA Suppl.* 34), P. Mountjoy, ed., London, pp. 167–194.

Faber, E.W., V. Kilikoglou, P.M. Day, and D.E. Wilson. 2002. "A Technological Study of Middle Minoan Polychrome Pottery from Knossos, Crete," in *Modern Trends in Scientific Studies on Ancient Ceramics. Papers Presented at the 5th European Meeting on Ancient Ceramics, Athens 1999* (*BAR-IS* 1011), V. Kilikoglou, A. Hein, and Y. Maniatis, eds., Oxford, pp. 129–141.

Fassoulas, C.G. 2000. *Field Guide to the Geology of Crete*, Herakleion.

Ferrence, S. 2008. "Pacheia Ammos Tombs Trench Notebook, Alatzomouri Hill," unpublished field notebook, archives, INSTAP Study Center for East Crete, Pacheia Ammos, Crete, Greece.

Ferrence, S.C., and E.B. Shank. 2006. "Evidence for Beekeeping," in Betancourt 2006, pp. 391–392.

Ferrence, S.C., C.P. Swann, and P.P. Betancourt. 2001. "PIXE Analysis of White Pigments on Pottery from Five Bronze Age Minoan Archaeological Sites," *Aegean Archaeology* 5 [2002], pp. 47–54.

Floyd, C.R. 1998. *Pseira III: The Plateia Building* (*University Museum Monograph* 102), P.P. Betancourt and C. Davaras, eds., Philadelphia.

———. 2009a. "Pottery from Block AF," in Betancourt and Davaras, eds., 2009, pp. 39–94.

———. 2009b. "Pottery Statistics," in Betancourt and Davaras, eds., 2009, pp. 171–222.

Forsdyke, E.J. 1925. *Catalog of the Greek and Etruscan Vases in the British Museum*, vol. I, pt. 1, *Prehistoric Aegean Pottery*, London.

Fortuin, A.R. 1977. *Stratigraphy and Sedimentary History of the Neogene Deposits in the Ierapetra Region, Eastern Crete* (*GUA Papers of Geology* 1°, 8), Utrecht.

Foster, K.P. 1978. "The Mount Holyoke Collection of Minoan Pottery," *TUAS* 3, pp. 1–29.

Georgotas, A. 2013. "Raw Materials—Hunting and Experimental Research on Minoan Ceramic Technology: A View from the Mirabello," *Kentro: The Newsletter of the INSTAP Study Center for East Crete* 16, pp. 11–13.

Graham, J.W. 1987. *The Palaces of Crete*, Princeton.

Greig, J.R.A., and P. Warren. 1974. "Early Bronze Age Agriculture in Western Crete," *Antiquity* 48, pp. 130–132.

Haggis, D.C. 2005. *Kavousi I: The Archaeological Survey of the Kavousi Region* (*Prehistory Monographs* 16), Philadelphia.

———. 2007. "Stylistic Diversity and Diacritical Feasting at Protopalatial Petras: A Preliminary Analysis of the Lakkos Deposit," *AJA* 111, pp. 715–775.

———. 2012. "The Lakkos Pottery and Middle Minoan IB Petras," in *Petras, Siteia: 25 Years of Excavations and Studies. Acts of a Two-Day Conference Held at the Danish Institute at Athens, 9–10 October 2010* (*Monographs of the Danish Institute at Athens* 16), Athens, pp. 191–204.

Haggis, D.C., and M.S. Mook. 1993. "The Kavousi Coarse Wares: A Bronze Age Chronology for Survey in the Mirabello Area, East Crete," *AJA* 97, pp. 265–293.

Hall, E.H. 1904–1905. "Early Painted Pottery from Gournia, Crete," *University of Pennsylvania Transactions of the Free Museum of Science and Art* 1, pp. 191–205.

———. 1912. *Excavations in Eastern Crete: Sphoungaras* (*University of Pennsylvania, the Museum Anthropological Publications* 3 [2]), Philadelphia.

Hawes, H.B., B.E. Williams, R.B. Seager, and E.H. Hall. 1908. *Gournia, Vasiliki and Other Prehistoric Sites on the Isthmus of Hierapetra, Crete*, Philadelphia.

Hayden, B. 2004. *Reports on the Vrokastro Area, Eastern Crete II: The Settlement History of the Vrokastro Area and Related Studies* (*University Museum Monograph* 119), Philadelphia.

Hitchcock, L.A., R. Laffineur, and J. Crowley, eds. 2008. *DAIS: The Aegean Feast. Proceedings of the 12th International Aegean Conference, University of Melbourne, Centre for Classics and Archaeology, 25–29 March 2008* (*Aegaeum* 29), Liège.

Hood, M.S.F. 1990. "Autochthons or Settlers? Evidence for Immigration at the Beginning of the Early Bronze Age in Crete," in Πεπραγμένα του ΣΤ' Διεθνούς Κρητολογικού Συνεδρίου Α' (1), Chania, pp. 367–375.

Hsu, K.J. 1983. *The Mediterranean Was a Desert: A Voyage of the Glomar Challenger*, Princeton.

Hurlbut, C.S., and C. Klein. 1977. *Manual of Mineralogy*, New York.

IGME, Institute for Geology and Mining Exploration. 1959. Geological Map of Greece: Ierapetra Quadrangle. 1:24,000, Athens.

I.G.S.R. 1959. Geological Map of Greece: Kato Chorion (Ierapetra) Sheet. 1:50,000, Athens.

Jones, G., and A. Schofield. 2006. "Evidence for the Use of Threshing Remains at the Early Minoan Metallurgical Workshop," in Betancourt 2006, pp. 153–154.

Kanta, A., and L. Rocchetti. 1989. "La Ceramica del Primo Edificio," in *Scavi a Nerokourou, Kydonias (Ricerche Greco-Italiane in Creta Occidentale* 1), Rome, pp. 101–279.

Karali, L. 1999. *Shells in Aegean Prehistory (BAR-IS* 761), Oxford.

Kemp, B.J., and G. Vogelsang-Eastwood. 2001. *The Ancient Textile Industry at Amarna (Egypt Exploration Fund Memoir* 68), London.

Koh, A. 2008. "The ARCHEM Project Turns Five," *Kentro: The Newsletter of the INSTAP Study Center for East Crete* 11, pp. 6–8.

Koh, A.J., and P.P. Betancourt. 2010. "Wine and Olive Oil from an Early Minoan I Hilltop Fort," *Mediterranean Archaeology and Archaeometry* 10 (2), pp. 15–23.

Kollmorgen Instrument Organization. 1992. *Munsell Color Charts*, Newburgh, NY.

Koutsoubas, D., S. Galinou-Mitsoudi, S. Katsanevakis, P. Leontarakis, A. Metaxato, and A. Zenetos. 2007. "Bivalve and Gastropod Molluscs of Commercial Interest for Human Consumption in the Hellenic Seas," in *State of the Hellenic Fisheries*, C. Papaconstantinou, V. Vassilopoulou, G. Tserpes, and A. Zenetos, eds., Athens, pp. 70–84.

Lambeck, K., and A. Purcell. 2005. "Sea-Level Change in the Mediterranean Sea since the LGM: Model Predictions for Tectonically Stable Areas," *Quaternary Science Reviews* 24, pp. 1969–1988.

Langford-Verstegen, L. 2015. *Hagios Charalambos: A Minoan Burial Cave in Crete* II. *The Pottery (Prehistory Monographs* 51), Philadelphia.

Larson, R.L. 1991. "Latest Pulse of Earth: Evidence for a Mid-Cretaceous Superplume," *Geology* 19, pp. 547–550. doi:10.1130/0091-7613(1991)019<0547:LPOEEF>2.3.CO;2

Leukadites, G. 1941. *Τὸ ψάρεμα στὰ ἑλληνικὰ ἀκρογιάλια: Τὰ σύνεργα, οἱ τρόποι, τὰ ψάρια*, Athens.

Levi, D. 1976. *Festòs e la civiltà minoica (Incunabula graeca* 60), vol. I, Rome.

MacGillivray, J.A., L.H. Sackett, J.M. Driessen, and S. Hemingway. 1992. "Excavations at Palaikastro 1991," *BSA* 87, pp. 121–152.

Maraghiannis, G., and G. Karo. 1907–1921. *Antiquités crétoises* I–III, Vienna.

Margaritis, E. 2013a. "Arboriculture at Aphrodite's Kephali," in Betancourt 2013, pp. 111–115.

———. 2013b. "Distinguishing Exploitation, Domestication, Cultivation, and Production: The Olive in the Third Millennium Aegean," *Antiquity* 87, pp. 746–757.

———. 2014. "Acts of Destruction and Acts of Preservation: Plants in the Ritual Landscape of Prehistoric Greece," in *Physis: L'environnement naturel et la relation homme-milieu dans le monde égéen protohistorique. Actes de la 14e Rencontre Égéenne internationale, Paris, Institut National d'Histoire de l'Art (INHA), 11–14 décembre 2012 (Aegaeum* 37), G. Touchais, R. Laffineur, and F. Rougemont, eds., Liège, pp. 279–285.

———. 2017. "The Plant Remains of the House Tombs at Petras: Acts of Destruction, Transformation and Preservation," in *Petras, Siteia: The Pre- and Protopalatial Petras Cemetery in Context (Monographs of the Danish Institute at Athens* 21), M. Tsipopoulou, ed., Athens, pp. 225–236.

Margaritis, E., and M.K. Jones. 2006. "Beyond Cereals: Crop Processing and *Vitis vinifera* L. Ethnography, Experiment and Charred Grape Remains from Hellenistic Greece," *JAS* 33, pp. 784–805.

———. 2008. "Crop Processing and *Olea europaea* L.: An Experimental Approach for the Interpretation of Archaeobotanical Olive Remains," *Vegetation History and Archaeobotany* 17, pp. 381–392.

Marinatos, S., and M. Hirmer. 1976. *Kreta, Thera und das mykenische Hellas*, Munich.

Martlew, H. 1988. "Domestic Coarse Pottery in Bronze Age Crete," in *Problems in Greek Prehistory*, E.B. French and K.A. Wardle, eds., Bristol, pp. 421–424.

Matson, F.R. 1984. "Physical Characteristics of the Fabric, Slip, and Paint," in Betancourt 1984, pp. 52–59.

McCoy, F.W. 2013. "Geology and Geologic History," in Betancourt 2013, pp. 15–33.

Melas, E.M. 1985. *The Islands of Karpathos, Saros and Kasos in the Neolithic and Bronze Age (SIMA* 68), Göteborg.

———. 1999. "The Ethnography of Minoan and Mycenaean Beekeeping," in *Meletemata. Studies in Aegean Archaeology Presented to Malcolm H. Wiener as He Enters His 65th Year* (*Aegaeum* 20), P.P. Betancourt, V. Karageorghis, R. Laffineur, and W.-D. Niemeier, eds., Liège, pp. 485–491.

Molloy, B., J. Day, S. Bridgford, V. Isaakidou, E. Nodarou, G. Kotzmani, M. Milic, T. Carter, P. Westlake, V. Klotza-Jaklova, and B.J. Hayden. 2014. "Life and Death of a Bronze Age House: Excavation of Early Minoan I Levels at Priniatikos Pyrgos," *AJA* 118, pp. 307–358.

Momigliano, N. 1991. "MM IA Pottery from Evans' Excavations at Knossos: A Reassessment," *BSA* 86, pp. 149–271.

Montelli, R., G. Nolet, F. Dahlen, and G. Masters. 2006. "A Catalogue of Deep Mantle Plumes: New Results from Finite-Frequency Tomography," *Geochemistry, Geophysics, Geosystems* 7 (11), doi:10.1029/2006GC001248

Mortzos, C.E. 1972. "Πάρτιρα: Μία πρώιμος μινωϊκὴ κεραμεικὴ ομὰς," *Ἐπετηρὶς Ἐπιστημονικῶν Ἐρευνῶν* 3, pp. 386–419.

Myer, G.H. 1984. "Ceramic Petrography," in Betancourt 1984, pp. 60–66.

Myer, G.H., K.G. McIntosh, and P.P. Betancourt. 1995. "Definition of Pottery Fabrics by Ceramic Petrography," in Betancourt and Davaras, eds., 1995, pp. 143–153.

Nodarou, E. 2007. "Exploring Patterns of Intra Regional Pottery Distribution in Late Minoan IIIA–B East Crete: The Evidence from the Petrographic Analysis of Three Ceramic Assemblages," in *Archaeometric and Archaeological Approaches to Ceramics. Papers Presented at EMAC '05, 8th European Meeting on Ancient Ceramics Lyon 2005* (*BAR-IS* 1691), C.Y. Waksman, ed., Oxford, pp. 75–83.

———. 2013a. "Analysis of Granodiorite Pottery of the Vrokastro Area from the Neolithic Period to Modern Times," *Kentro: The Newsletter of the INSTAP Study Center for East Crete* 16, pp. 10–11.

———. 2013b. "Petrographic Analysis of the Pottery," in Betancourt 2013, pp. 151–175.

Nodarou, E., and J. Moody. 2014. "Mirabello" Fabric(s) Forever: An Analytical Study of the Granodiorite Pottery of the Vrokastro Area from the Final Neolithic Period to Modern Times," in *A Cretan Landscape through Time: Survey of Priniatikos Pyrgos and Environs* (*BAR-IS* 2634), B.P.C. Molloy and C. Duckworth, eds., Oxford, pp. 91–98.

Panagiotaki, M. 1999. *The Central Palace Sanctuary at Knossos* (*BSA Suppl.* 31), London.

Papadopoulos, G.A. 2011. *A Seismic History of Crete: The Hellenic Arc and Trench*, Athens.

Pendlebury, J.D.S. 1939. *The Archaeology of Crete*, London.

Pendlebury, J.D.S., H.W. Pendlebury, and M.B. Money-Coutts. 1935–1936. "Excavations in the Plain of Lasithi. I. The Cave of Trapeza," *BSA* 36, pp. 5–131.

Platon, N. 1960. "Ἀνασκαφαὶ Κάτω Ζάκρου," *Prakt* 115 [1966], pp. 294–307.

———. 1961. "Ἀνασκαφαὶ Κάτω Ζάκρου," *Prakt* 116 [1964], pp. 216–224.

———. 1963. "Ἀνασκαφαὶ Κάτω Ζάκρου," *Prakt* 118 [1966], pp. 160–188.

———. 1965. "Ἀνασκαφαὶ περιοχῆς Πραισοῦ," *Prakt* 120 [1967], pp. 167–224.

———. 1974. "The New-Palace Minoan Period," in *History of the Hellenic World: Prehistory and Protohistory*, G.A. Christopoulos, ed., University Park, PA, pp. 174–219.

Postma, G., A.R. Fortuin, and W.A. van Wamel. 1994. "Basin-Fill Patterns Controlled by Tectonics and Climate: The Neogene 'Fore-Arc' Basins of Eastern Crete as a Case History," in *Tectonic Controls and Signatures in Sedimentary Successions*, L.E. Frostick and R.J. Steel, eds., Oxford. doi:10.1002/9781444304053.ch18

Potter, P.E., and P. Szatmari. 2009. "Global Miocene Tectonics and the Modern World," *Earth-Science Reviews* 96, pp. 279–295.

Poursat, J.-C. 1984. "Poissons Minoens à Mallia," in *Aux origins de l'Hellénisme: La Crète et la Grèce. Hommage à Henri van Effenterre présenté par le Centre G Glotz.* (*Histoire Ancienne et Médiéval* 15), Paris, pp. 25–28.

Reese, D.S. 1995a. "The Faunal Remains," in Betancourt and Davaras, eds. 1995, pp. 11, 45–46, 56–57, 83.

———. 1995b. "The Minoan Fauna: 5. The Marine Invertebrates," in *Kommos* I: *The Kommos Region and Houses of the Minoan Town.* Part 1: *The Kommos Region, Ecology, and Minoan Industries*, J.W. Shaw and M.C. Shaw, eds., Princeton, pp. 240–273.

———. 1998. "The Faunal Remains," in C.R. Floyd 1998, pp. 131–144.

———. 1999. "The Faunal Remains," in Betancourt and Davaras, eds., 1999, pp. 36–37, 80, 99, 136, 162–164, 184, 282–283.

———. 2009. "The Faunal Remains from Block AF," in Betancourt and Davaras, eds., 2009, pp. 131–142.

Reese, D.S., D. Mylona, J. Bending, A. Sarpaki, W.H. Schoch, and M. Ntinou. 2004. "Fauna and Flora," in *Mochlos* IC: *Period III. Neopalatial Settlement on the Coast: The Artisans' Quarter and the Farmhouse at Chalinomouri. The Small Finds* (*Prehistory Monographs* 9), J.S. Soles and C. Davaras, eds., Philadelphia, pp. 117–138.

Rehak, P. 1995. "The Use and Destruction of Minoan Stone Bull's Head Rhyta," in POLITEIA: *Society and State in the Aegean Bronze Age. Proceedings of the 5th International Aegean Conference, University of Heidelberg, Archäologisches Institut, 10–13 April 1994* (*Aegaeum* 12), R. Laffineur and W.-D. Niemeier, eds., Liège, pp. 435–460.

Renfrew, C. 2012. *Cognitive Archaeology from Theory to Practice: The Early Cycladic Sanctuary at Keros* (*The Annual Balzan Lecture* 3), Florence.

Renfrew, J. 1972. "The Plant Remains," in Warren 1972 pp. 315–317.

Roveri, M., A. Bertini, D. Cosentino, A. Di Stefano, R. Gennari, E. Gliozzi, F. Grossi, S.M. Iaccarino, S. Lugli, V. Manzi, and M. Taviani. 2008. "A High-Resolution Stratigraphic Framework for the Latest Messinian Events in the Mediterranean Area," *Stratigraphy* 5, pp. 323–342.

Ruscillo, D. 2006. "Faunal Remains and Murex Dye Production," in *Kommos V: The Monumental Minoan Buildings at Kommos*, J.W. Shaw and M.C. Shaw, eds., Princeton, pp. 776–844.

Ryan, W.B.F., and M.B. Cita. 1978. "The Nature and Distribution of Messinian Erosional Surfaces: Indicators of a Several-Kilometer-Deep Mediterranean in the Miocene," *Marine Geology* 27, pp. 193–230.

Sakellarakis, Y., and E. Sapouna-Sakellaraki. 1997. *Archanes: Minoan Crete in a New Light*, Athens.

Seager, R.B. 1904–1905. "Excavations at Vasiliki, 1904," *University of Pennsylvania Transactions of the Free Museum of Science and Art* 1, pp. 207–221.

———. 1906–1907. "Report of Excavations at Vasiliki, Crete, in 1906," *University of Pennsylvania Transactions of the Free Museum of Science and Art* 2, pp. 111–132.

———. 1908. "Excavations at Vasiliki," in *Gournia, Vasiliki, and Other Prehistoric Sites on the Isthmus of Ierapetra, Crete: Excavations of the Wells-Houston-Cramp Expeditions, 1901, 1903, 1904*, H.A. Boyd Hawes, B.E. Williams, R.B. Seager, and E.H. Hall, Philadelphia, pp. 49–50.

———. 1909. "Excavations on the Island of Mochlos, Crete, in 1908," *AJA* 13, pp. 273–303.

———. 1910. *Excavations on the Island of Pseira, Crete*, Philadelphia.

———. 1912. *Explorations in the Island of Mochlos*, Boston.

———. 1916. *The Cemetery of Pachyammos, Crete* (*University of Pennsylvania, The University Museum Anthropological Publications* 7 [1]), Philadelphia.

Shank, E. 2005. "New Evidence for Anatolian Relations with Crete in EM I–IIA," in EMPORIA: *Aegeans in the Central and Eastern Mediterranean. Proceedings of the 10th International Aegean Conference, Italian School of Archaeology in Athens, 14–18 April 2004* (*Aegaeum* 25), R. Laffineur and E. Greco, eds., Liège, pp. 103–106.

Shaw, J.W. 1973. *Minoan Architecture: Materials and Techniques* (*ASAtene* 49 [n.s. 33, 1971]), Rome.

———. 2009. *Minoan Architecture: Materials and Techniques* (*Studi di Archeologia Cretese* 7), Padua.

Soles, J.S. 1992. *The Prepalatial Cemeteries at Mochlos and Gournia and the House Tombs of Bronze Age Crete* (*Hesperia Suppl.* 24), Princeton.

Sotirakopoulou, P. 1997. "Κυκλάδες και Βόρειο Αιγαίο: Οι σχέσεις τους κατά το δεύτερο ήμισυ της 3ης χλιετίας π. Χ.," in *Η Πολιόχνη και η πρώιμη εποχή του χαλκού στο βόρειο Αιγαίο/Poliochni e l'antica età del bronzo nell'Egeo settentrionale*, C.G. Doumas and V. La Rosa, eds., Athens, pp. 522–542.

Swann, C.P., S. Ferrence, and P.P. Betancourt. 2000. "Analysis of Minoan White Pigments Used on Pottery from Palaikastro," *Nuclear Instruments and Methods in Physics Research* B 161–163, pp. 714–717.

Syrides, G. Forthcoming. "Sea Shells from the Sanctuary of Poseidon at Kalaureia," in *Physical Environment and Daily Life in the Sanctuary of Poseidon at Kalaureia (Poros): The Swedish Institute Excavataions in 2003–2005* (*ActaAth* 4°, 56), D. Mylona and A. Penttinen, eds., Stockholm.

ten Veen, J.H., and K.L. Kleinspehn. 2003. "Incipient Continental Collision and Plate-Boundary Curvature: Late Pliocene-Holocene Transtensional Hellenic Forearc, Crete, Greece," *Journal of the Geological Society* 160, pp. 161–181.

Theodoropoulou, T. 2007. *L'exploitation des faunes aquatiques en Égée septentrionale aux périodes pré- et protohistoriques*, Ph.D. diss., Université de Sorbonne I, Panthéon-Sorbonne.

Tod, M.N. 1902–1903. "Excavations at Palaikastro II. 10–Hagios Nikolaos," *BSA* 9, pp. 336–343.

Todaro, S. 2001. "Nuove prospettive sulla produzione in stile Pyrgos nella Creta meridionale: il caso della pisside e della coppa su base ad anello," *Creta Antica* 2, pp. 11–28.

Tsikouras, B., H. Dierckx, and K. Hatzipanagiotou. 2008. "Geological and Mineral-Petrographic Study of Dioritic-Granodioritic Rocks from East Crete, Aiming to the Investigation of Provenance of Stone Implements and Tempers in Ceramics of Minoan Age from the Area of Mirabello Bay," unpublished research project, Department of Geology, University of Patras, Greece.

Tsipopoulou, M. 2012a. "Defining the End of the Prepalatial Period at Petras," in *Petras, Siteia: 25 Years of Excavations and Studies. Acts of a Two-Day Conference Held at the Danish Institute at Athens, 9–10 October 2010* (*Monographs of the Danish Institute at Athens* 16), M. Tsipopoulou, ed., Athens, pp. 179–190.

———. 2012b. "The Prepalatial–Early Protopalatial Cemetery at Petras, Siteia: A Diachronic Symbol of Social Coherence," in *Petras, Siteia—25 Years of Excavations and Studies. Acts of a Two-Day Conference Held at the Danish Institute at Athens, 9–10 October 2010* (*Monographs of the Danish Institute at Athens* 16), M. Tsipopoulou, ed., Athens, pp. 117–129.

Tzedakis, Y., and H. Martlew, eds. 1999. *Minoans and Mycenaeans: Flavours of Their Time,* Athens.

van Hinsbergen, D.J.J., and J.E. Meulenkamp. 2006. "Neogene Supradetachment Basin Development on Crete (Greece) during Exhumation of the South Aegean Core Complex," *Basin Research* 8, pp. 103–124. doi:10.1111/j.1356-2117.2005.00282.x

Vaughan, S.J. 2002. "Petrographic Analysis of Fabrics from the Pseira Cemetery," in Betancourt and Davaras, eds., 2002, pp. 147–165.

Vogeikoff-Brogan, N. 2012. "Hellenistic and Roman Periods: Expansion of the Isthmus in an International Era," in Watrous et al. 2012, pp. 81–93.

Warren, P.M. 1965. "The First Minoan Stone Vases and Early Minoan Chronology," *CretChron* 19, pp. 7–43.

———. 1972. *Myrtos: An Early Bronze Age Settlement in Crete* (*BSA Suppl.* 7), Oxford.

———. 2004. "Part II. The Contents of the Tombs," in *The Early Minoan Tombs of Lebena, Southern Crete* (*SIMA* 30), S. Alexiou and P. Warren, Sävedalen, pp. 25–198.

Watrous, L.V. 1992. *Kommos III: The Late Bronze Age Pottery,* Princeton.

———. 1996. *The Cave Sanctuary of Zeus at Psychro: A Study of Extra-Urban Sanctuaries in Minoan and Early Iron Age Crete* (*Aegaeum* 15), Liège and Austin.

———. 2012. "Catalog of Sites" and "Conclusions," in Watrous et al. 2012, pp. 105–133, 97–102.

Watrous, L.V., D.M. Buell, J.C. McEnroe, J.G. Younger, L.A. Turner, B.S. Kunkel, K. Glowacki, S. Gallimore, A. Smith, P.A. Pantou, A. Chapin, and E. Margaritis. 2015. "Excavations at Gournia, 2010–2012," *Hesperia* 84, pp. 397–465.

Watrous, L.V., D. Haggis, K. Nowicki, N. Vogeikoff-Brogan, and M. Schultz. 2012. *An Archaeological Survey of the Gournia Landscape: A Regional History of the Mirabello Bay, Crete, in Antiquity* (*Prehistory Monographs* 37), Philadelphia.

Watrous, L.V., and A. Heimroth. 2011. "Household Industries of Late Minoan IB Gournia and the Socioeconomic Status of the Town," in ΣΤΕΓΑ: *The Archaeology of Houses and Households in Ancient Crete* (*Hesperia Suppl.* 44), K. Glowacki and N. Vogeikoff-Brogan, eds., Princeton, pp. 199–212.

Watrous, L.V., and M. Schultz. 2012a. "Early Minoan I–II Periods: Emergence of a Stratified Society," in Watrous et al. 2012, pp. 21–31.

———. 2012b. "Middle Minoan IB–II Periods: Growth of Regional Factions and Conflict," in Watrous et al. 2012, pp. 41–50.

———. 2012c. "Middle Minoan III–Late Minoan I Periods: The Rise of a Regional State," in Watrous et al. 2012, pp. 51–63.

Wentworth, C.K. 1922. "A Scale of Grade and Class Terms for Clastic Sediments," *Journal of Geology* 30, pp. 377–392

Whitelaw, T., P.M. Day, E. Kiriatzi, V. Kilikoglou, and D.E. Wilson. 1997. "Ceramic Traditions at EM IIB Myrtos, Fournou Korifi," in ΤΕΧΝΗ: *Craftsmen, Craftswomen and Craftsmanship in the Aegean Bronze Age/Artisanat et artisans en Égée à l'âge du Bronze. Proceedings of the 6th International Aegean Conference, Philadelphia, Temple University, 18–21 April 1996* (*Aegaeum* 16), R. Laffineur and P.P. Betancourt, eds., Liège, pp. 265–274.

Xanthoudides, S. 1918. "Μέγας πρωτομινωϊκὸς τάφος Πύργου," *ArchDelt* 4, pp. 136–170.

———. 1924. *The Vaulted Tombs of Mesara,* London.

Zachariasse, W.J., D.J.J. van Hinsbergen, and A.R. Fortuin. 2008. "Mass Wasting and Uplift on Crete and Karpathos during the Early Pliocene Related to Initiation of South Aegean Left-Lateral, Strike-Slip Tectonics," *Geological Society of America Bulletin* 120, pp. 976–993. doi:10.1130/B26175.1

Zois, A.A. 1968a. "'Υπάρχει ΠΜ III εποχή;" in *Πεπραγμένα του Β' Διεθνούς Κρητολογικού Συνεδρίου* Α', Athens, pp. 141–156.

———. 1968b. "Ἔρευνα περὶ τῆς μινωϊκῆς κεραμεικῆς," *Ἐπετηρὶς Ἐπιστημονικῶν Ἐρευνῶν* 1967–1968, pp. 703–731.

———. 1980. "Ἀνασκαφὴ εἰς Βασιλικὴν Ἱεραπέτρας," *Prakt* 136 (1982), pp. 331–336.

———. 1991. "Βασιλικὴ Ἱεράπετρας," *Ergon* 38 (1992), pp. 100–103.

———. 2007. *Βασιλική Ιεράπετρας: Ο μινωικός οικισμός στη θέση Κεφάλι. Αρχαιολογικός και ιστορικός οδηγός*, Athens.

Concordance A

Hagios Nikolaos Museum Numbers and Catalog Numbers

HN 13500 **506**	HN 14432 **49**	HN 14450 **203**	HN 14468 **299**
HN 13538 **334**	HN 14433 **183**	HN 14451 **250**	HN 14469 **300**
HN 13556 **336**	HN 14434 **182**	HN 14452 **249**	HN 14470 **302**
HN 13612 **355**	HN 14435 **169**	HN 14453 **290**	HN 14471 **301**
HN 13636 **206**	HN 14436 **384**	HN 14454 **377**	HN 14472 **303**
HN 13729 **105**	HN 14437 **180**	HN 14455 **238**	HN 14473 **304**
HN 14335 **42**	HN 14438 **15**	HN 14456 **345**	HN 14474 **306**
HN 14421 **330**	HN 14439 **165**	HN 14457 **39**	HN 14475 **307**
HN 14422 **168**	HN 14440 **279**	HN 14458 **278**	HN 14476 **308**
HN 14423 **167**	HN 14441 **51**	HN 14459 **277**	HN 14477 **309**
HN 14424 **138**	HN 14442 **374**	HN 14460 **274**	HN 14478 **310**
HN 14425 **326**	HN 14443 **375**	HN 14461 **84**	HN 14479 **320**
HN 14426 **16**	HN 14444 **292**	HN 14462 **293**	HN 14480 **318**
HN 14427 **101**	HN 14445 **196**	HN 14463 **251**	HN 14481 **319**
HN 14428 **33**	HN 14446 **147**	HN 14464 **296**	HN 14482 **317**
HN 14429 **325**	HN 14447 **291**	HN 14465 **294**	HN 14483 **316**
HN 14430 **322**	HN 14448 **266**	HN 14466 **329**	HN 14484 **314**
HN 14431 **321**	HN 14449 **242**	HN 14467 **298**	HN 14485 **315**

HN 14486	**313**	HN 14529	**22**	HN 14571	**373**	HN 14613	**259**
HN 14487	**166**	HN 14530	**132**	HN 14572	**141**	HN 14614	**88**
HN 14488	**312**	HN 14531	**35**	HN 14573	**264**	HN 14615	**378**
HN 14489	**27**	HN 14532	**131**	HN 14574	**275**	HN 14616	**10**
HN 14490	**118**	HN 14533	**328**	HN 14575	**18**	HN 14617	**265**
HN 14491	**110**	HN 14534	**3**	HN 14576	**273**	HN 14618	**64**
HN 14492	**172**	HN 14535	**83**	HN 14577	**140**	HN 14619	**93**
HN 14493	**117**	HN 14536	**31**	HN 14578	**371**	HN 14620	**156**
HN 14494	**502**	HN 14537	**243**	HN 14579	**372**	HN 14621	**122**
HN 14495	**115**	HN 14539	**331**	HN 14580	**361**	HN 14622	**260**
HN 14496	**90**	HN 14540	**107**	HN 14581	**360**	HN 14623	**54**
HN 14497	**17**	HN 14541	**333**	HN 14582	**364**	HN 14624	**154**
HN 14498	**503**	HN 14543	**63**	HN 14583	**98**	HN 14625	**1**
HN 14499	**44**	HN 14544	**65**	HN 14584	**362**	HN 14626	**155**
HN 14501	**20**	HN 14545	**70**	HN 14585	**386**	HN 14627	**139**
HN 14504	**129**	HN 14546	**66**	HN 14586	**385**	HN 14628	**356**
HN 14505	**504**	HN 14547	**339**	HN 14587	**348**	HN 14629	**288**
HN 14506	**507**	HN 14548	**45**	HN 14588	**247**	HN 14630	**128**
HN 14507	**505**	HN 14549	**149**	HN 14589	**343**	HN 14631	**286**
HN 14508	**508**	HN 14550	**179**	HN 14590	**221**	HN 14632	**287**
HN 14509	**510**	HN 14551	**188**	HN 14591	**363**	HN 14633	**353**
HN 14510	**134**	HN 14552	**58**	HN 14592	**280**	HN 14634	**173**
HN 14511	**511**	HN 14553	**390**	HN 14593	**268**	HN 14637	**74**
HN 14512	**106**	HN 14554	**369**	HN 14594	**359**	HN 14638	**282**
HN 14513	**36**	HN 14555	**335**	HN 14595	**289**	HN 14639	**281**
HN 14514	**514**	HN 14557	**337**	HN 14596	**283**	HN 14640	**199**
HN 14515	**111**	HN 14558	**338**	HN 14597	**193**	HN 14641	**9**
HN 14516	**29**	HN 14559	**19**	HN 14598	**269**	HN 14642	**127**
HN 14517	**32**	HN 14560	**144**	HN 14599	**358**	HN 14644	**216**
HN 14518	**513**	HN 14561	**332**	HN 14600	**28**	HN 14645	**114**
HN 14520	**21**	HN 14562	**163**	HN 14601	**342**	HN 14646	**235**
HN 14521	**34**	HN 14563	**14**	HN 14602	**357**	HN 14647	**24**
HN 14522	**191**	HN 14564	**143**	HN 14603	**158**	HN 14648	**25**
HN 14523	**37**	HN 14565	**261**	HN 14604	**267**	HN 14649	**86**
HN 14524	**109**	HN 14566	**184**	HN 14605	**157**	HN 14650	**23**
HN 14525	**7**	HN 14567	**164**	HN 14606	**354**	HN 14651	**125**
HN 14526	**60**	HN 14568	**75**	HN 14608	**89**	HN 14652	**124**
HN 14527	**186**	HN 14569	**112**	HN 14610	**217**	HN 14653	**229**
HN 14528	**133**	HN 14570	**43**	HN 14611	**174**	HN 14654	**228**

CONCORDANCE A

HN 14655 **152**	HN 14697 **368**	HN 14737 **327**	HN 14778 **376**
HN 14656 **151**	HN 14698 **176**	HN 14738 **59**	HN 14779 **130**
HN 14657 **189**	HN 14699 **162**	HN 14739 **148**	HN 14780 **344**
HN 14658 **381**	HN 14700 **272**	HN 14740 **77**	HN 14781 **30**
HN 14659 **391**	HN 14701 **237**	HN 14741 **205**	HN 14782 **104**
HN 14660 **231**	HN 14702 **367**	HN 14743 **255**	HN 14783 **350**
HN 14661 **87**	HN 14703 **244**	HN 14744 **347**	HN 14785 **341**
HN 14663 **126**	HN 14704 **153**	HN 14745 **232**	HN 14786 **241**
HN 14664 **71**	HN 14705 **389**	HN 14747 **187**	HN 14787 **81**
HN 14665 **91**	HN 14706 **56**	HN 14748 **46**	HN 14789 **352**
HN 14666 **512**	HN 14707 **41**	HN 14749 **509**	HN 14790 **13**
HN 14667 **208**	HN 14708 **248**	HN 14750 **515**	HN 14791 **94**
HN 14668 **47**	HN 14709 **2**	HN 14751 **501**	HN 14792 **198**
HN 14669 **192**	HN 14710 **146**	HN 14752 **8**	HN 14793 **170**
HN 14670 **214**	HN 14711 **270**	HN 14753 **52**	HN 14794 **245**
HN 14671 **208**	HN 14712 **161**	HN 14754 **61**	HN 14796 **226**
HN 14673 **97**	HN 14713 **256**	HN 14755 **100**	HN 14797 **227**
HN 14674 **96**	HN 14714 **388**	HN 14756 **380**	HN 14798 **50**
HN 14675 **57**	HN 14715 **142**	HN 14757 **79**	HN 14799 **121**
HN 14676 **80**	HN 14716 **222**	HN 14758 **349**	HN 14800 **76**
HN 14678 **69**	HN 14717 **40**	HN 14759 **78**	HN 14801 **254**
HN 14679 **382**	HN 14718 **370**	HN 14760 **181**	HN 14802 **150**
HN 14680 **230**	HN 14719 **387**	HN 14761 **12**	HN 14803 **48**
HN 14681 **213**	HN 14720 **5**	HN 14762 **108**	HN 14804 **225**
HN 14682 **252**	HN 14721 **159**	HN 14763 **185**	HN 14805 **53**
HN 14683 **209**	HN 14722 **4**	HN 14764 **383**	HN 14806 **201**
HN 14684 **211**	HN 14723 **253**	HN 14765 **62**	HN 14807 **190**
HN 14685 **135**	HN 14724 **123**	HN 14766 **285**	HN 14808 **145**
HN 14686 **136**	HN 14725 **284**	HN 14767 **305**	HN 14809 **175**
HN 14687 **236**	HN 14726 **103**	HN 14768 **92**	HN 14810 **223**
HN 14688 **233**	HN 14727 **365**	HN 14769 **178**	HN 14811 **73**
HN 14689 **224**	HN 14728 **102**	HN 14770 **258**	HN 14812 **295**
HN 14690 **26**	HN 14730 **68**	HN 14771 **177**	HN 14813 **340**
HN 14691 **215**	HN 14731 **263**	HN 14772 **113**	HN 14814 **55**
HN 14692 **194**	HN 14732 **160**	HN 14773 **366**	HN 14815 **219**
HN 14693 **202**	HN 14733 **257**	HN 14774 **85**	HN 14816 **99**
HN 14694 **239**	HN 14734 **204**	HN 14775 **324**	HN 14817 **346**
HN 14695 **120**	HN 14735 **197**	HN 14776 **351**	HN 14818 **246**
HN 14696 **271**	HN 14736 **82**	HN 14777 **67**	HN 14819 **171**

HN 14820	**220**	HN 14848	**470**	HN 14876	**431**	HN 14904	**486**
HN 14821	**200**	HN 14849	**439**	HN 14877	**488**	HN 18137	**411**
HN 14822	**262**	HN 14850	**428**	HN 14878	**449**	HN 18138	**413**
HN 14823	**323**	HN 14851	**466**	HN 14879	**473**	HN 18139	**415**
HN 14824	**297**	HN 14852	**420**	HN 14880	**450**	HN 18140	**418**
HN 14825	**240**	HN 14853	**457**	HN 14881	**478**	HN 18141	**419**
HN 14826	**218**	HN 14854	**468**	HN 14882	**474**	HN 18142	**423**
HN 14827	**95**	HN 14855	**416**	HN 14883	**436**	HN 18143	**425**
HN 14828	**116**	HN 14856	**463**	HN 14884	**453**	HN 18144	**427**
HN 14829	**11**	HN 14857	**479**	HN 14885	**476**	HN 18145	**429**
HN 14830	**195**	HN 14858	**464**	HN 14886	**475**	HN 18146	**430**
HN 14831	**119**	HN 14859	**442**	HN 14887	**443**	HN 18147	**432**
HN 14832	**276**	HN 14860	**414**	HN 14888	**489**	HN 18148	**433**
HN 14833	**207**	HN 14861	**460**	HN 14889	**492**	HN 18149	**435**
HN 14834	**6**	HN 14862	**452**	HN 14890	**494**	HN 18150	**437**
HN 14835	**72**	HN 14863	**434**	HN 14892	**490**	HN 18151	**438**
HN 14836	**38**	HN 14864	**441**	HN 14893	**491**	HN 18152	**445**
HN 14837	**311**	HN 14865	**465**	HN 14894	**484**	HN 18153	**451**
HN 14838	**481**	HN 14866	**462**	HN 14895	**497**	HN 18154	**458**
HN 14839	**417**	HN 14867	**469**	HN 14896	**487**	HN 18155	**459**
HN 14840	**446**	HN 14868	**471**	HN 14897	**409**	HN 18156	**461**
HN 14841	**493**	HN 14869	**422**	HN 14898	**495**	HN 18157	**472**
HN 14842	**448**	HN 14870	**424**	HN 14899	**500**	HN 18158	**477**
HN 14843	**447**	HN 14871	**456**	HN 14900	**444**	HN 18159	**480**
HN 14844	**426**	HN 14872	**410**	HN 14901	**455**	HN 18160	**482**
HN 14845	**421**	HN 14873	**440**	HN 14902	**496**	HN 18161	**483**
HN 14846	**412**	HN 14874	**467**	HN 14903	**498**	HN 18162	**499**
HN 14847	**454**	HN 14875	**485**				

Concordance B

Field and Catalog Numbers

The field numbers written on the artifacts were preceded by the abbreviation PAR (for Pacheia Ammos Rock Shelter).

PAR 1	**127**	PAR 17	**200**	PAR 33	**259**	PAR 49	**267**
PAR 2	**9**	PAR 18	**73**	PAR 34	**88**	PAR 50	**157**
PAR 3	**349**	PAR 19	**206**	PAR 35	**217**	PAR 51	**269**
PAR 4	**198**	PAR 20	**173**	PAR 36	**10**	PAR 53	**193**
PAR 5	**241**	PAR 21	**100**	PAR 37	**265**	PAR 54	**289**
PAR 6	**199**	PAR 22	**353**	PAR 38	**260**	PAR 55	**354**
PAR 7	**281**	PAR 23	**286**	PAR 39	**154**	PAR 56	**158**
PAR 8	**350**	PAR 24	**95**	PAR 40	**211**	PAR 57	**357**
PAR 9	**85**	PAR 25	**287**	PAR 41	**64**	PAR 58	**342**
PAR 10	**351**	PAR 26	**128**	PAR 42	**93**	PAR 59	**28**
PAR 11	**42**	PAR 27	**288**	PAR 43	**156**	PAR 60	**283**
PAR 12	**352**	PAR 28	**356**	PAR 44	**122**	PAR 61	**358**
PAR 13	**155**	PAR 29	**347**	PAR 45	**54**	PAR 62	**359**
PAR 14	**344**	PAR 30	**139**	PAR 46	**1**	PAR 63	**11**
PAR 15	**282**	PAR 31	**174**	PAR 47	**89**	PAR 64	**268**
PAR 16	**74**	PAR 32	**355**	PAR 48	**378**	PAR 65	**362**

PAR 66	**98**	PAR 105	**24**	PAR 144	**153**	PAR 184	**371**
PAR 67	**360**	PAR 106	**229**	PAR 145	**244**	PAR 185	**372**
PAR 68	**361**	PAR 107	**46**	PAR 146	**367**	PAR 186	**273**
PAR 69	**280**	PAR 108	**151**	PAR 147	**341**	PAR 187	**18**
PAR 70	**363**	PAR 109	**230**	PAR 148	**99**	PAR 188	**69**
PAR 71	**113**	PAR 110	**126**	PAR 149	**237**	PAR 189	**140**
PAR 72	**221**	PAR 111	**366**	PAR 150	**202**	PAR 190	**275**
PAR 73	**343**	PAR 112	**220**	PAR 151	**162**	PAR 191	**264**
PAR 74	**247**	PAR 113	**125**	PAR 152	**232**	PAR 192	**219**
PAR 75	**12**	PAR 114	**231**	PAR 153	**176**	PAR 193	**274**
PAR 76	**348**	PAR 115	**47**	PAR 154	**271**	PAR 194	**373**
PAR 77	**385**	PAR 116	**8**	PAR 155	**368**	PAR 195	**57**
PAR 78	**386**	PAR 117	**222**	PAR 156	**120**	PAR 196	**223**
PAR 79	**218**	PAR 118	**370**	PAR 157	**194**	PAR 198	**67**
PAR 80	**263**	PAR 119	**40**	PAR 158	**239**	PAR 199	**141**
PAR 81	**13**	PAR 120	**119**	PAR 159	**78**	PAR 200	**112**
PAR 82	**105**	PAR 121	**72**	PAR 160	**177**	PAR 201	**75**
PAR 83	**102**	PAR 122	**256**	PAR 161	**246**	PAR 202	**43**
PAR 84	**365**	PAR 123	**92**	PAR 162	**390**	PAR 203	**195**
PAR 85	**253**	PAR 124	**387**	PAR 163	**178**	PAR 204	**51**
PAR 86	**160**	PAR 125	**161**	PAR 164	**369**	PAR 205	**374**
PAR 87	**257**	PAR 126	**6**	PAR 165	**26**	PAR 206	**53**
PAR 88	**68**	PAR 127	**388**	PAR 166	**58**	PAR 207	**375**
PAR 89	**213**	PAR 128	**364**	PAR 167	**179**	PAR 208	**207**
PAR 90	**175**	PAR 129	**270**	PAR 168	**240**	PAR 209	**224**
PAR 91	**55**	PAR 130	**389**	PAR 169	**188**	PAR 211	**279**
PAR 92	**159**	PAR 131	**142**	PAR 170	**170**	PAR 212	**152**
PAR 93	**38**	PAR 132	**146**	PAR 171	**262**	PAR 213	**376**
PAR 94	**123**	PAR 133	**30**	PAR 172	**14**	PAR 214	**165**
PAR 95	**103**	PAR 134	**2**	PAR 173	**235**	PAR 215	**15**
PAR 96	**284**	PAR 135	**56**	PAR 174	**143**	PAR 216	**180**
PAR 97	**204**	PAR 136	**48**	PAR 175	**276**	PAR 217	**277**
PAR 98	**94**	PAR 137	**97**	PAR 176	**332**	PAR 218	**84**
PAR 99	**5**	PAR 138	**248**	PAR 177	**163**	PAR 219	**39**
PAR 100	**4**	PAR 139	**41**	PAR 178	**164**	PAR 220	**278**
PAR 101	**96**	PAR 140	**330**	PAR 179	**145**	PAR 221	**238**
PAR 102	**71**	PAR 141	**272**	PAR 180	**144**	PAR 222	**252**
PAR 103	**23**	PAR 142	**245**	PAR 182	**184**	PAR 223	**345**
PAR 104	**124**	PAR 143	**201**	PAR 183	**261**	PAR 224	**79**

CONCORDANCE B

PAR 225	**377**	PAR 264	**315**	PAR 303	**466**	PAR 342	**190**
PAR 226	**203**	PAR 265	**316**	PAR 304	**33**	PAR 343	**104**
PAR 227	**250**	PAR 266	**317**	PAR 305	**101**	PAR 344	**479**
PAR 228	**249**	PAR 267	**318**	PAR 306	**16**	PAR 345	**471**
PAR 229	**266**	PAR 268	**319**	PAR 307	**454**	PAR 346	**150**
PAR 230	**242**	PAR 269	**320**	PAR 308	**442**	PAR 347	**475**
PAR 231	**290**	PAR 270	**321**	PAR 309	**456**	PAR 348	**476**
PAR 232	**147**	PAR 271	**322**	PAR 310	**420**	PAR 349	**474**
PAR 233	**291**	PAR 272	**323**	PAR 311	**447**	PAR 350	**481**
PAR 234	**196**	PAR 273	**324**	PAR 312	**410**	PAR 351	**473**
PAR 235	**292**	PAR 274	**325**	PAR 313	**409**	PAR 352	**439**
PAR 236	**181**	PAR 275	**406**	PAR 314	**431**	PAR 353	**450**
PAR 237	**293**	PAR 276	**395**	PAR 315	**460**	PAR 354	**449**
PAR 238	**294**	PAR 277	**400**	PAR 316	**457**	PAR 355	**493**
PAR 239	**295**	PAR 278	**396**	PAR 317	**446**	PAR 356	**185**
PAR 240	**296**	PAR 279	**407**	PAR 318	**464**	PAR 357	**168**
PAR 241	**251**	PAR 280	**393**	PAR 319	**455**	PAR 358	**189**
PAR 242	**255**	PAR 281	**392**	PAR 320	**436**	PAR 359	**25**
PAR 243	**297**	PAR 282	**398**	PAR 321	**434**	PAR 360	**87**
PAR 244	**298**	PAR 283	**397**	PAR 322	**428**	PAR 362B	**489**
PAR 245	**329**	PAR 284	**401**	PAR 323	**440**	PAR 363	**487**
PAR 246	**299**	PAR 285	**394**	PAR 324	**416**	PAR 364	**495**
PAR 247	**285**	PAR 286	**402**	PAR 325	**462**	PAR 365	**498**
PAR 248	**300**	PAR 287	**399**	PAR 326	**422**	PAR 366	**500**
PAR 249	**301**	PAR 288	**403**	PAR 327	**452**	PAR 367	**496**
PAR 250	**302**	PAR 289	**408**	PAR 328	**414**	PAR 368	**497**
PAR 251	**303**	PAR 290	**404**	PAR 329	**478**	PAR 369	**484**
PAR 252	**304**	PAR 291	**405**	PAR 330	**444**	PAR 370	**494**
PAR 253	**305**	PAR 292	**468**	PAR 331	**326**	PAR 371	**486**
PAR 254	**306**	PAR 293	**469**	PAR 332	**233**	PAR 372	**488**
PAR 255	**307**	PAR 294	**470**	PAR 333	**465**	PAR 373	**485**
PAR 256	**308**	PAR 295	**443**	PAR 334	**467**	PAR 374	**490**
PAR 257	**309**	PAR 296	**424**	PAR 335	**453**	PAR 375	**492**
PAR 258	**310**	PAR 297	**421**	PAR 336	**138**	PAR 376	**491**
PAR 259	**311**	PAR 298	**426**	PAR 337	**419**	PAR 377	**225**
PAR 260	**312**	PAR 299	**448**	PAR 338	**463**	PAR 378	**86**
PAR 261	**166**	PAR 300	**417**	PAR 339	**383**	PAR 379	**135**
PAR 262	**313**	PAR 301	**441**	PAR 340	**258**	PAR 380	**381**
PAR 263	**314**	PAR 302	**412**	PAR 341	**167**	PAR 381	**382**

PAR 382	**121**	PAR 418	**501**	PAR 466	**335**	PAR 506	**333**
PAR 383	**346**	PAR 419	**502**	PAR 467	**513**	PAR 507	**334**
PAR 384	**148**	PAR 420	**503**	PAR 468	**514**	PAR 508	**149**
PAR 385	**254**	PAR 421	**504**	PAR 469	**32**	PAR 509	**107**
PAR 386	**379**	PAR 422	**510**	PAR 470	**338**	PAR 510	**331**
PAR 387	**80**	PAR 423	**505**	PAR 471	**197**	PAR 511	**243**
PAR 388	**384**	PAR 424	**509**	PAR 472	**111**	PAR 512	**66**
PAR 389	**169**	PAR 425	**506**	PAR 473	**37**	PAR 513	**31**
PAR 390	**182**	PAR 426	**507**	PAR 474	**29**	PAR 514	**63**
PAR 391	**76**	PAR 427	**508**	PAR 475	**336**	PAR 515	**215**
PAR 392	**183**	PAR 428	**511**	PAR 476	**21**	PAR 516	**339**
PAR 393	**171**	PAR 429	**512**	PAR 477	**186**	PAR 547	**418**
PAR 394	**49**	PAR 430	**515**	PAR 478	**133**	PAR 548	**423**
PAR 395	**136**	PAR 431	**340**	PAR 479	**132**	PAR 549	**425**
PAR 396	**91**	PAR 433	**214**	PAR 480	**131**	PAR 550	**432**
PAR 397	**227**	PAR 436	**210**	PAR 481	**35**	PAR 551	**429**
PAR 398	**90**	PAR 437	**380**	PAR 482	**22**	PAR 552	**415**
PAR 399	**17**	PAR 438	**226**	PAR 484	**59**	PAR 553	**413**
PAR 400	**192**	PAR 440	**234**	PAR 485	**205**	PAR 554	**427**
PAR 401	**52**	PAR 444	**236**	PAR 486	**109**	PAR 555	**445**
PAR 402	**137**	PAR 445	**228**	PAR 487	**60**	PAR 556	**411**
PAR 403	**391**	PAR 447	**GRAVE 1A**	PAR 488	**191**	PAR 557	**499**
PAR 404	**212**	PAR 448	**GRAVE 1B**	PAR 489	**34**	PAR 558	**433**
PAR 405	**50**	PAR 449	**36**	PAR 490	**3**	PAR 559	**437**
PAR 406	**208**	PAR 451	**337**	PAR 491	**134**	PAR 560	**430**
PAR 407	**209**	PAR 452	**106**	PAR 492	**328**	PAR 561	**438**
PAR 408	**115**	PAR 453	**81**	PAR 493	**45**	PAR 562	**472**
PAR 409	**116**	PAR 454	**44**	PAR 495	**61**	PAR 563	**451**
PAR 410	**117**	PAR 455	**108**	PAR 496	**216**	PAR 564	**482**
PAR 411	**77**	PAR 456	**82**	PAR 499	**7**	PAR 565	**480**
PAR 412	**110**	PAR 457	**27**	PAR 501	**83**	PAR 566	**461**
PAR 413	**172**	PAR 458	**19**	PAR 502	**70**	PAR 567	**477**
PAR 414	**114**	PAR 459	**327**	PAR 503	**65**	PAR 568	**483**
PAR 415	**118**	PAR 460	**20**	PAR 504	**187**	PAR 569	**435**
PAR 416	**458**	PAR 461	**129**	PAR 505	**62**		
PAR 417	**459**	PAR 462	**130**				

Index

acid weight-loss technique, 4
Alatzomouri Rock Shelter, 64
Alatzomouri settlement, 10, 37, 38
Alatzomouri Survey, site 17, 10, 38, 111
Ammoudares Formation, 5
Anatolia, 58
andesite, 71, 72, 76
Aphrodite's Kephali, 90
archaeobotanical remains, main discussion, 89–90
Archanes, 49, 83
ASPIS, 90

basalt, 100
basin with interior scoring, 51
baskets, 1
bast fibers, 83
beehives, 51
biotite, discredited mineral, 93, 96
blocking wall at doorway, 14, 17, 111, 112
breakage of objects, 2, 14, 26, 35

calcarenites, 5–6
calcareous breccia, 3–5

calcite, 76
cenotaphs, 113
Center for Textile Research, 81
chert, 100
chipped and ground stone tools, main discussion, 69–79
Chrysokamino, 41, 42, 45, 47, 50, 51, 52, 62, 90, 110
compositional system for White-on-Dark Ware, 109
conglomerate, 71, 77, 78, 100
Cyclades, 49

date of the deposit, 96
debitage, 70
Debla, 90
definitions of pottery wares and styles, 38
Demenagaki, 70
desiccation of the Mediterranean Sea, 3, 4, 6
Dictaean Mountain Range, 9
differential global positioning system, 3
discovery and survey of the site, 10–11
dolerite, 100
drain fragment, main discussion, 91–92
Drakones, 52, 83

early pottery, *see* heirlooms
East Cretan White-on-Dark Ware
    analysis of white paints, 109
    definition of the ware, 38
    main discussions, 38, 93, 99–105
Egypt, 83
elongated spout development, 49
Epano Zakros, 64
Eurasian convergent orogenic belt, 4
Evraika Field, 47

fiber-wetting bowls, 51, 52, 81, 83, 96
Finikia Group, 5
flax, 52

gabbro, 100
Gournia
    Gournia Moon site, 47
    Gournia Survey, site 17, 10, 37
    Gournia Survey, site 81, 47
    granitic-dioritic rock formations at Gournia, 38
    Minoan settlement, 5, 9, 10, 35, 37, 38, 42, 45, 46, 47,
        48, 49, 53, 54, 64, 90, 92, 93, 99, 104, 105, 108,
        109, 110
grapes, 89–90
ground stone tools, *see* chipped and ground stone tools
gypsum, 3, 4

Hagia Photeini church, 10
Hagia Photia, 42
Hagia Triada, 39
Hagios Charalambos cave, 45, 46
Hagios Demetrios church, 10
Hagios Nikolaos, 39
Hagios Onouphrios Style, 109
Halepa, 10
halite, 4
Harris Matrix, 11
heirlooms in the deposit, 38–41
Hellenikon Group, 5

Ierapetra Fault Zone, 7
Ierapetra graben, 5
importance of the deposit, 37
interpretations of the evidence, 112–114
Isthmus of Ierapetra, 5, 9

Kalamavka Formation, 5–6
Kamares Ware, 1
Karpathos, 51
Kasos island, 51
Kato Syme, 51

Kato Zakros, 64
Kavousi, 10, 38
Keros, 114
knapping not present, 70
Knossos, 39, 72, 83, 110, 114
Kommos, 51, 52, 64, 72, 83, 85
Koumasa Style, 109
Koutsokeras, 39–40

Lasithi White-on-Dark Style, 41, 45
Lebena, 40
limestone, 71–79, 3–8, 100, 102–105
limpet gathering, 87
linen thread, 83
Livari, 90
locus system for excavation, 11
loomweights, 1, 2, 11, 17, 23, 24, 27, 37, 82

Malia, 41, 42, 43, 44, 45, 46, 47, 49, 64, 83
marls, 3–8, 100, 101
marly chalks, 7
Melos, 70
Messinian Salinity Crisis, 4, 6, 9
meta-andesite, 71, 72
metacarbonate, 71, 72, 76, 77
Miocene–Pliocene oceans, 3, 4
Mirabello Fabric
    absence of centralized production, 104, 109
    main discussions, 38, 93, 99–105
    sub-groups of Mirabello Fabric, 100–105
Mochlos, 10, 42, 43, 44, 45, 46, 48, 49, 52, 53, 72, 83,
    85, 90, 109, 112
Modified Mercelli scale, 5
murex, 87
Myrsini, 72
Myrtos Phournou Korifi, 38, 49, 52, 53, 62, 64, 82, 83,
    90, 92, 108, 109
Myrtos Pyrgos, 46

Nerokourou, 51
nicking on limpets, 86
North Trench deposit, 109, 110

objects from the grave, main discussion, 39
obsidian, *see* chipped and ground stone tools
olives, 89–90
organization of the deposit, 35

Pacheia Ammos
    cemetery, 10, 54, 57, 113
    settlement, 9

INDEX

Pacheia Ammos Formation, 5
Palaikastro, 41, 42, 43, 45, 46, 47, 48, 49, 58, 83, 109, 110
Papadiokampos, 86
Partira, 40
pebbles as possible offerings, 87
Pefka, 71–83
    Pefka Layer, 6–9
Pera Alatzomouri, 9, 38, 41, 42, 47, 109, 111
Petras, 83, 90
Phaistos, 52, 83
Phournou Korifi, *see* Myrtos Phournou Korifi
Phrouzi, 105
phytoliths, 69
Poseidonia, 87
potter's batts, 92
pottery, main discussions, 37–67
pottery, shape catalog entries for the deposit
    bridge-spouted jar, 47, 48, 57
    closed vessel, 49, 65–67
    collared jar, 48, 49, 57
    conical bowl, 42–43
    conical bowl or basin, 50
    conical bowl with frying pan handle, 43, 44
    conical bowl without spout, 42
    cooking dish, 62, 63, 64
    cup, 39
    cup, cylindrical, 46
    cup, straight-sided or rounded, 45
    cup, with straight sides, 52, 53
    hole-mouthed jar, 57, 58
    jar, 45
    jug or jar, 48
    jug, 40, 41, 53
    jug, wide-mouthed, 43, 54
    lid, flat, 54
    pithos, 54, 55
    pithos or jar, 55, 56
    pyxis lid, 39
    rounded cup, 46
    rounded cup with tiny lugs or handles, 46
    shallow bowl, 40
    shallow conical bowl and basin, 41, 42
    spinning bowl, 51, 52
    spouted conical bowl, 44–45
    spouted jar, 56
    spouted two-handled cup, 46, 47
    teapot, 49
    tripod cooking pot, 58, 59, 60, 61, 62
    vat, 64
    vessel with elliptical base, 49
    vessel with knobs, 57

Priniatikos Pyrgos, 10, 38, 45, 52, 90, 93, 104–105
Prophetes Elias Tourtoulon, 64
Pseira, 42, 44, 45, 46, 47, 49, 5, 62, 72, 85, 109, 112
Psychro Cave, 47
pumice, 7, 71, 72, 76
Pyrgos Cave, 40, 42
Pyrgos Style pottery, 40

quarries for obsidian, 70
quartzite, 71, 76, 77

Roman Pacheia Ammos, 10
roof of the rock shelter, 6, 8, 27
rubble wall blocking the entrance, 11

sandstone, 3–6, 71, 75, 77, 78
Santorini, 7
schist, 75–78
scoring on pottery, 51
sea pebbles, 7
sea shells, 85–88
sequence of construction and deposition, 34, 35
Siteia, 73
size of the pottery deposit, 93
South Coast Fabric, 99
south coast jug, 14–16, 18, 20, 21
Sphoungaras, 38, 42, 45, 46, 47, 48, 50, 52, 112, 113
spindle whorl, *see* textile tools
spinning bowls, *see* fiber wetting bowls
sponge spicules, 6
Sta Nychia, 70
statistics on the deposit, 93–97
Strait of Gibraltar, 4
super-plumes, 4
Syme, *see* Kato Syme

teapot development, 49
Tefeli Group, 5–6
Tethys Ocean, 4
textile tools, main discussion, 81–84
Thriphti Mountain Range, 9
tools of stone, *see* chipped and ground stone tools
Trapeza Cave, 45, 50
triangle of clay, 92
turntables, *see* potter's batts

Vasiliki Ware,
    analyzed, 103, 108, 109
    definition of, 38

Vasiliki, 10, 41, 42, 44, 46, 47, 48, 49, 50, 53, 83, 90, 108, 109
Vathypetro, 26
Venetian Pacheia Ammos, 9
VGI, 90
Vrokastro, 100, 102, 103, 105
Vronda, 72

warp-wighted loom, 81
White-on-Dark Ware, *see* East Cretan White-on-Dark Ware
workshop and craft activities in the deposit, 95–96
World War II bunkers, 8, 9
worm burrows, 6

Figures

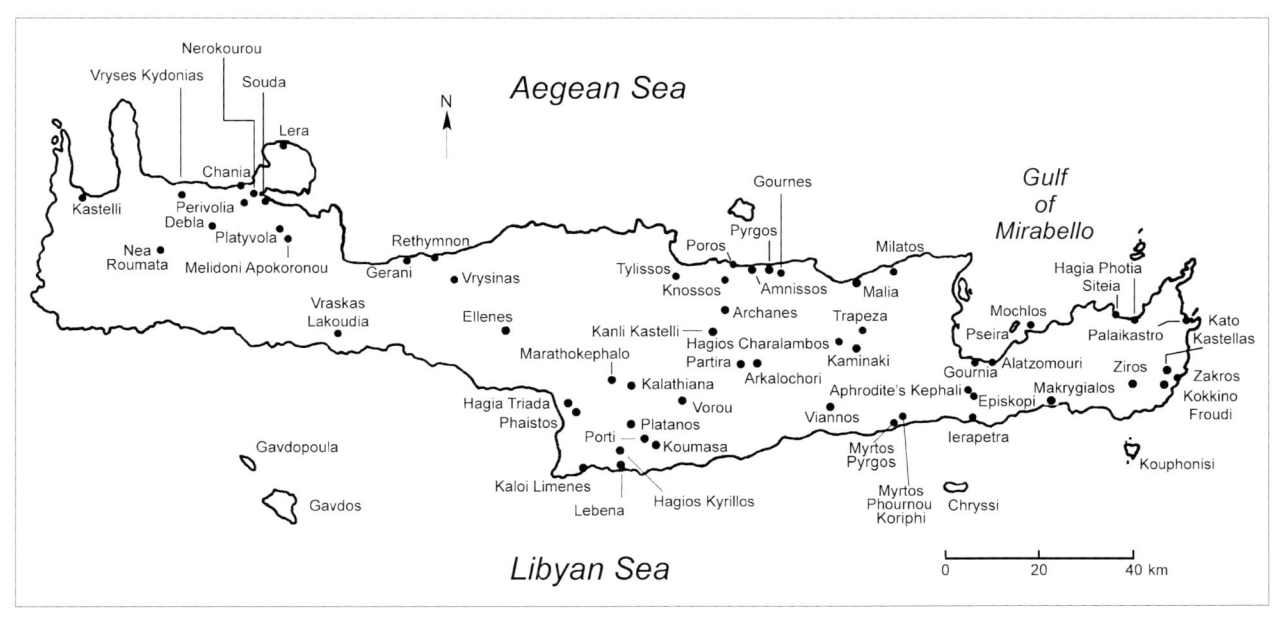

Figure 1. Map of Crete.

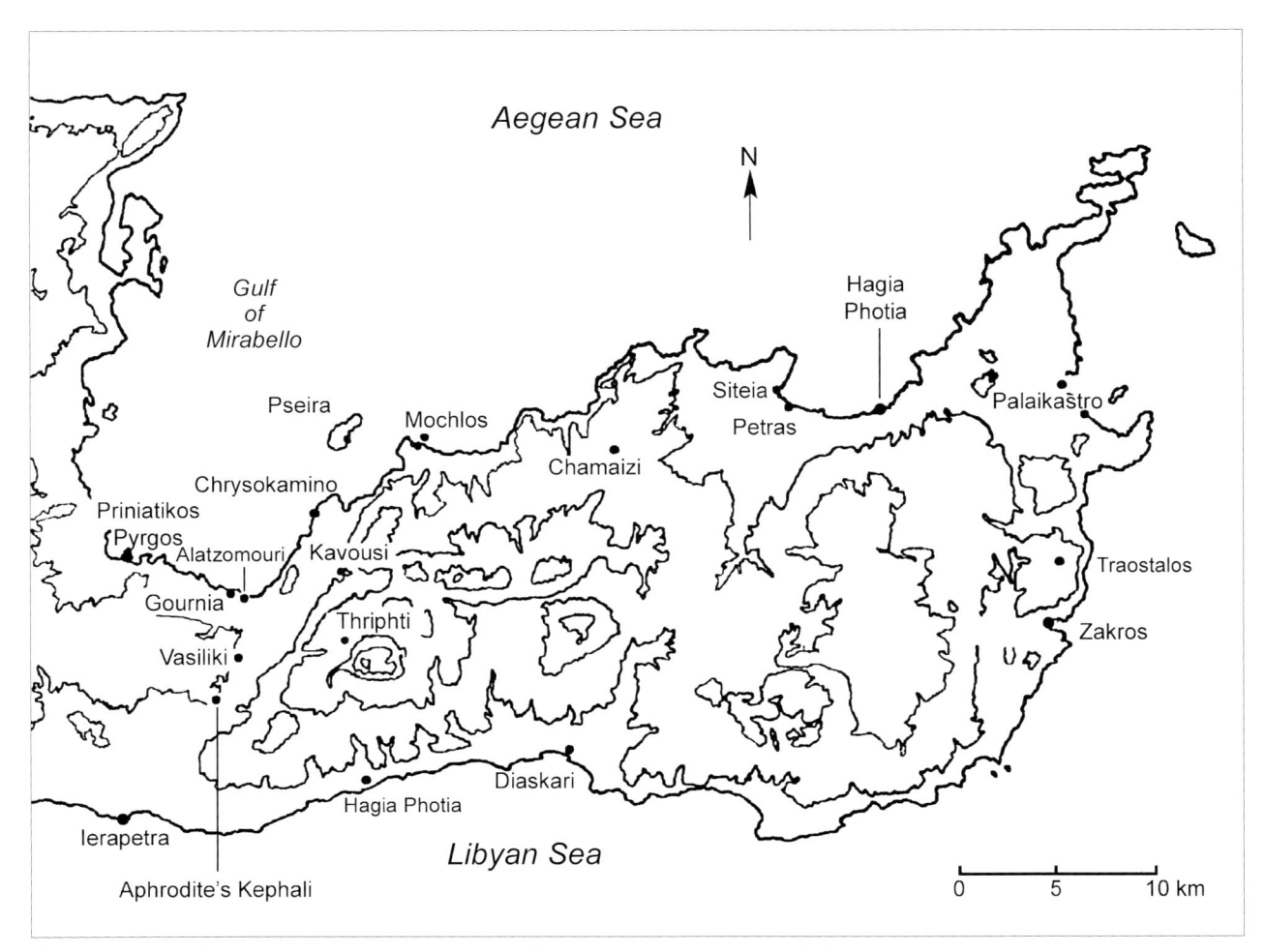

Figure 2. Map of East Crete. Contour interval 300 m. Map A. Insua and P. Betancourt.

FIGURE 3

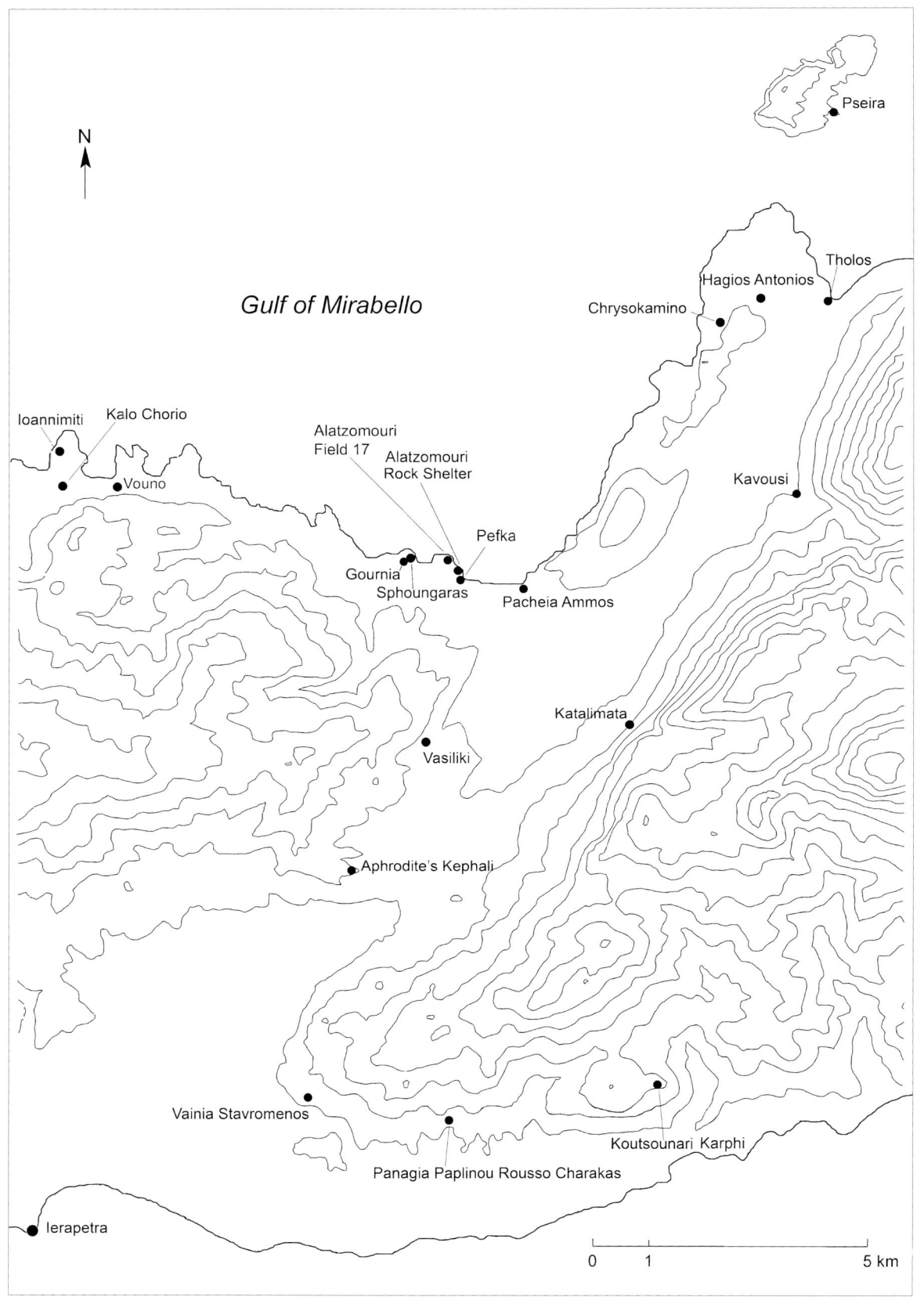

Figure 3. Map of the Ierapetra Isthmus. Contour interval 100 m.

FIGURE 4

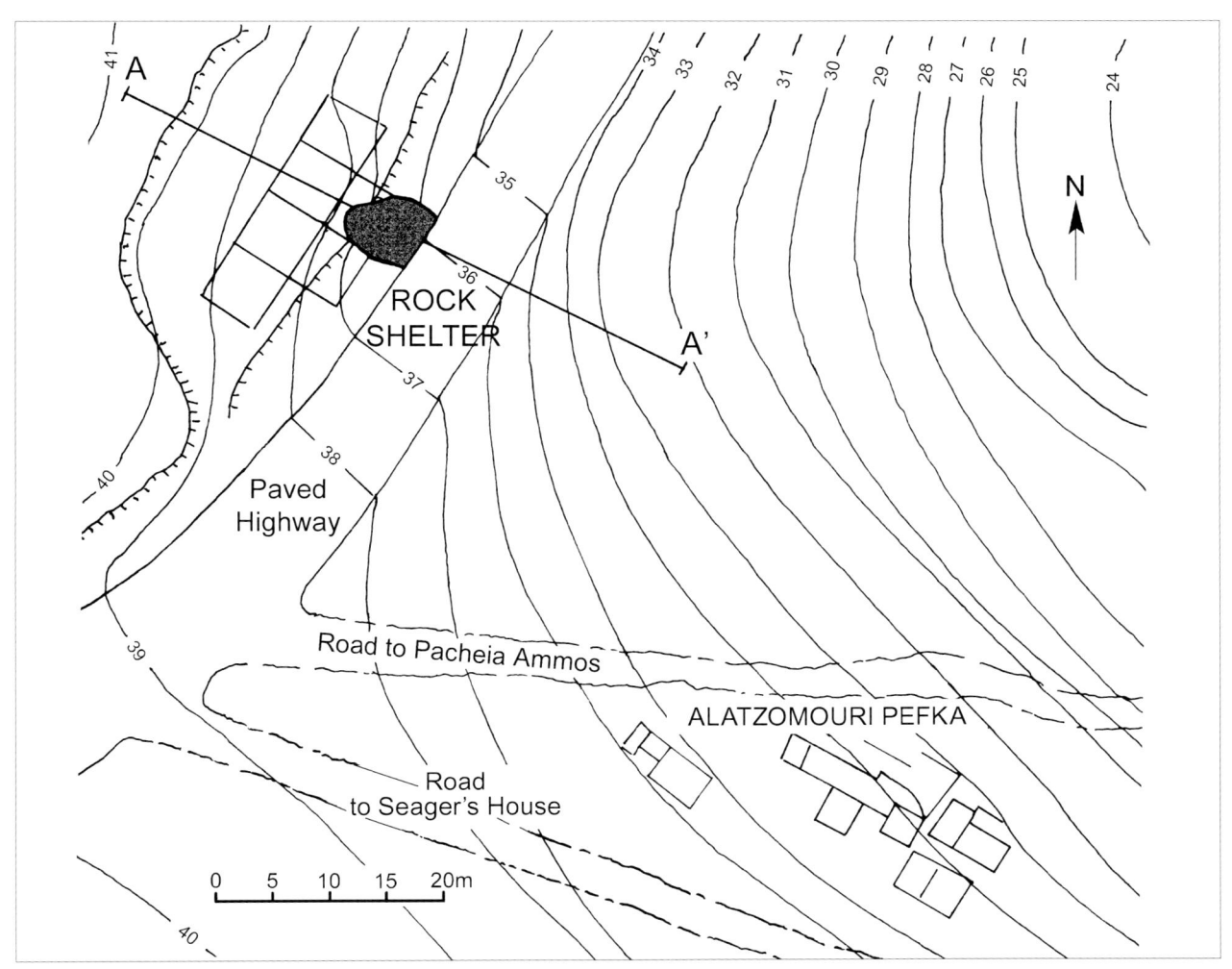

Figure 4. Topographic map of the southeastern slope of Alatzomouri Hill and surrounding area showing position and approximate outline of the Alatzomouri Rock Shelter (solid black) and archaeological field grids (2007–2008). Contour interval is 1 meter. Escarpments are depicted by lines with hatchures drawn on the lower topographic portion of the feature. The national paved highway cuts across the map; tertiary unpaved roads are indicated by wavy irregular lines. Outlined in the lower right is a portion of the Alatzomouri Industrial Area (Pefka). The position of cross-section A–A' (Fig. 6) is shown. Topographic data are from: (1) surveys during excavations at the Alatzomouri Rock Shelter, (2) a field survey of Alatzomouri Hill, (3) ground penetrating radar surveys east of the Alatzomouri Rock Shelter (right, off map), and (4) topographic surveys using dGPS. Map A. Insua, F. McCoy, A. Stamos, and D. Faulmann.

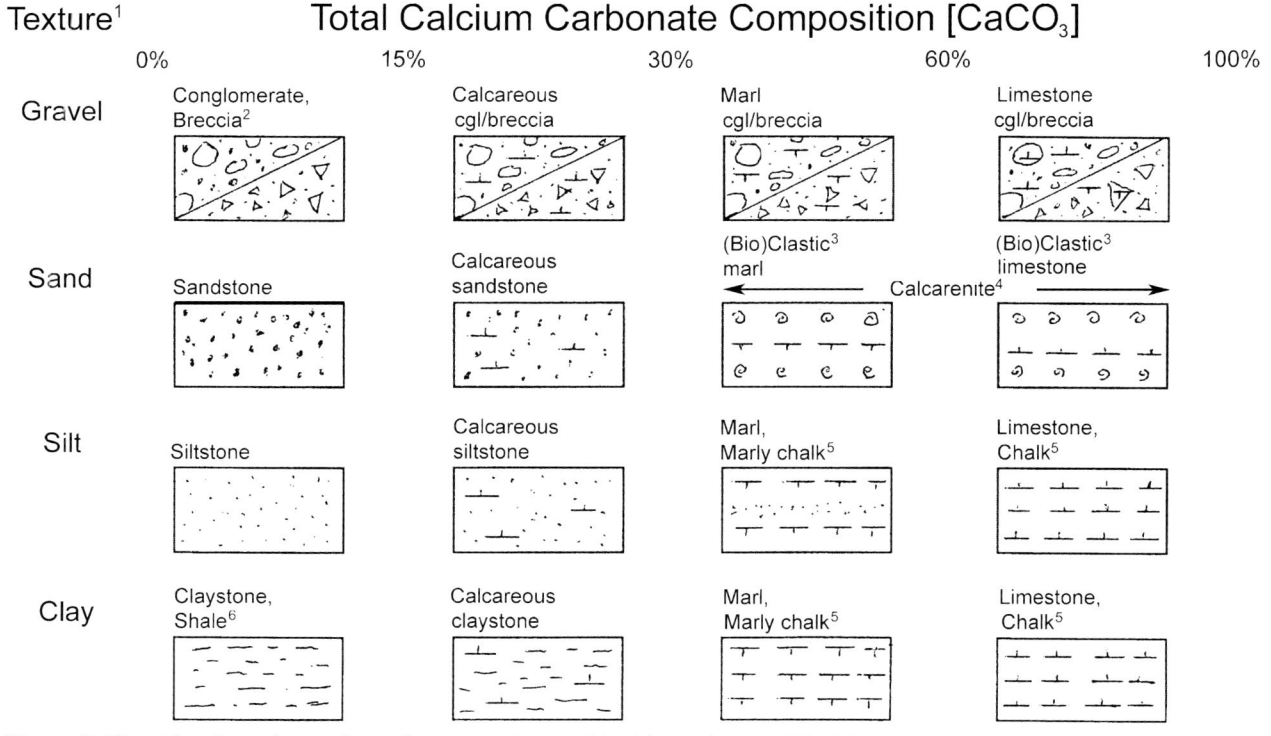

Figure 5. Classification scheme for carbonate rocks used in this study (modified from McCoy 2013, 29, table 3.2). Cgl = conglomerate. [1]Mean diameter of the dominant size of particles forming the sediment, following the standard Wentworth (1922) grade scale: gravel >2 mm; sand 2–1/16 mm [62 μm]; silt 62–64 μm; clay <4 μm. [2]The dominant particles in breccias are angular; those in conglomerates are semi- to well-rounded. [3]Clastic refers to sand-sized particles of either nonbiogenic (such as limestone fragments) or biogenic origin (e.g., shell fragments) and can be composed of calcium carbonate or silica (if the latter then the modifier in the rock name would be "siliceous"). [4]Alternate designator often used for clastic carbonate-rich rocks, with >50% clastic sand particles composed of calcium carbonate, thus they could be fragments of eroded carbonate rocks or biogenic material (e.g., shells). [5]Chalk refers to a less-indurated, softer, friable limestone that is usually white and easily crumbles, as differentiated from limestone, which can be very hard and well-indurated; colors vary from white to black. [6]Claystone is massive without fissility (splitting easily into thin slabs along closely spaced roughly parallel surfaces such as bedding planes), whereas shale has marked fissility.

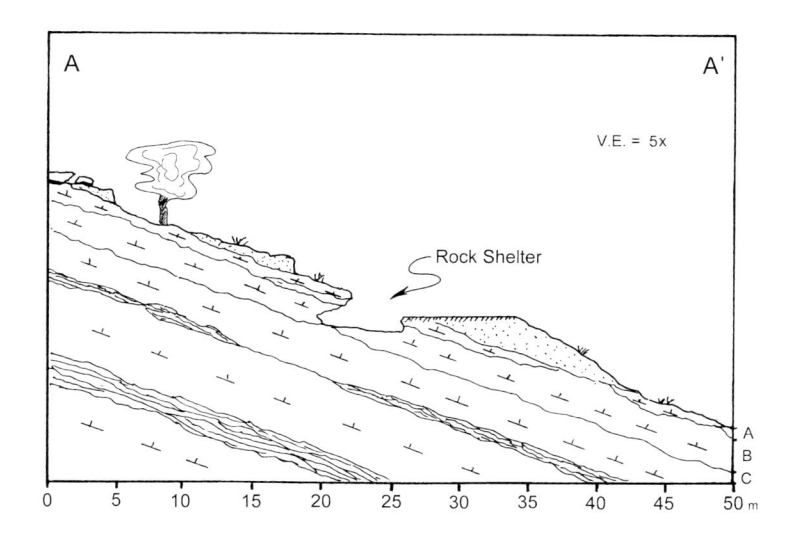

Figure 6. Stratigraphic section A–A' in Figure 4 along the southeastern slope of Alatzomouri Hill. The Alatzomouri Rock Shelter within this geological framework is indicated; note the use of local geologic factors for constructing this shelter with the softer sedimentary rock excavated to create the shelter (layer marked as B) and the lithified stronger layers utilized as floor and ceiling (layers marked as A [Pefka layer] and C). Lithologic symbols are those defined in Figure 5. Dips on layers are 9°–17° E. V.E. = vertical exaggeration. Drawing F. McCoy and A. Parisky.

FIGURE 7

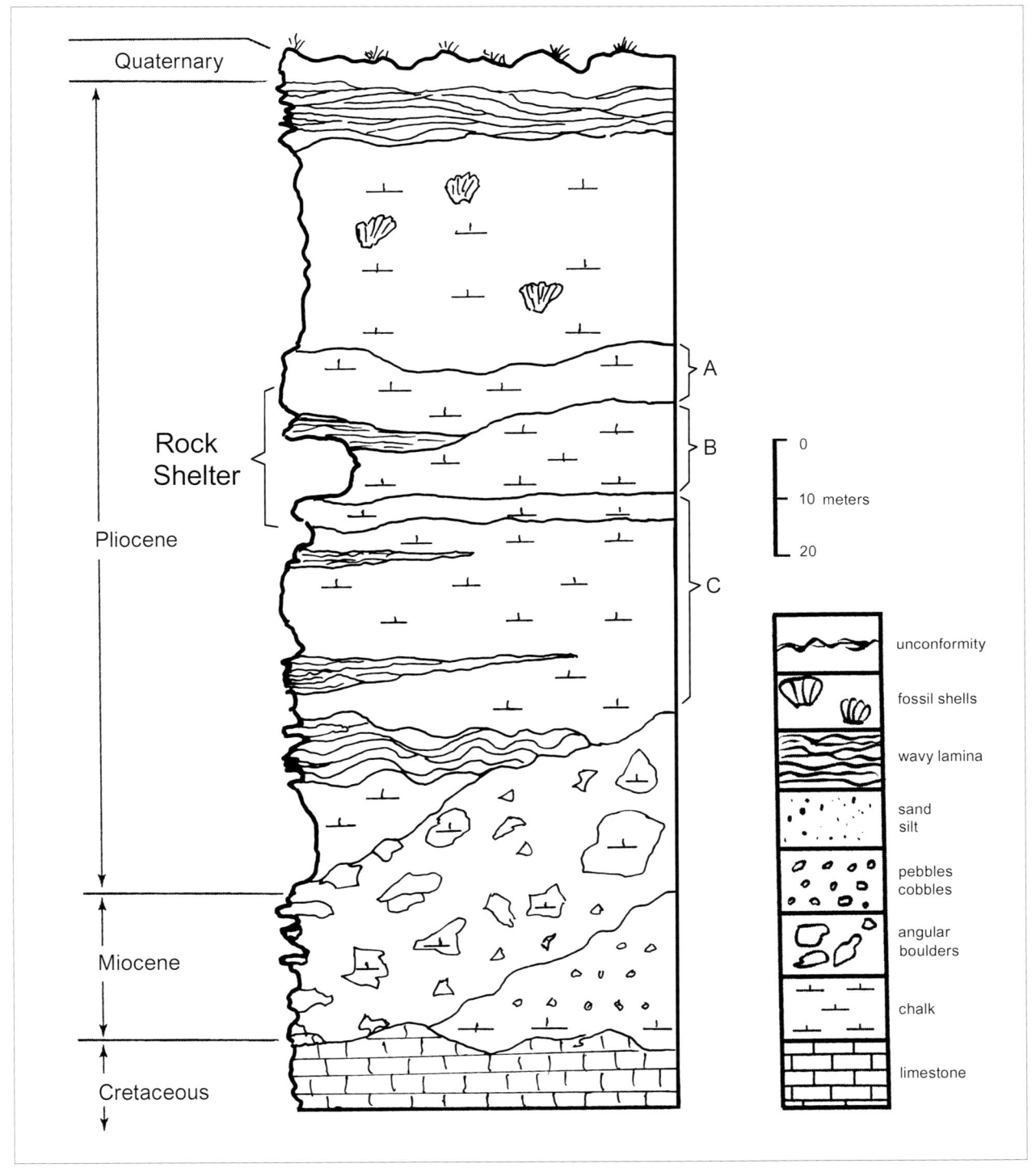

Figure 7. Generalized lithologic and stratigraphic section of the Pacheia Ammos Formation at Alatzomouri Hill. Data are from geologic mapping of the hill, extrapolated to create this section. The relationship of the Alatzomouri Rock Shelter within this regional stratigraphic section is noted. Layers marked A, B, and C are those noted in Figure 6. Relative induration of rocks, thus hardness and resistance to erosion, are indicated along the left margin by protruding ledges and undercut areas. Patterns for identifying lithologic types are those defined in Figure 5; the exception is the pattern used for the basal lithologic unit, which is an Upper Cretaceous limestone. Closely spaced wavy lines depict fine laminations; wavy lines with wider spacing indicate soft sediment deformation structures such as slumps and folds. Breccia is indicated by large angular clasts within the host rock. Fossiliferous zones are indicated by the shell symbol. Drawing F. McCoy and A. Parisky.

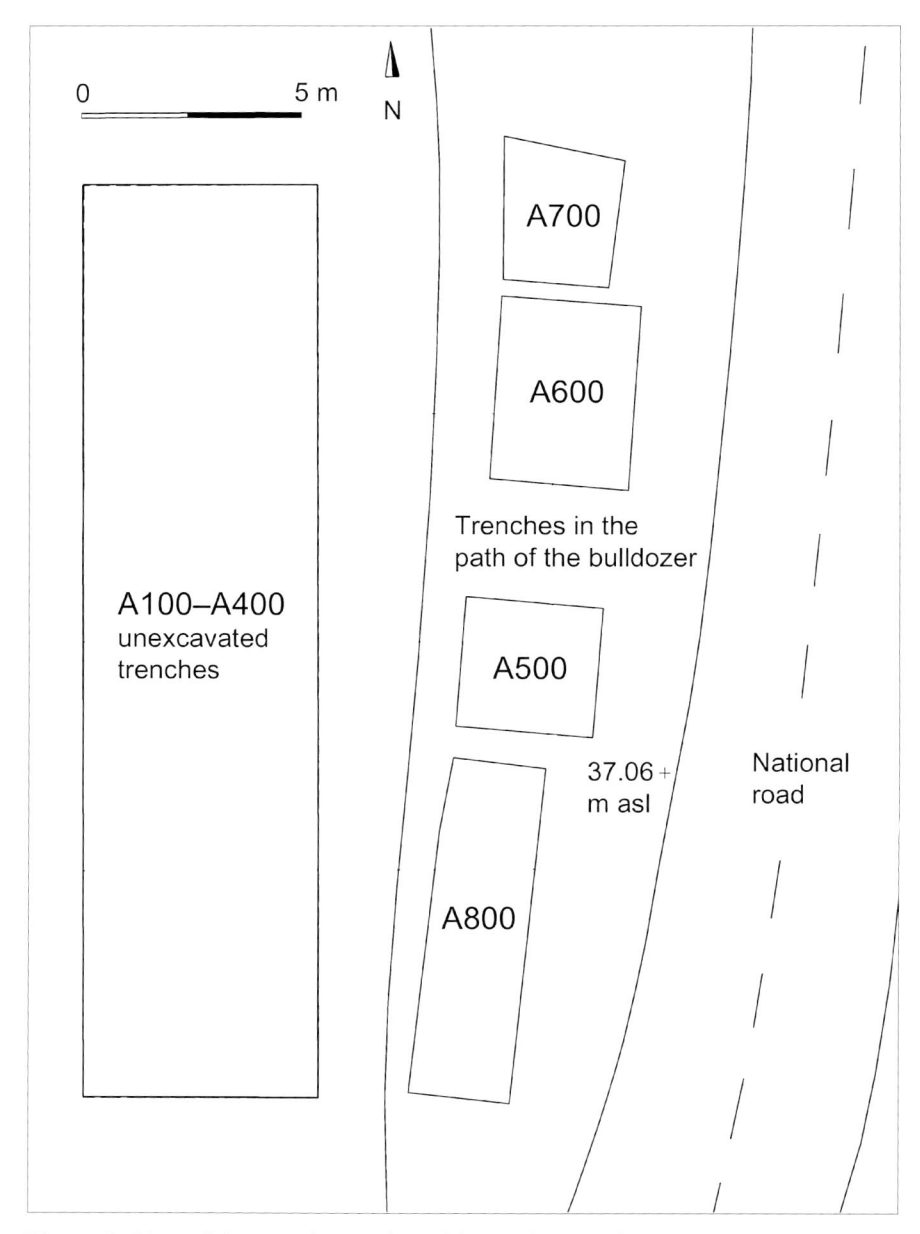

Figure 8. Plan of the trenches and road located west of Pacheia Ammos. Plan G. Damaskanakis and K. Chalikias.

FIGURE 9

	S 36.57 m						N
1			A600.1			A600.2	
2			A600.1	A600.1/2/3/4/5/6 (clay vessels)		A602.1 (rock shelter)	
3	A604.2/3/4/5 (clay vessels)		A601.7	"			
4	"		A604.1	"			
5	A604.14 (clay vessel)	"	A604.8	"	A604.7	A604.6	A604.15/16 (pottery scatters)
6	"	"	A604.10	"	A604.9 (trial trench)		"
7	"	"	A604.11 (S and W part)	"			"
8	"	A604.2	A604.13		A601.2/3	A604.12	A602.2
9	"	A604.18 (S part)	A604.17 (W–NW part)		A601.8		A602.3
10	"	A604.20		the wall A605	A604.19		A604.25/26 (clay vessels)
11	"	A604.23/24/30/31/32/33/34/35 (clay vessels)	A604.22		A604.21		"
12	"	A604.36/37/38/41/42/43/44/45 (clay vessels)	A604.40 A604.27 (W part) A604.28 (E part)		A604.29		A604.39
13	A604.50/54 (clay vessels)		A604.47 (S part)		A604.48/51/52 (clay vessels)		A604.46
14	"		A604.53		"		A604.49
15			A604.55 (S part)				
16			A604.57			A604.56	A604.58
17	A604.61 (SE corner)	A604.60 (clay vessel)	A604.59				
18			A604.62				
19	A604.64 (pottery scatter)		A604.63				
20	A604.66 (SE corner)	A604.68/69 (clay vessels)	A604.65				
21	"		A604.67				

35.72 m

Figure 9. Schematic south–north section of the rock shelter stratigraphy showing loci and finds by layer/level. Image K. Chalikias.

FIGURE 10

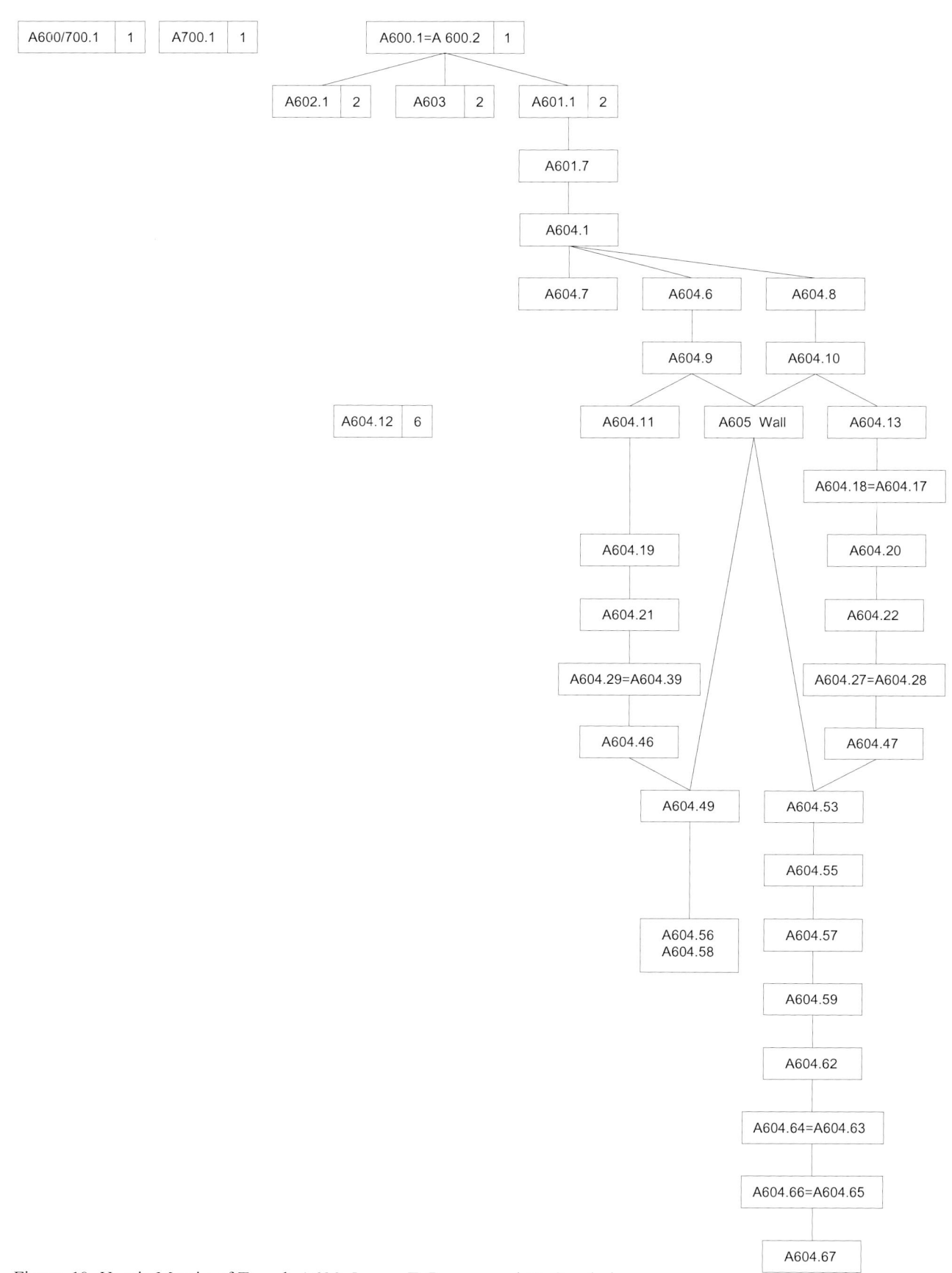

Figure 10. Harris Matrix of Trench A600. Image T. Brogan and H. Goodwin.

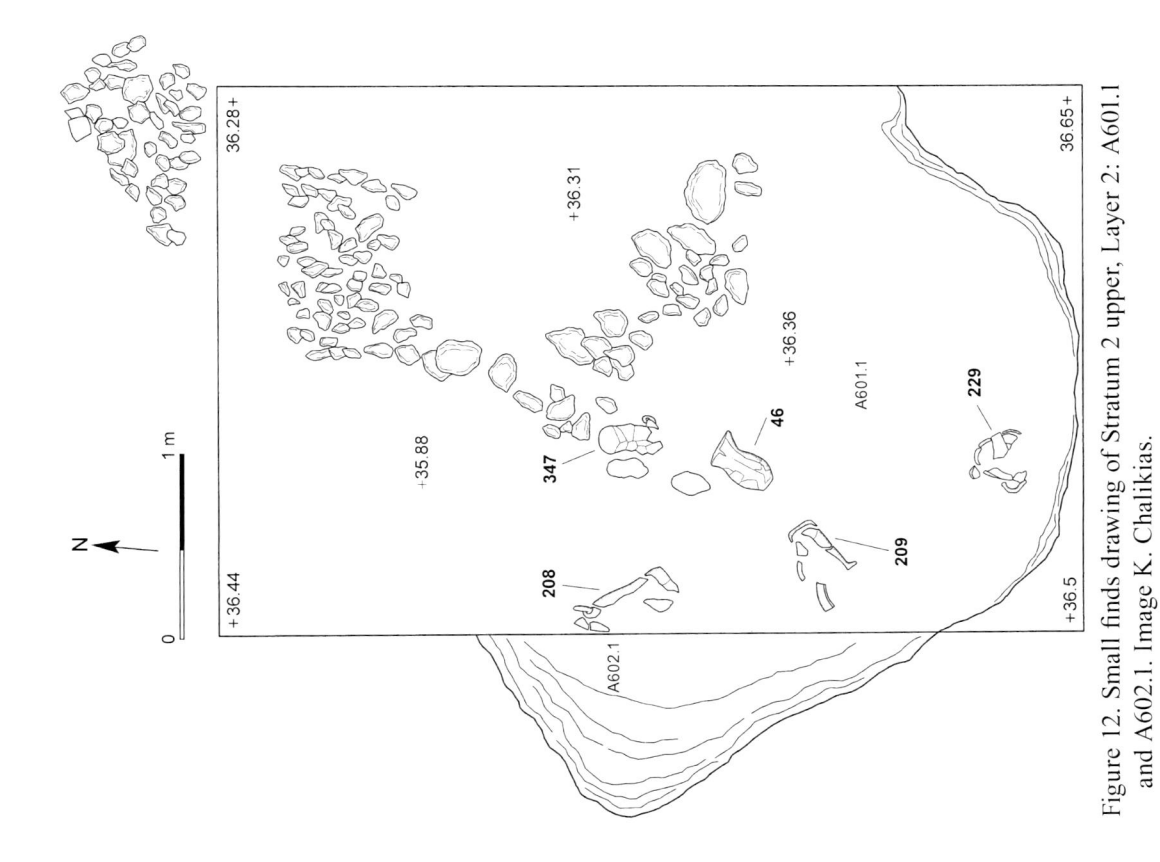

Figure 12. Small finds drawing of Stratum 2 upper, Layer 2: A601.1 and A602.1. Image K. Chalikias.

Figure 11. Small finds drawing of Stratum 1, Layer 1: A600.1. Image K. Chalikias.

FIGURES 13 AND 14

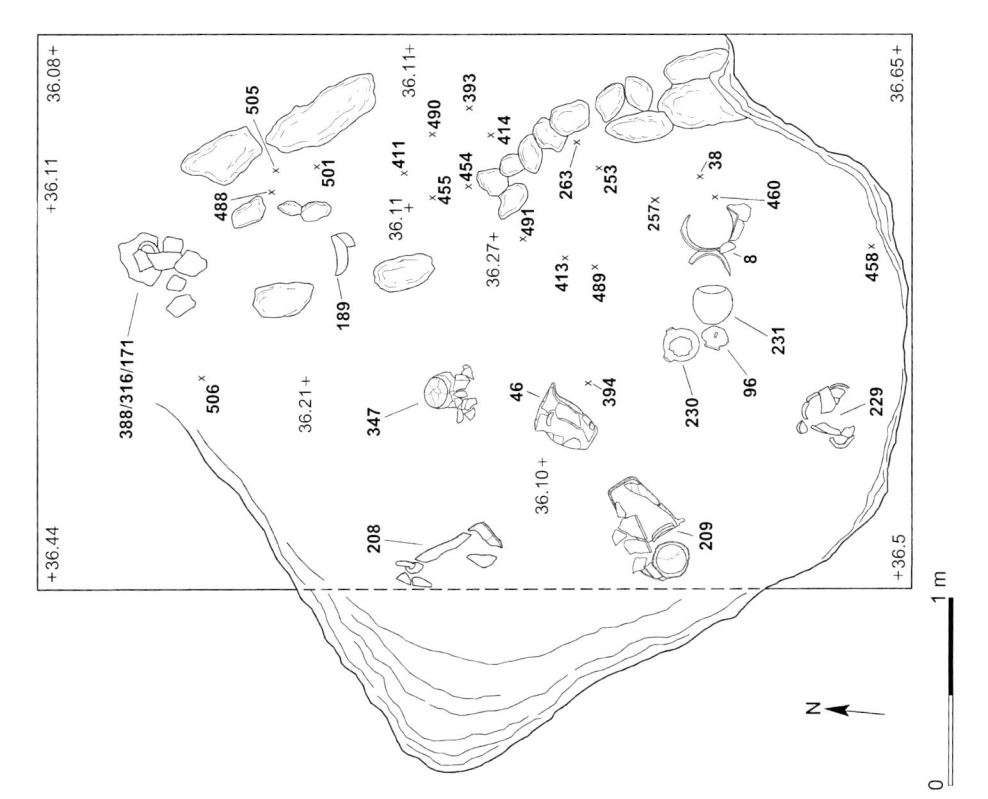

Figure 14. Small finds drawing of upper portion of Stratum 2 interior and exterior: objects from Layers 2–9 and stones from Layers 4–9. Image K. Chalikias.

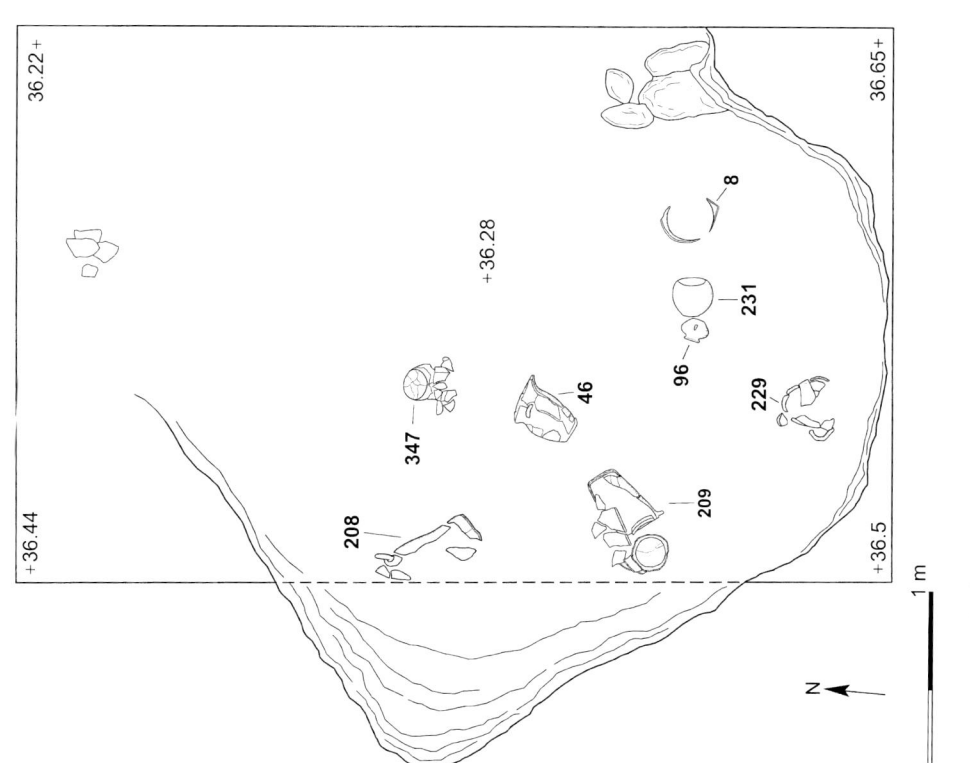

Figure 13. Small finds drawing of upper portion of Stratum 2 upper: Layers 2–4. Image K. Chalikias.

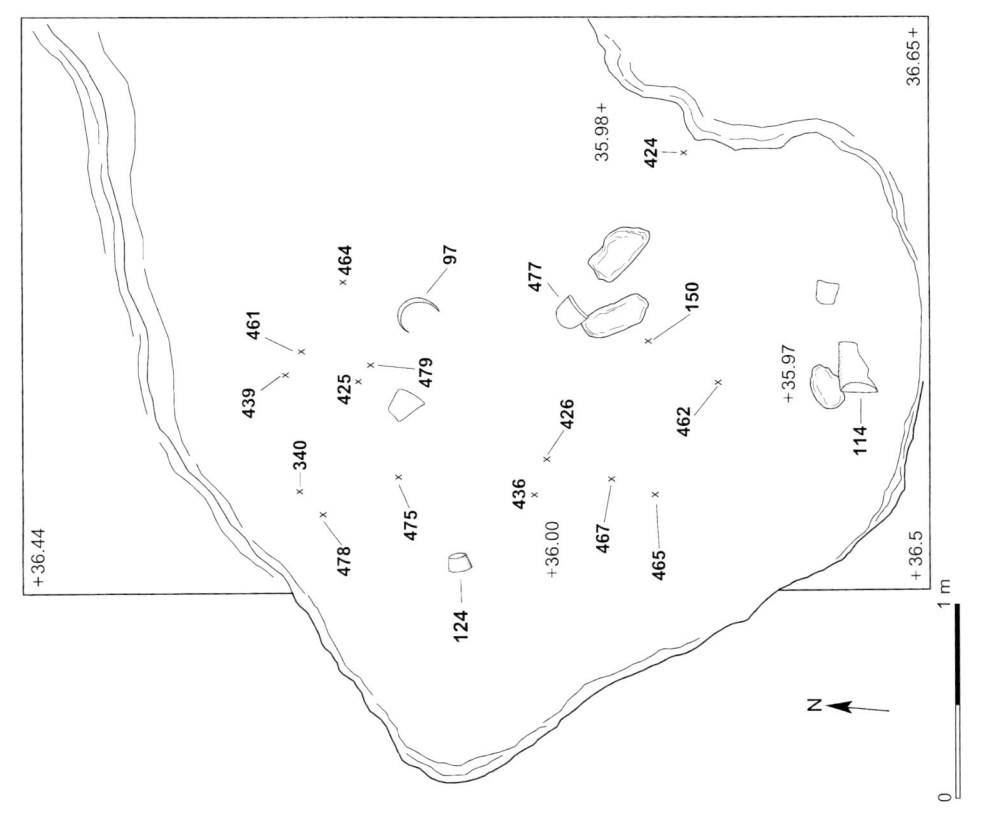

Figure 16. Small finds drawing of Stratum 3a: Layers 13–14. Image K. Chalikias.

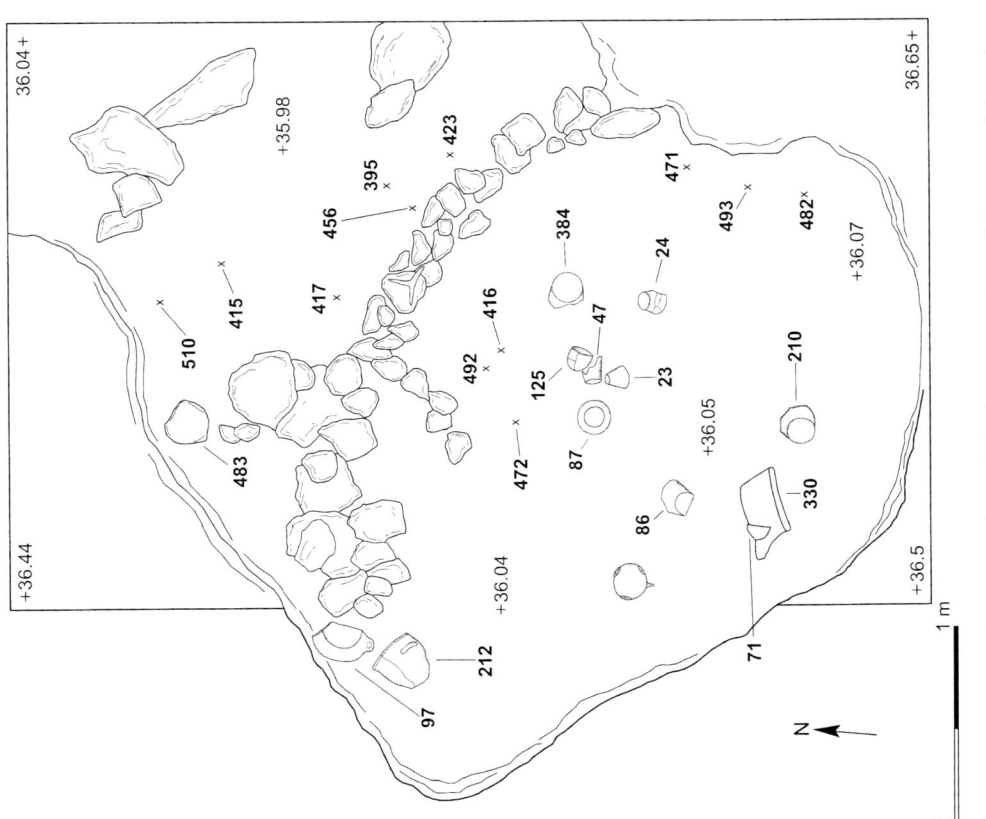

Figure 15. Small finds drawing of the lower portion of Stratum 2 interior and exterior: Layers 10–12. Image K. Chalikias.

FIGURES 17 AND 18

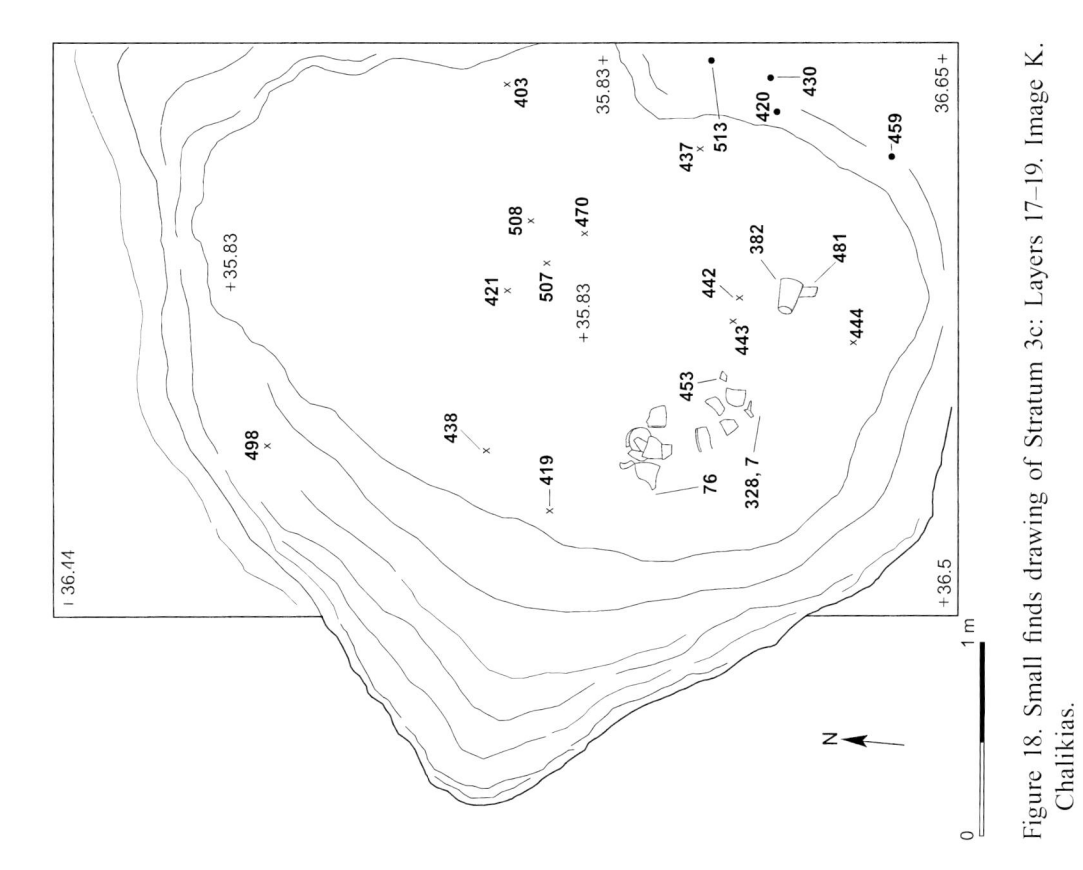

Figure 18. Small finds drawing of Stratum 3c: Layers 17–19. Image K. Chalikias.

Figure 17. Small finds drawing of Stratum 3b: Layers 15–16. Image K. Chalikias.

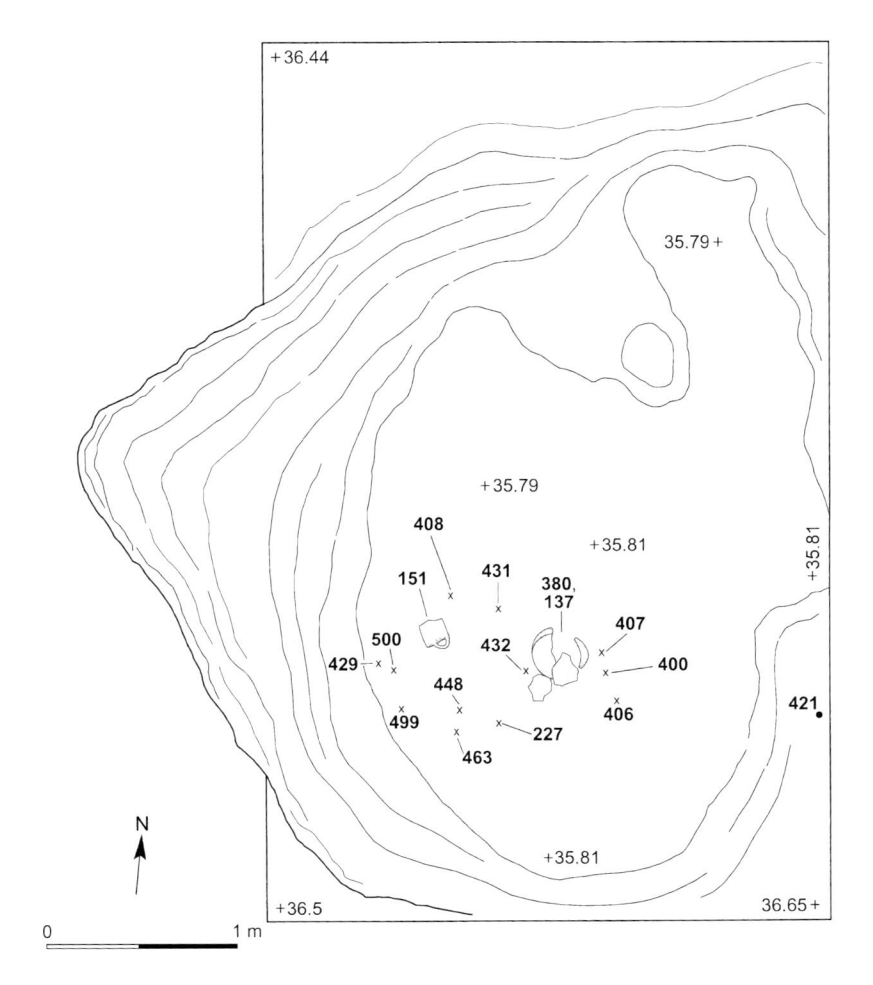

Figure 19. Small finds drawing of Stratum 3d: Layers 20–21. Image K. Chalikias.

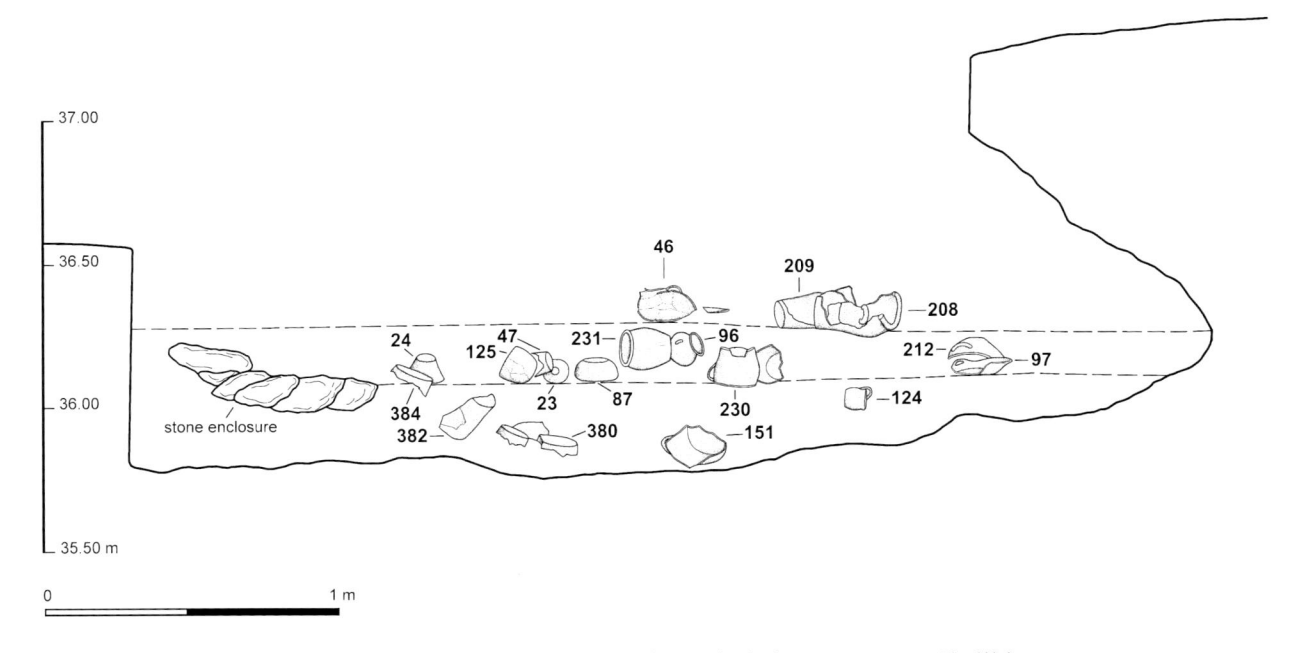

Figure 20. Stratigraphic section running E–W through the rock shelter. Image K. Chalikias.

FIGURE 21

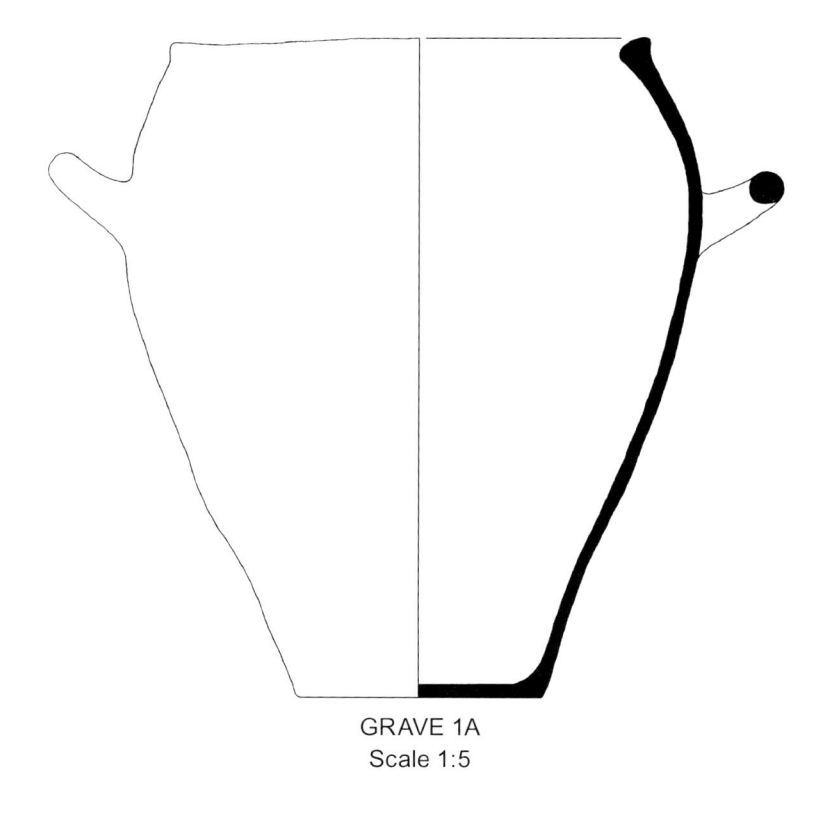

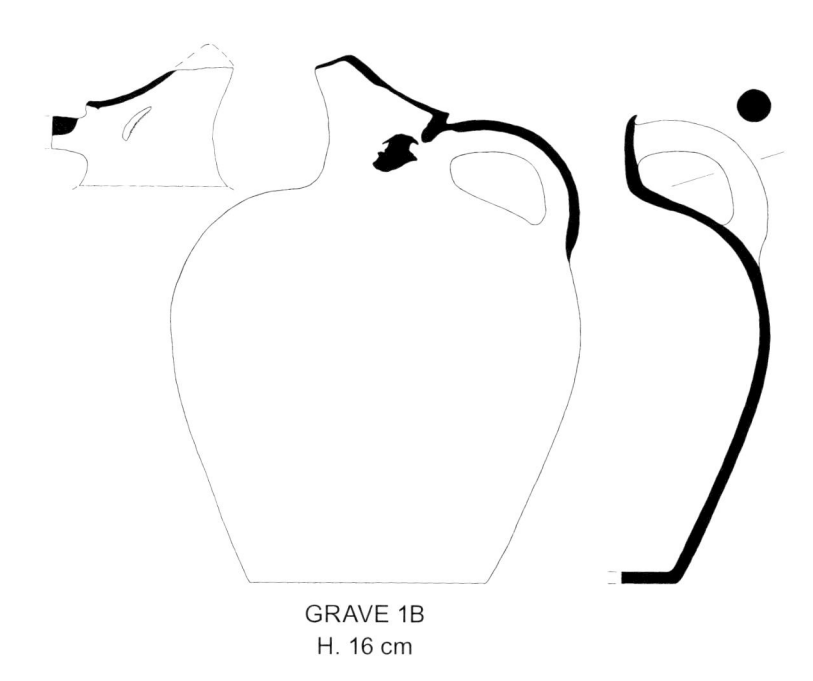

GRAVE 1A
Scale 1:5

GRAVE 1B
H. 16 cm

Figure 21. Plain pottery and pottery with slight decoration found outside the cave. Scale as marked. Drawings D. Faulmann.

FIGURE 22

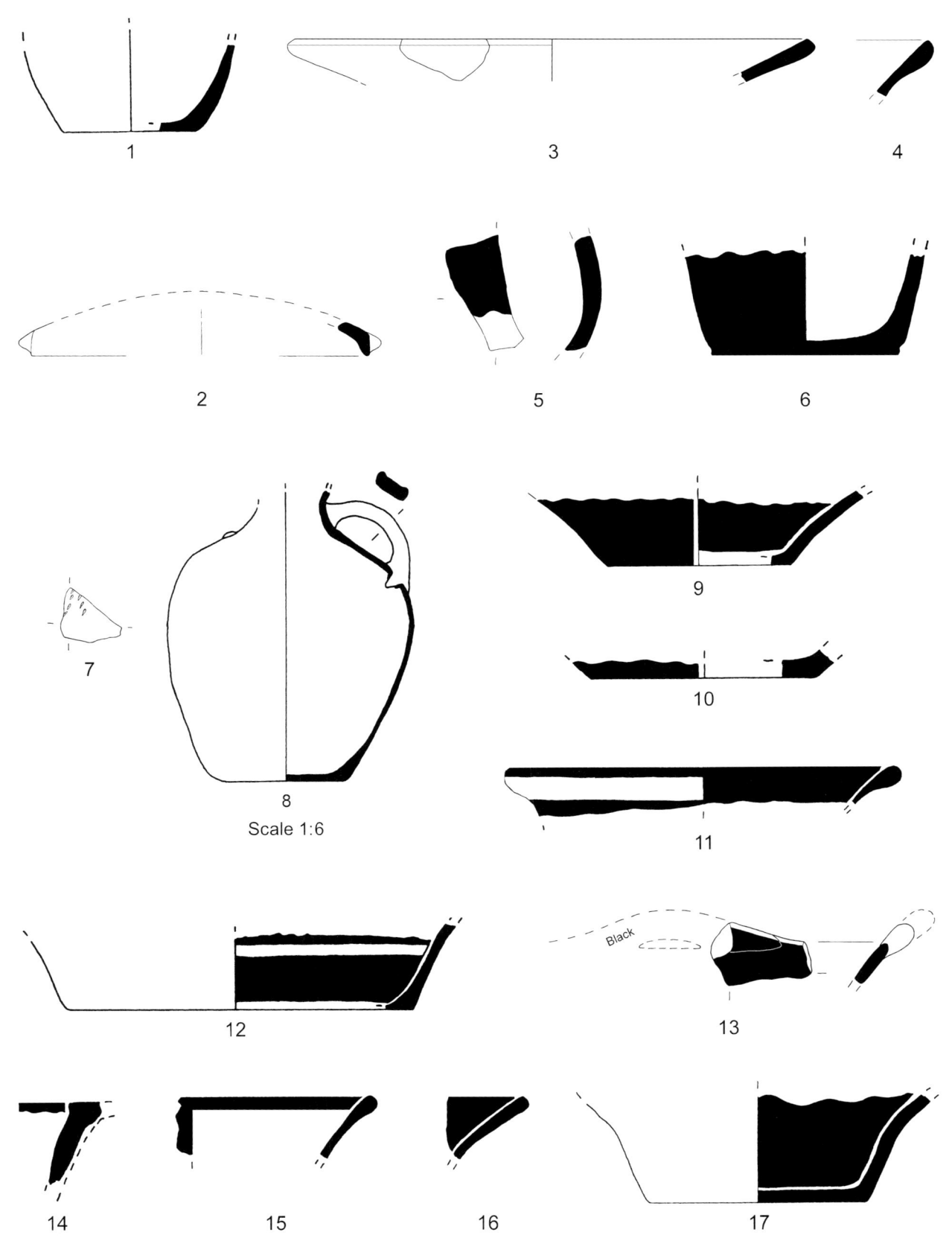

Figure 22. Final Neolithic–EM I cup (**1**); EM I pyxis lid (**2**); EM IIB Vasiliki Ware: shallow bowls (**3**, **4**) and jugs (**5**–**7**); EM IIB South Coast jug (**8**); and EM III East Cretan White-on-Dark Ware shallow bowls and basins (**9**–**17**). Scale 1:3 except for **8** (scale 1:6). Drawings D. Faulmann, L. Bonga, K. Chalikias, and P. Betancourt.

FIGURE 23

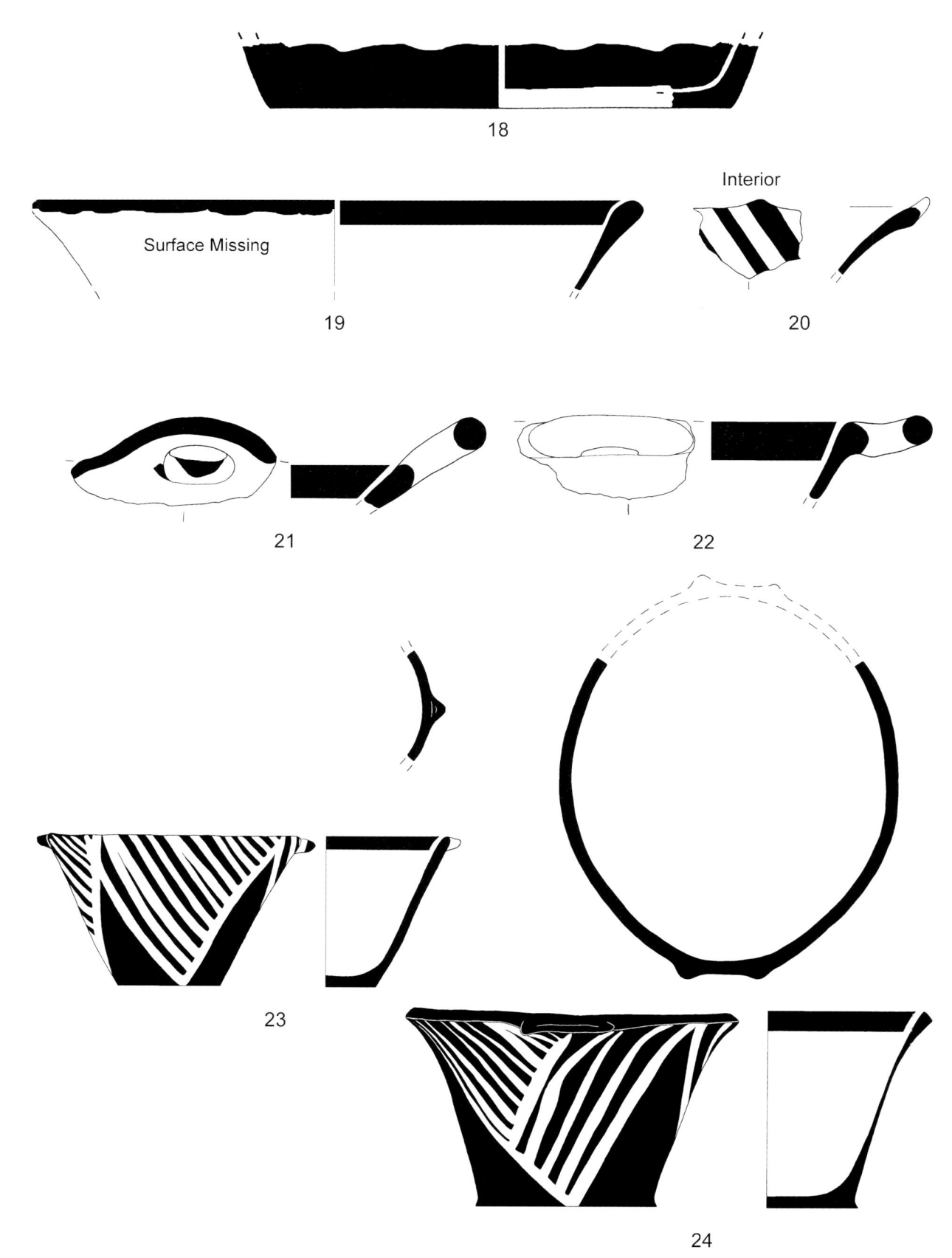

Figure 23. East Cretan White-on-Dark Ware: shallow bowls and basins (**18–22**) and conical bowls (**23**, **24**). Scale 1:3. Drawings D. Faulmann, L. Bonga, K. Chalikias, and P. Betancourt.

FIGURE 24

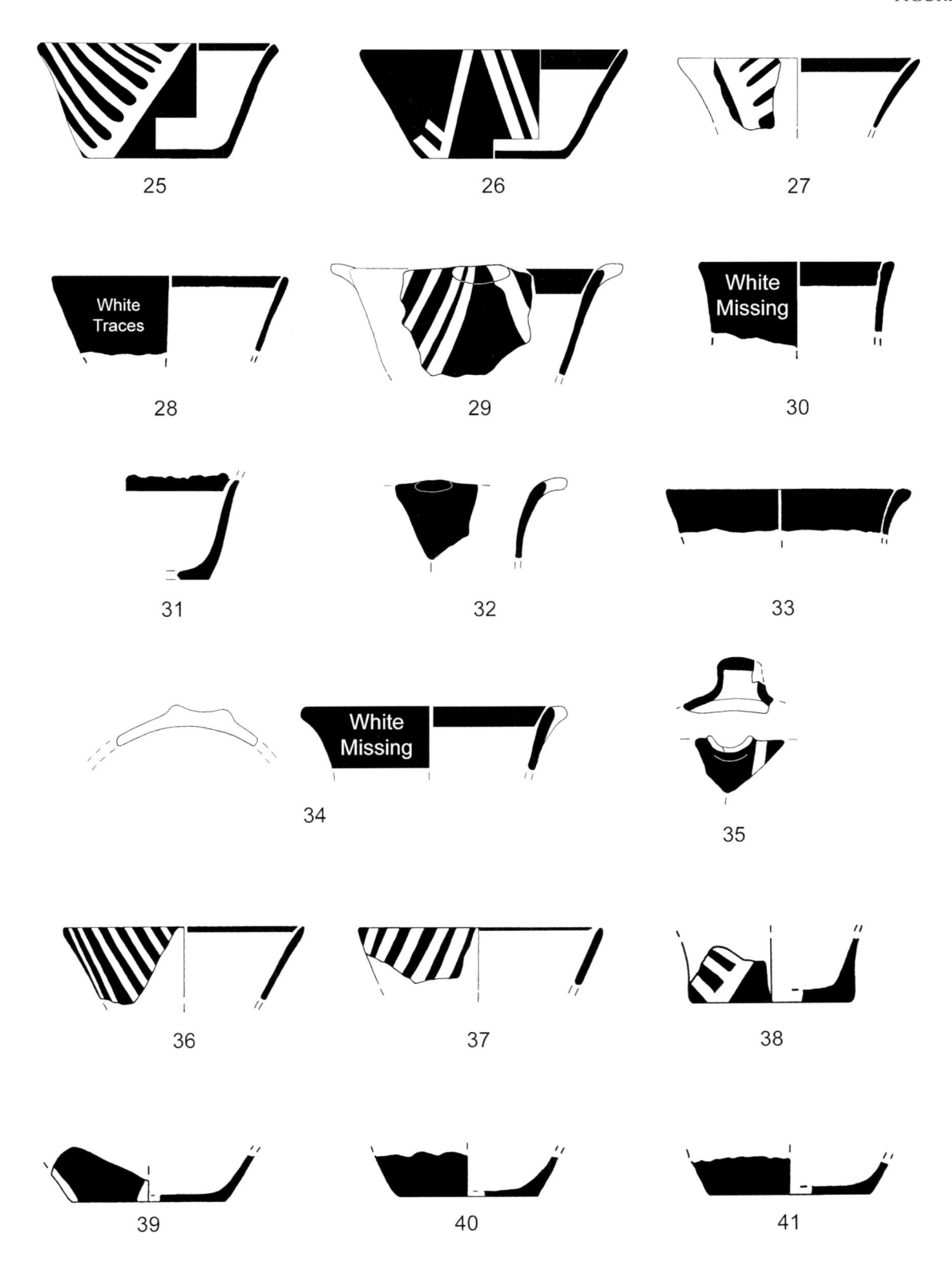

Figure 24. East Cretan White-on-Dark Ware: conical bowls (**25–41**). Scale 1:3. Drawings D. Faulmann, L. Bonga, K. Chalikias, and P. Betancourt.

FIGURE 25

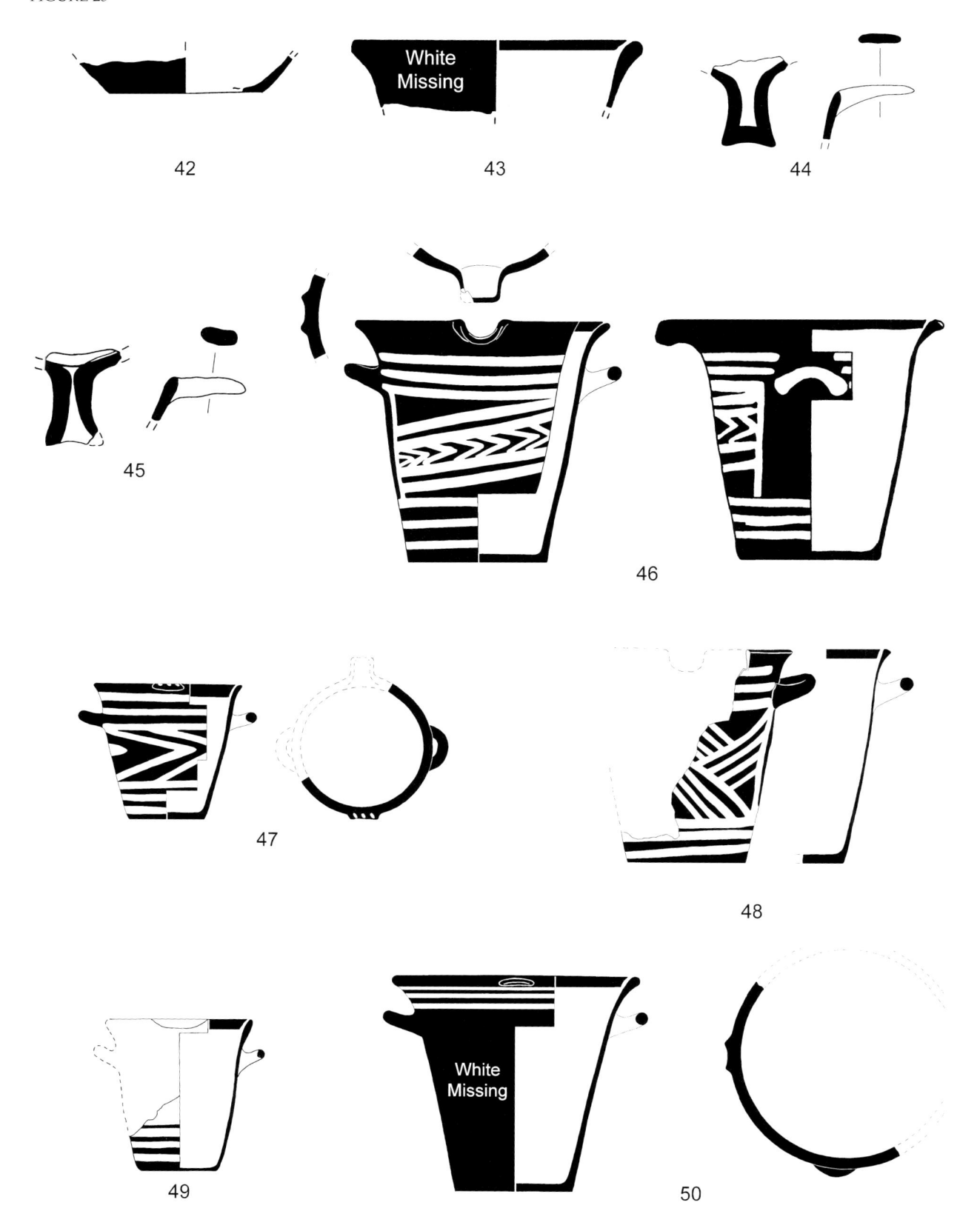

Figure 25. East Cretan White-on-Dark Ware: conical bowls (**42**, **43**), conical bowls with frying pan handles (**44**, **45**), and spouted conical jars (**46**–**50**). Scales 1:3 (**42**–**45**) and 1:6 (**46**–**50**). Drawings D. Faulmann, L. Bonga, K. Chalikias, and P. Betancourt.

FIGURE 26

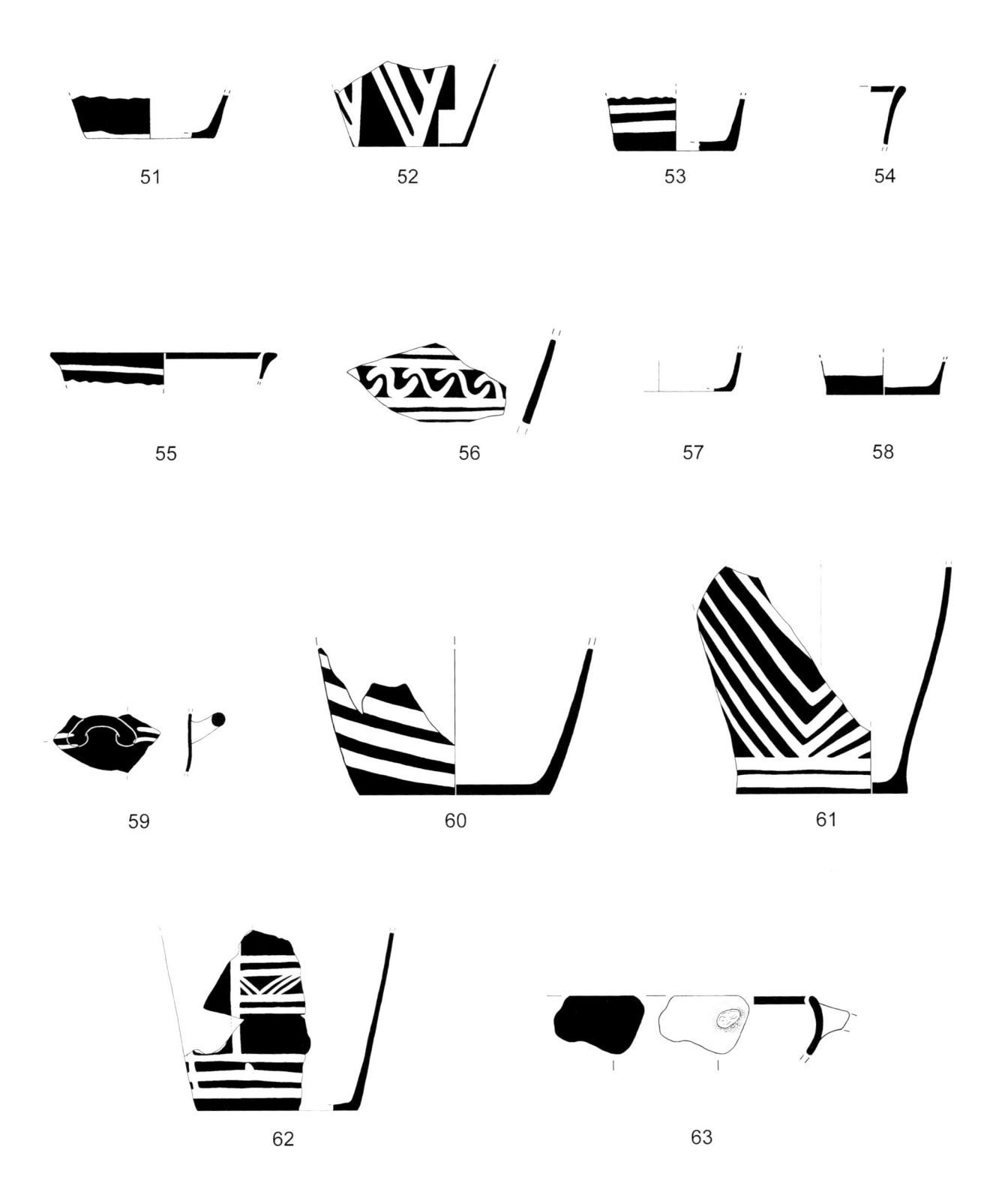

Figure 26. East Cretan White-on-Dark Ware: spouted conical jars (**51–60**), jars (**61**, **62**), and straight-sided or rounded cup (**63**). Scales 1:3 (**56**, **60**, **63**) and 1:6 (**51–55**, **57–59**, **61**, **62**). Drawings D. Faulmann, L. Bonga, K. Chalikias, and P. Betancourt.

FIGURE 27

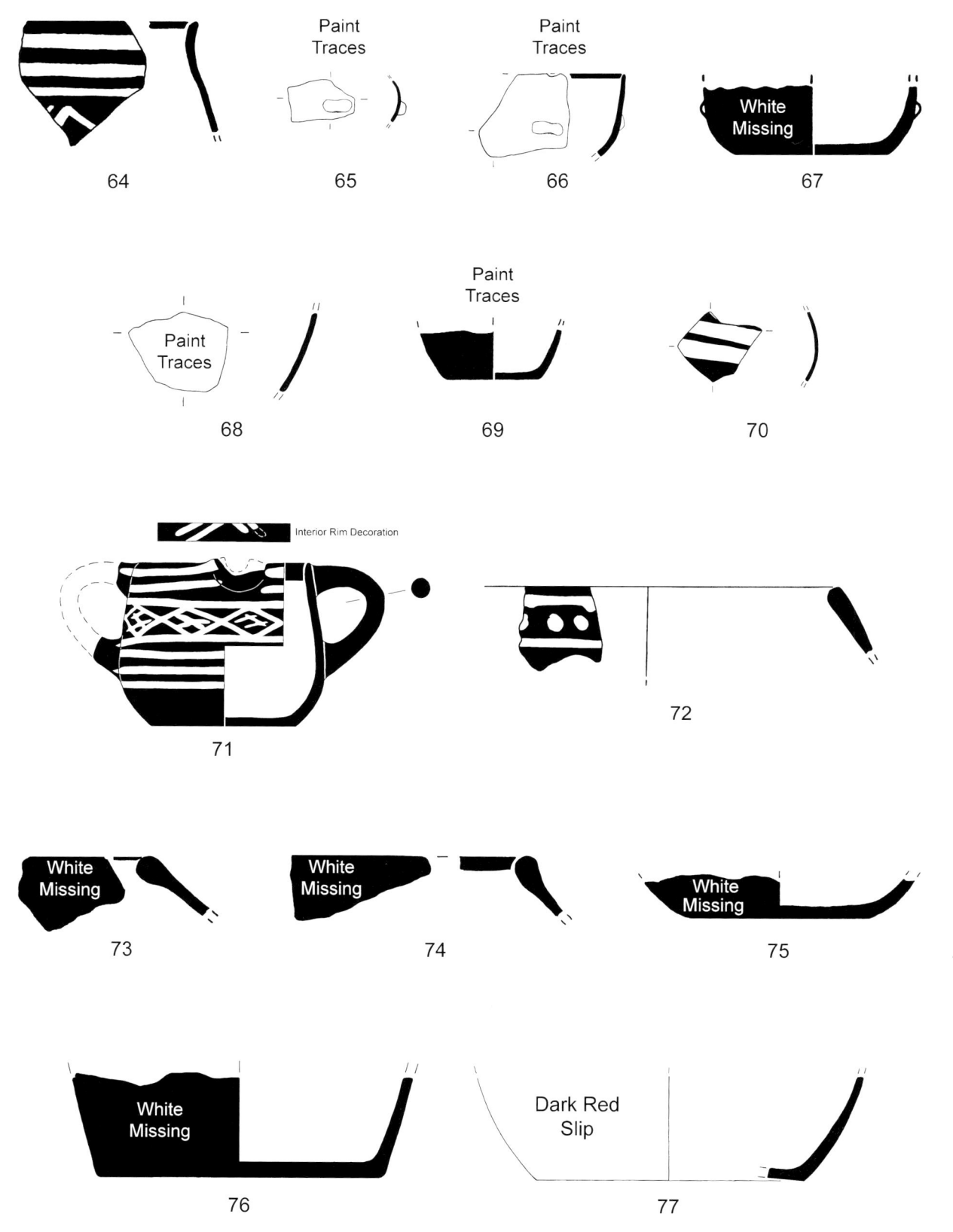

Figure 27. East Cretan White-on-Dark Ware: cylindrical cup (**64**), rounded cup with tiny lugs or handles (**65–67**), rounded cups (**68–70**), spouted cup (**71**), and bridge-spouted jars (**72–77**). Scale 1:3. Drawings D. Faulmann, L. Bonga, K. Chalikias, and P. Betancourt.

FIGURE 28

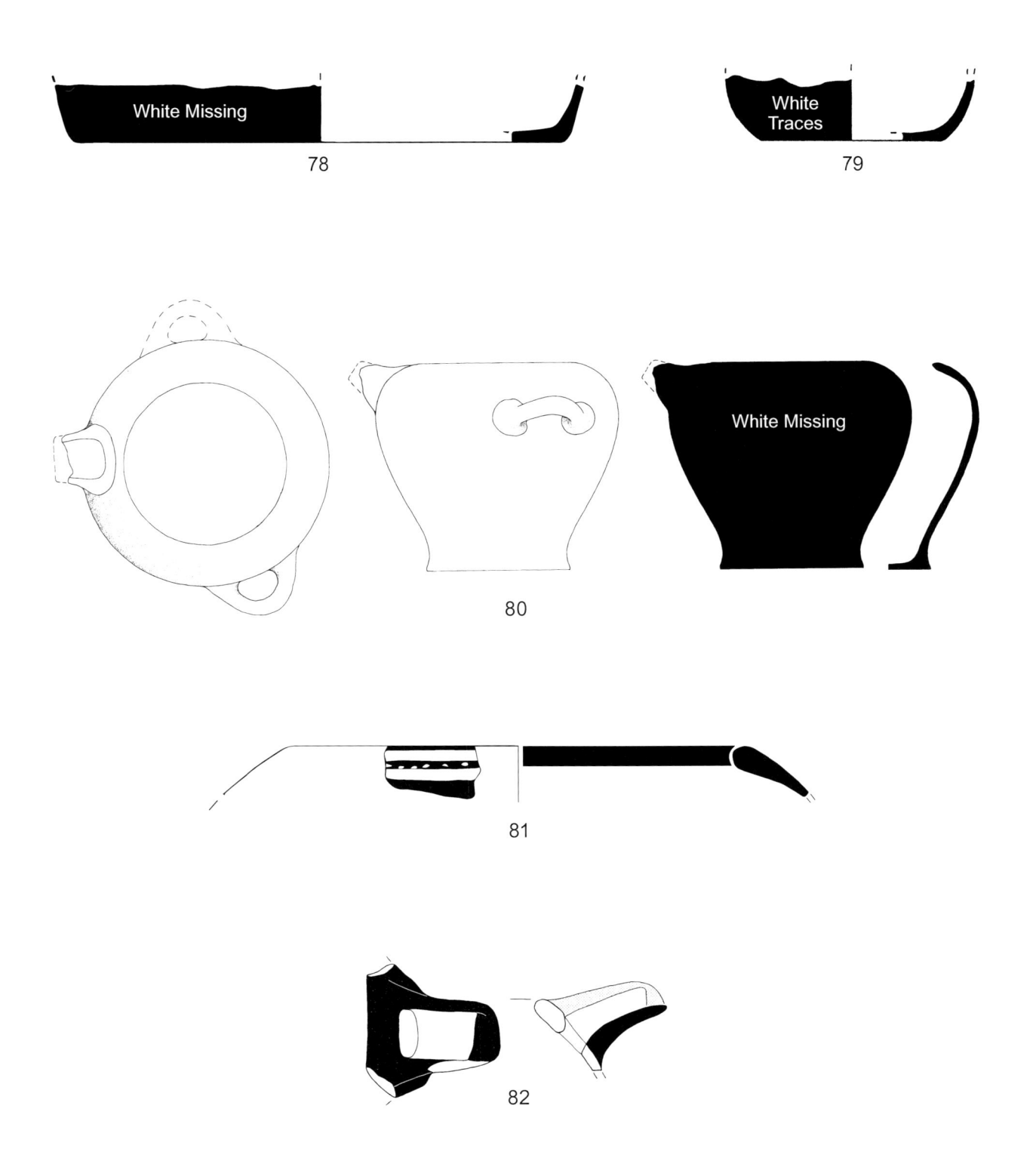

Figure 28. East Cretan White-on-Dark Ware: bridge-spouted jars (**78**–**82**). Scales 1:3 (**78**, **79**, **81**, **82**) and 1:6 (**80**). Drawings D. Faulmann, L. Bonga, K. Chalikias, and P. Betancourt.

FIGURE 29

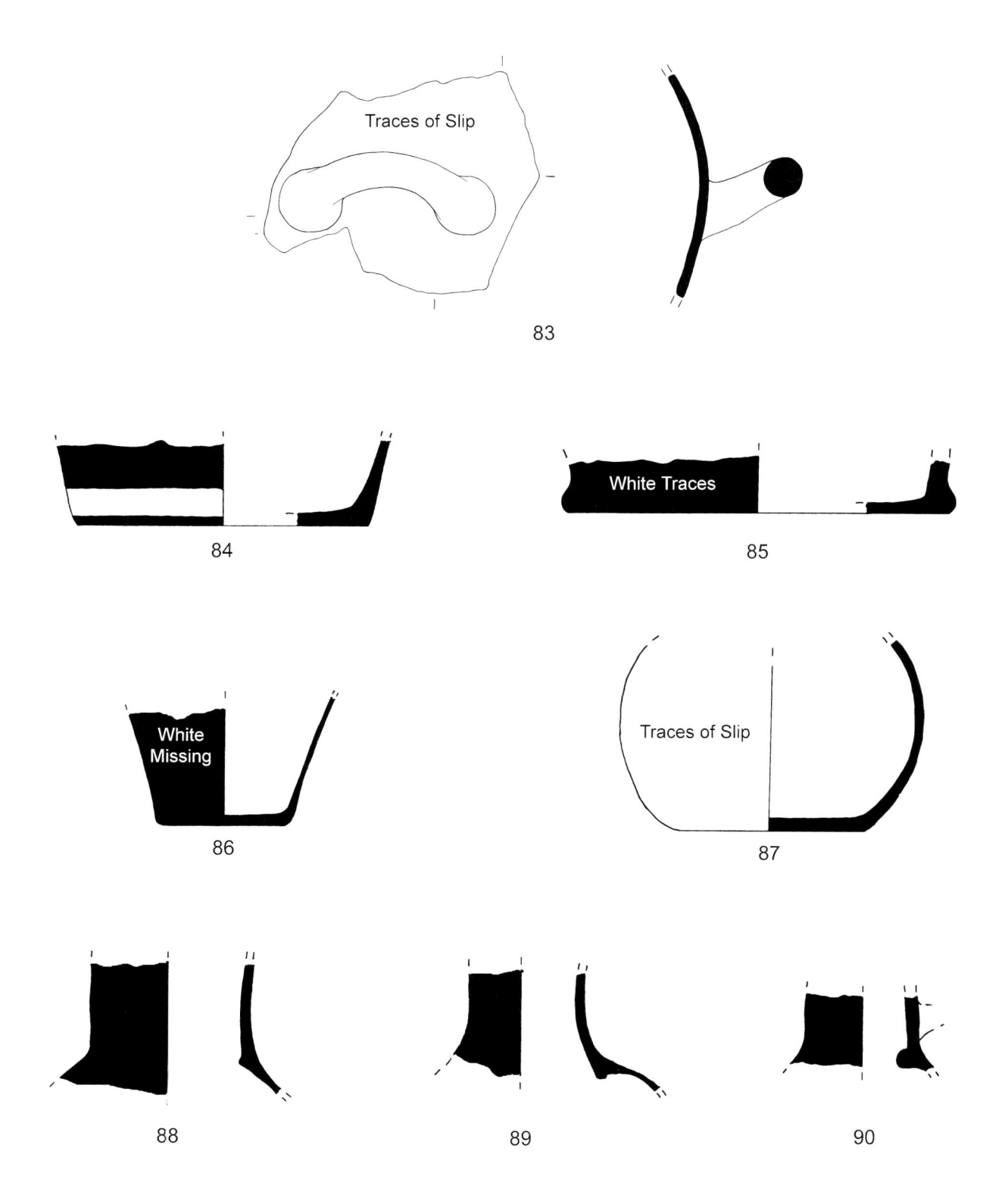

Figure 29. East Cretan White-on-Dark Ware: bridge-spouted jar (**83**), jugs or jars (**84**–**86**), and jugs (**87**–**90**). Scales 1:3 (**83**–**85**, **87**–**90**) and 1:6 (**86**). Drawings D. Faulmann, L. Bonga, K. Chalikias, and P. Betancourt.

FIGURE 30

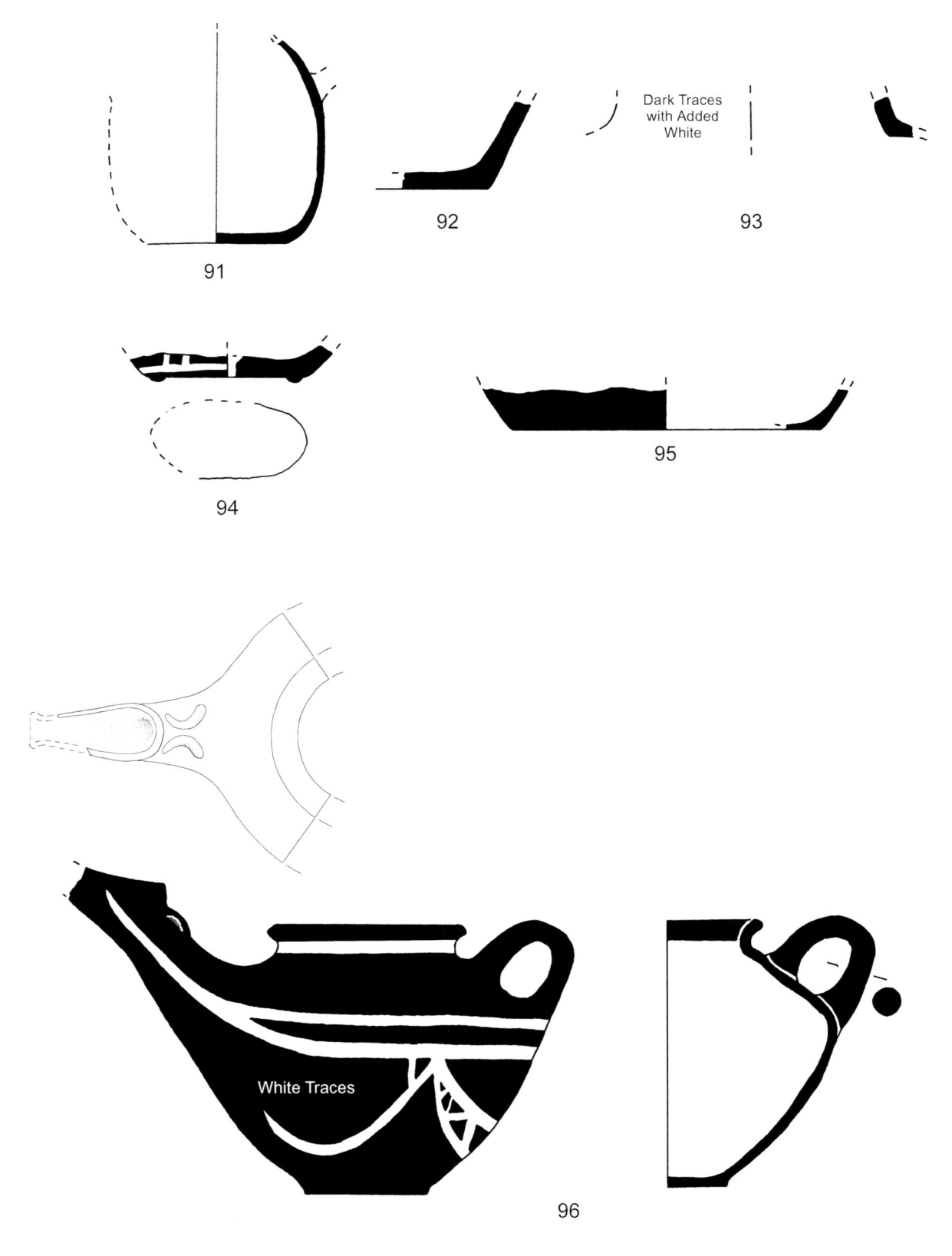

Figure 30. East Cretan White-on-Dark Ware: jug (**91**), jug or jar (**92**), collared jar (**93**), vessel with elliptical base (**94**), closed vessel (**95**), and teapot (**96**). Scales 1:3 (**91**–**95**) and 1:6 (**96**). Drawings D. Faulmann, L. Bonga, K. Chalikias, and P. Betancourt.

FIGURE 31

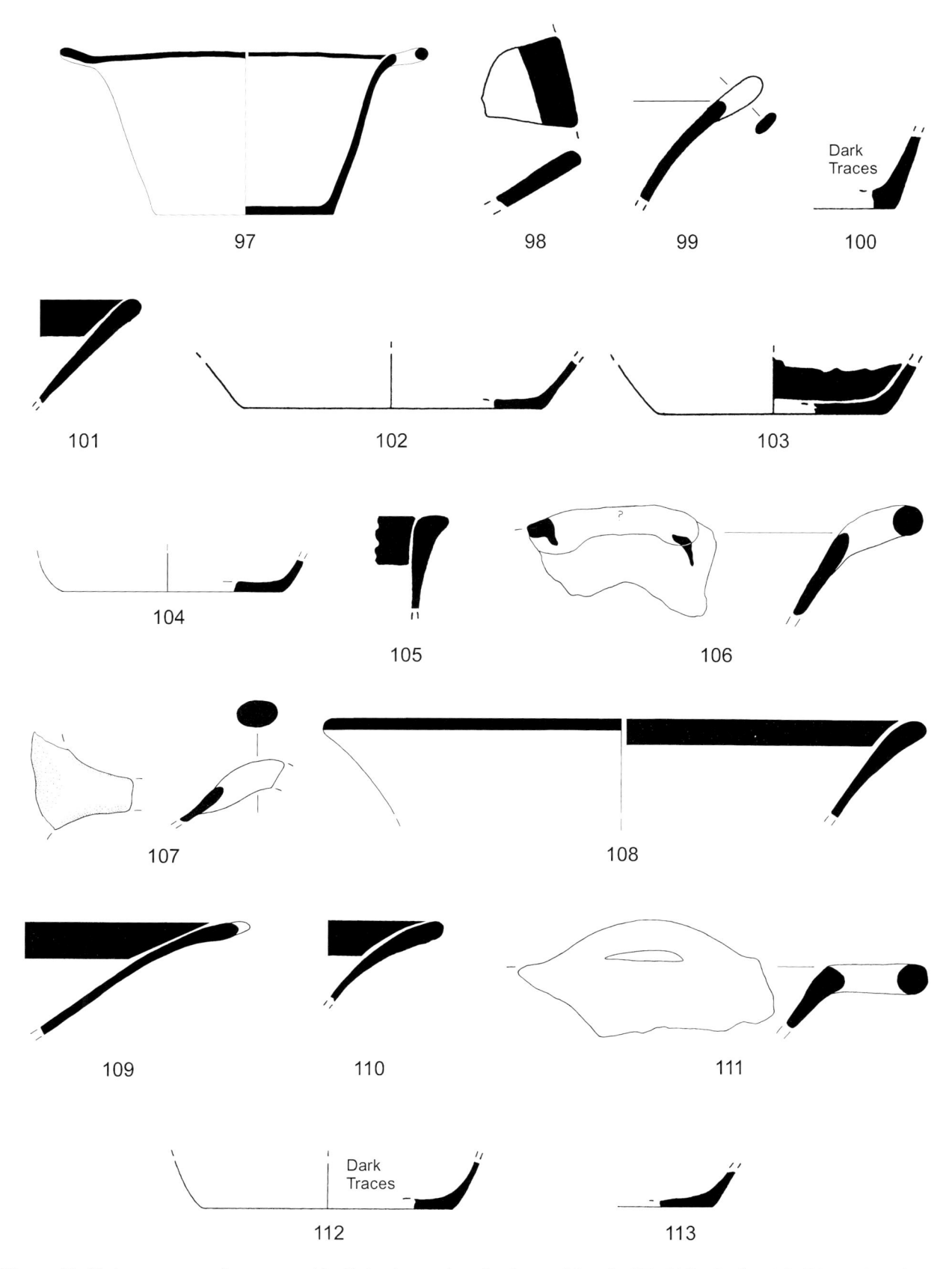

Figure 31. Plain pottery and pottery with slight decoration: basins and bowls (**97–113**). Scales 1:3 (**98–113**) and 1:6 (**97**). Drawings D. Faulmann, L. Bonga, K. Chalikias, and P. Betancourt.

FIGURE 32

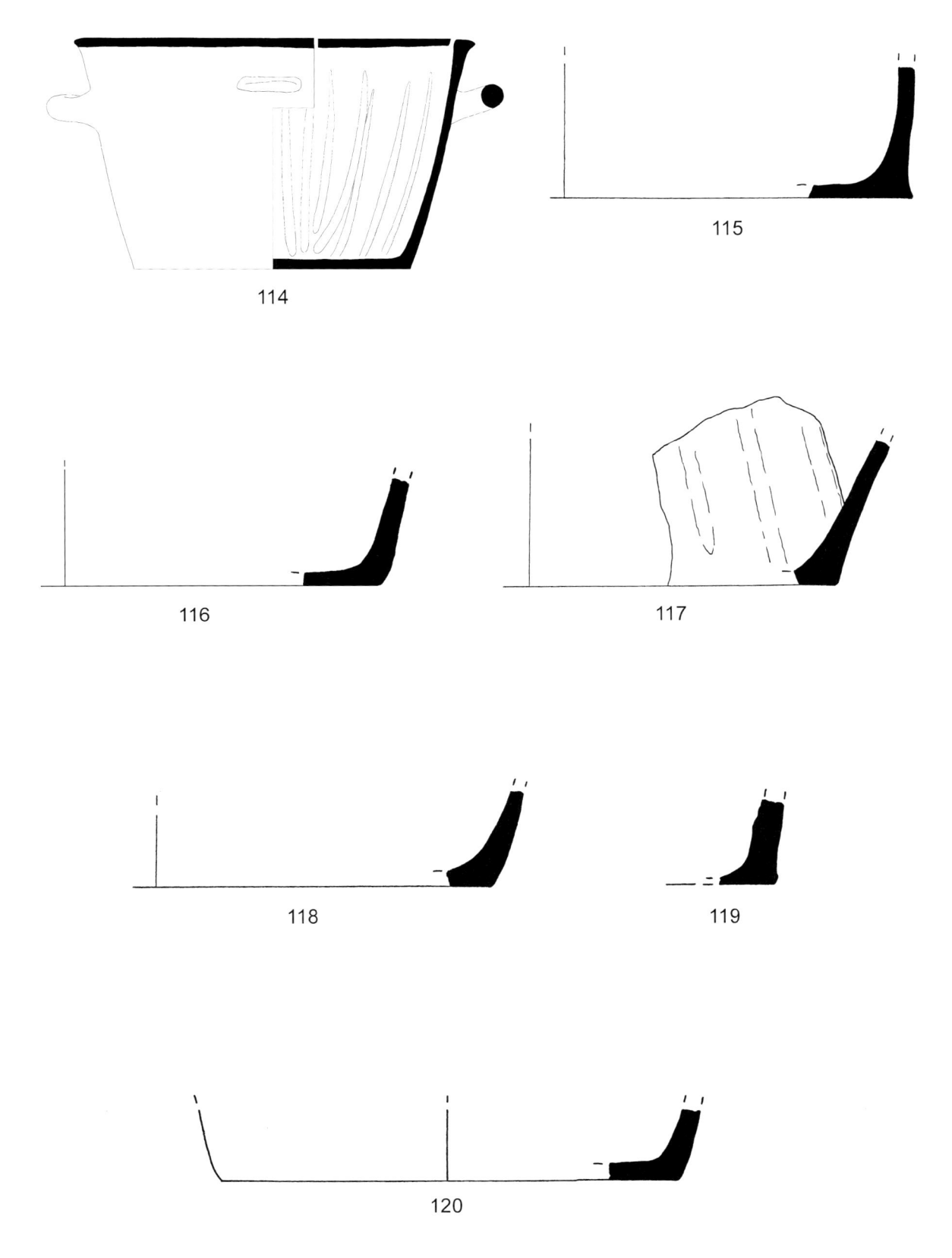

114

115

116

117

118

119

120

Figure 32. Plain pottery and pottery with slight decoration: basins with scoring inside (**114–120**). Scales 1:3 (**115–120**) and 1:6 (**114**). Drawings D. Faulmann, L. Bonga, K. Chalikias, and P. Betancourt.

FIGURE 33

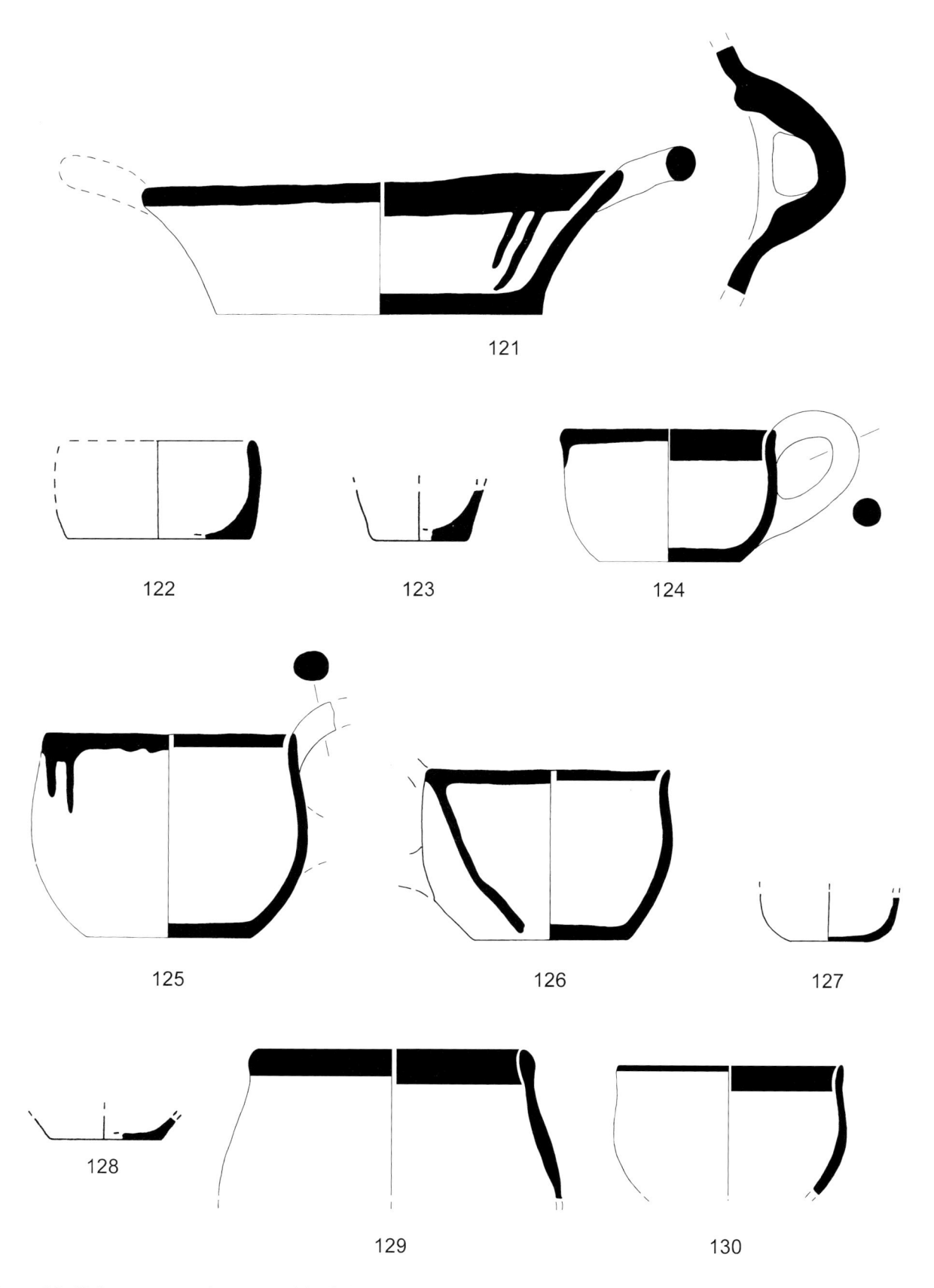

Figure 33. Plain pottery and pottery with slight decoration: spinning bowl (**121**) and cups (**122**–**130**). Scale 1:3. Drawings D. Faulmann, L. Bonga, K. Chalikias, and P. Betancourt.

FIGURE 34

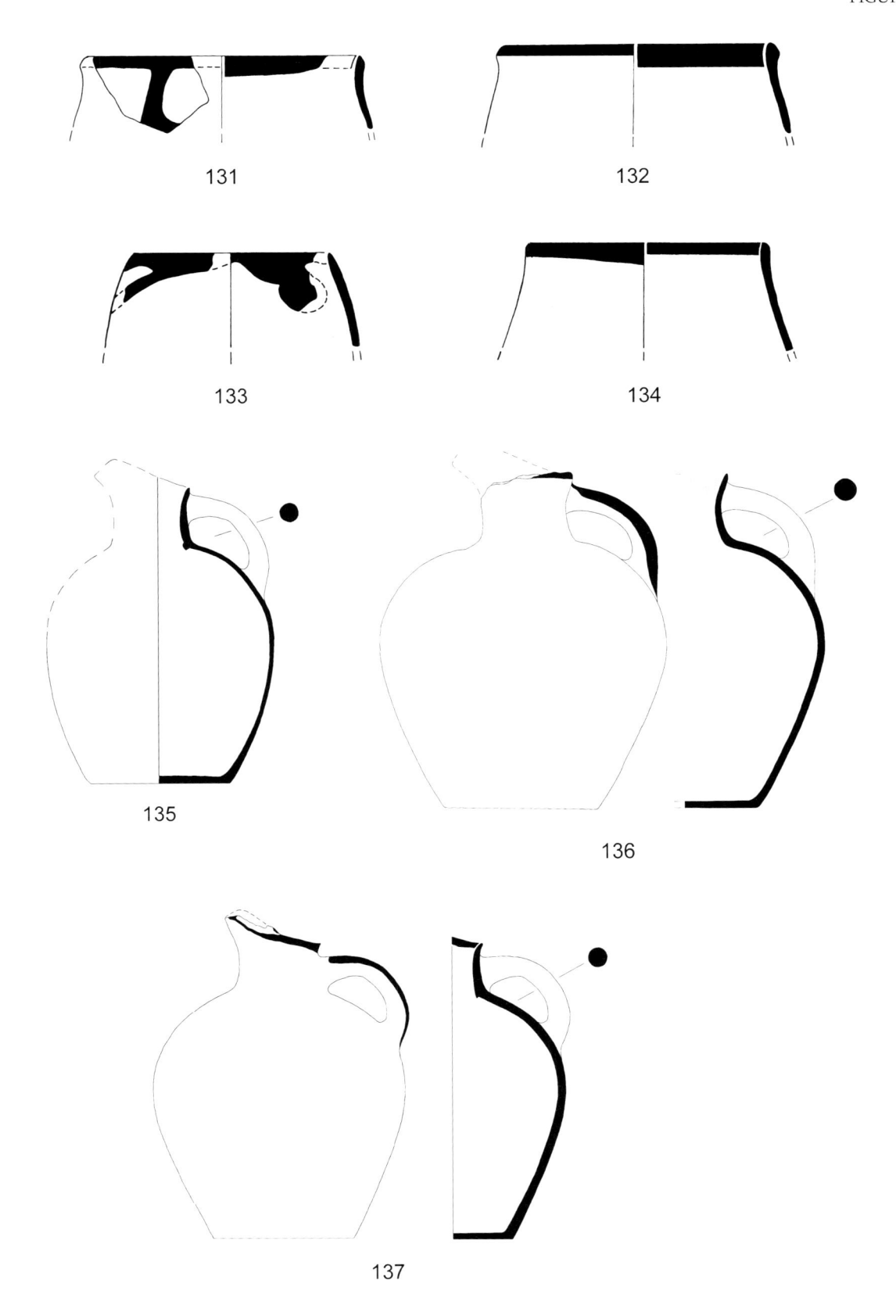

Figure 34. Plain pottery and pottery with slight decoration: cups (**131–134**) and jugs (**135–137**). Scales 1:3 (**131–134**) and 1:6 (**135–137**). Drawings D. Faulmann, L. Bonga, K. Chalikias, and P. Betancourt.

FIGURE 35

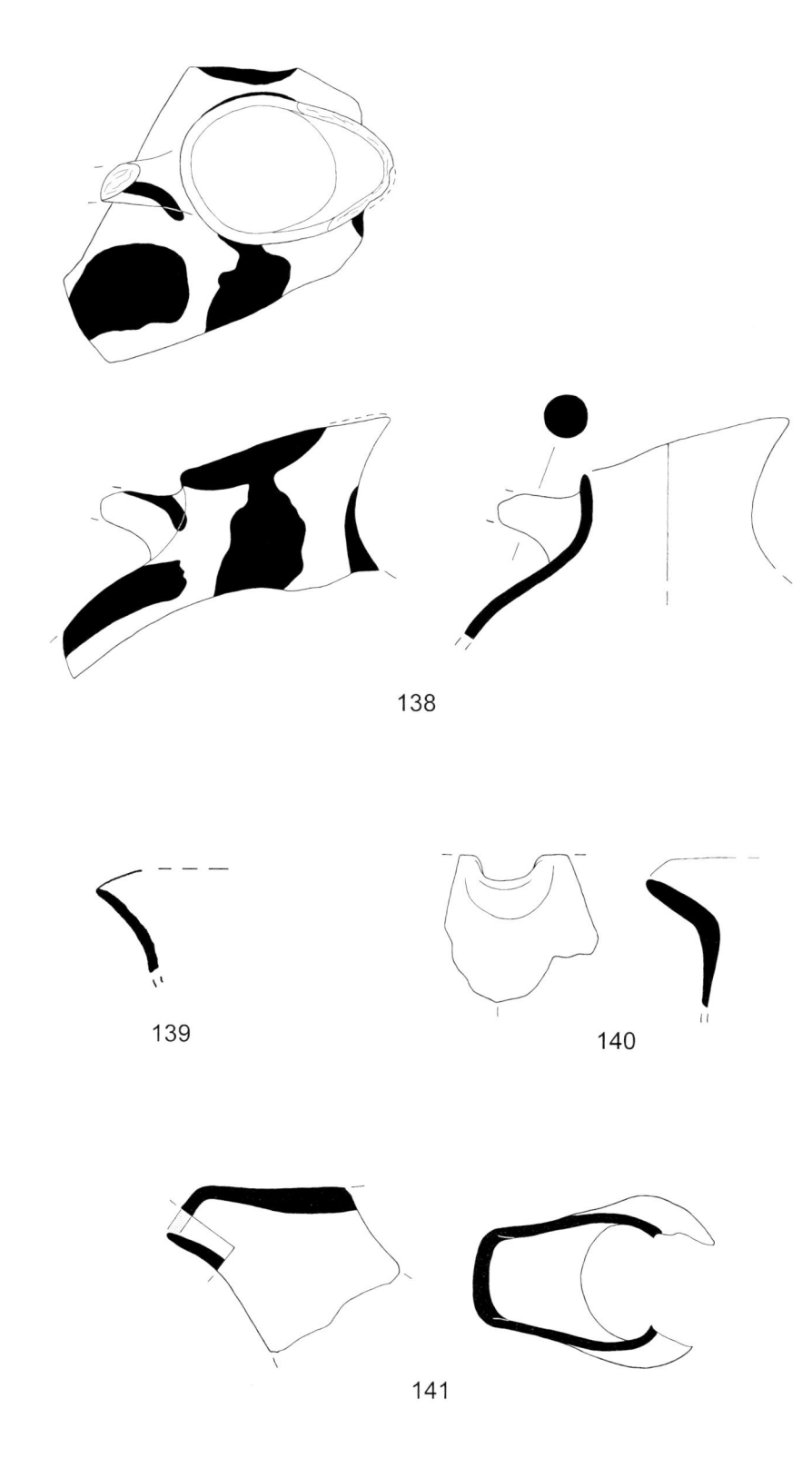

138

139

140

141

Figure 35. Plain pottery and pottery with slight decoration: jugs (**138–141**). Scale 1:3. Drawings D. Faulmann, L. Bonga, K. Chalikias, and P. Betancourt.

FIGURE 36

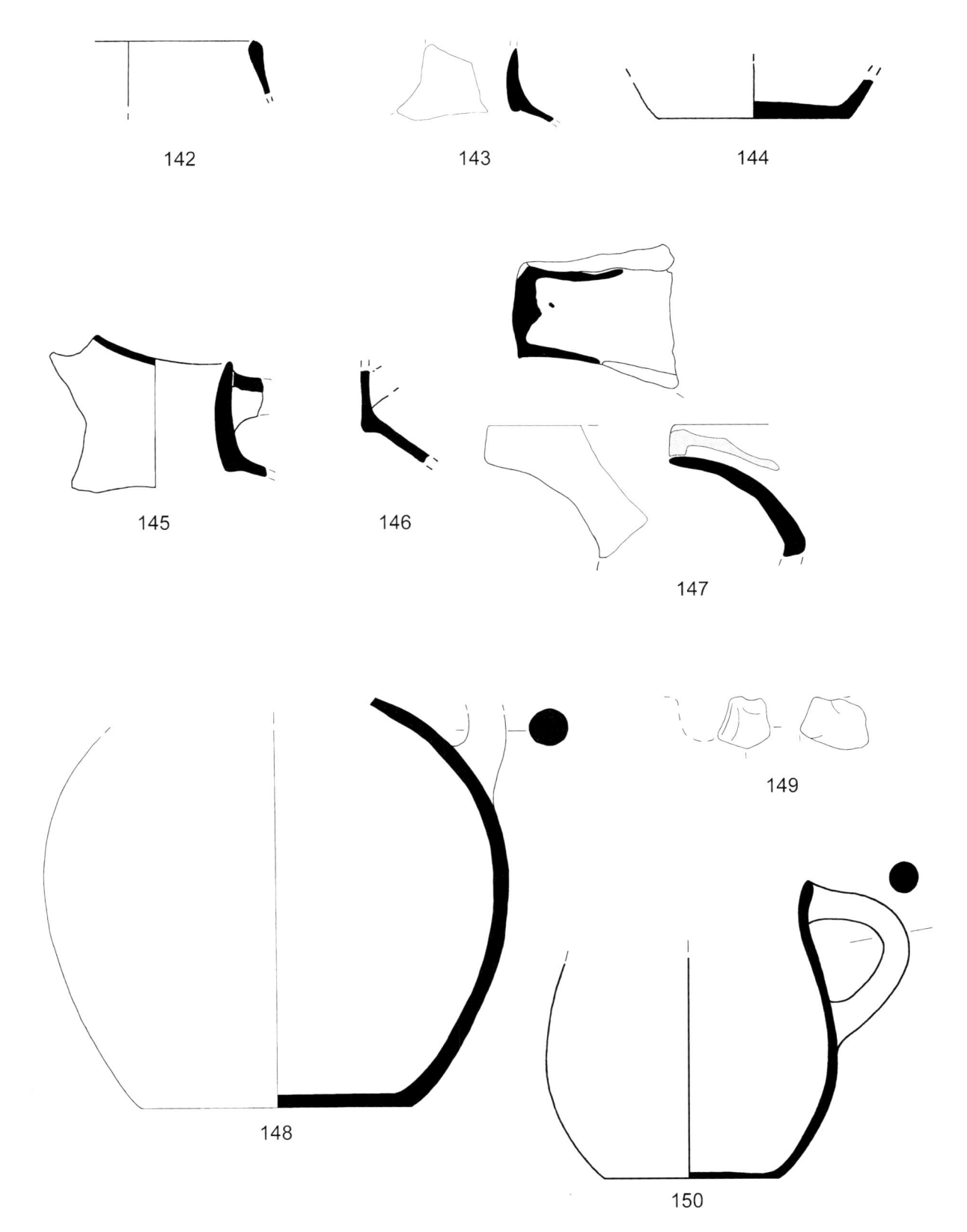

142 143 144

145 146 147

148 149 150

Figure 36. Plain pottery and pottery with slight decoration: jugs (**142–150**). Scale 1:3. Drawings D. Faulmann, L. Bonga, K. Chalikias, and P. Betancourt.

FIGURE 37

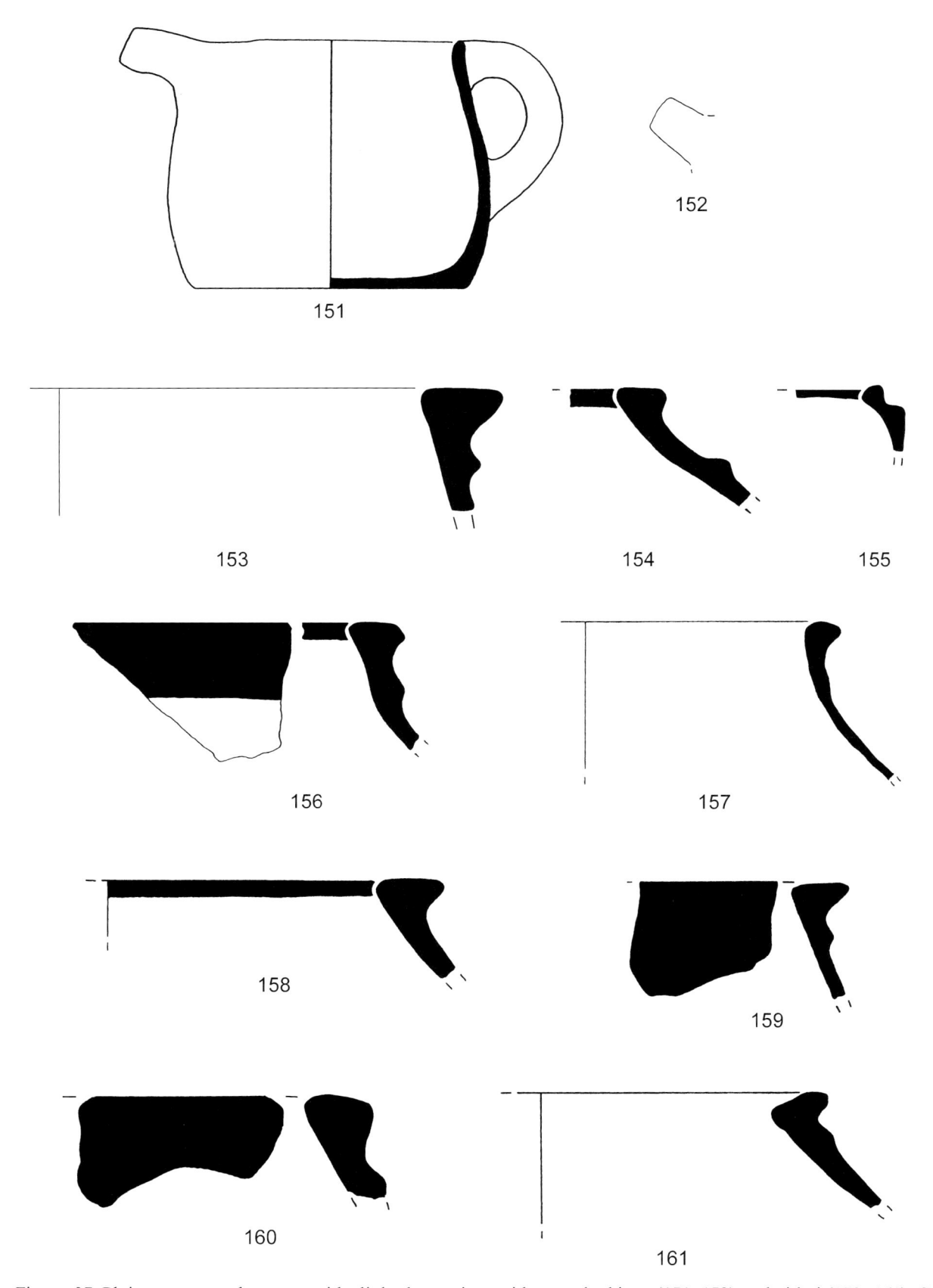

151
152
153
154
155
156
157
158
159
160
161

Figure 37. Plain pottery and pottery with slight decoration: wide-mouthed jugs (**151**, **152**) and pithoi (**153**–**161**). Scale 1:3. Drawings D. Faulmann, L. Bonga, K. Chalikias, and P. Betancourt.

FIGURE 38

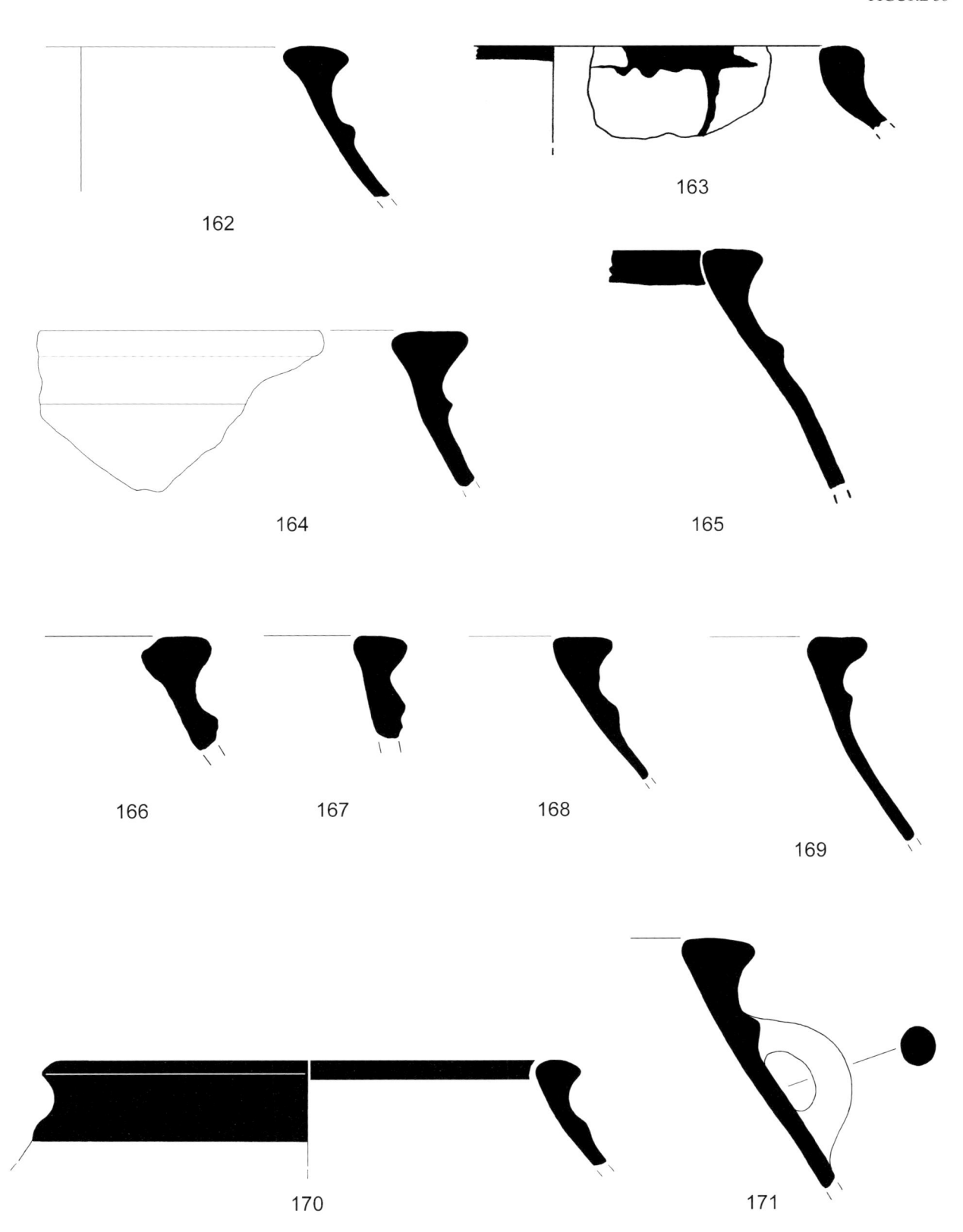

162
163
164
165
166
167
168
169
170
171

Figure 38. Plain pottery and pottery with slight decoration: pithoi (**162**–**171**). Scales 1:3 (**162**–**170**) and 1:6 (**171**). Drawings D. Faulmann, L. Bonga, K. Chalikias, and P. Betancourt.

FIGURE 39

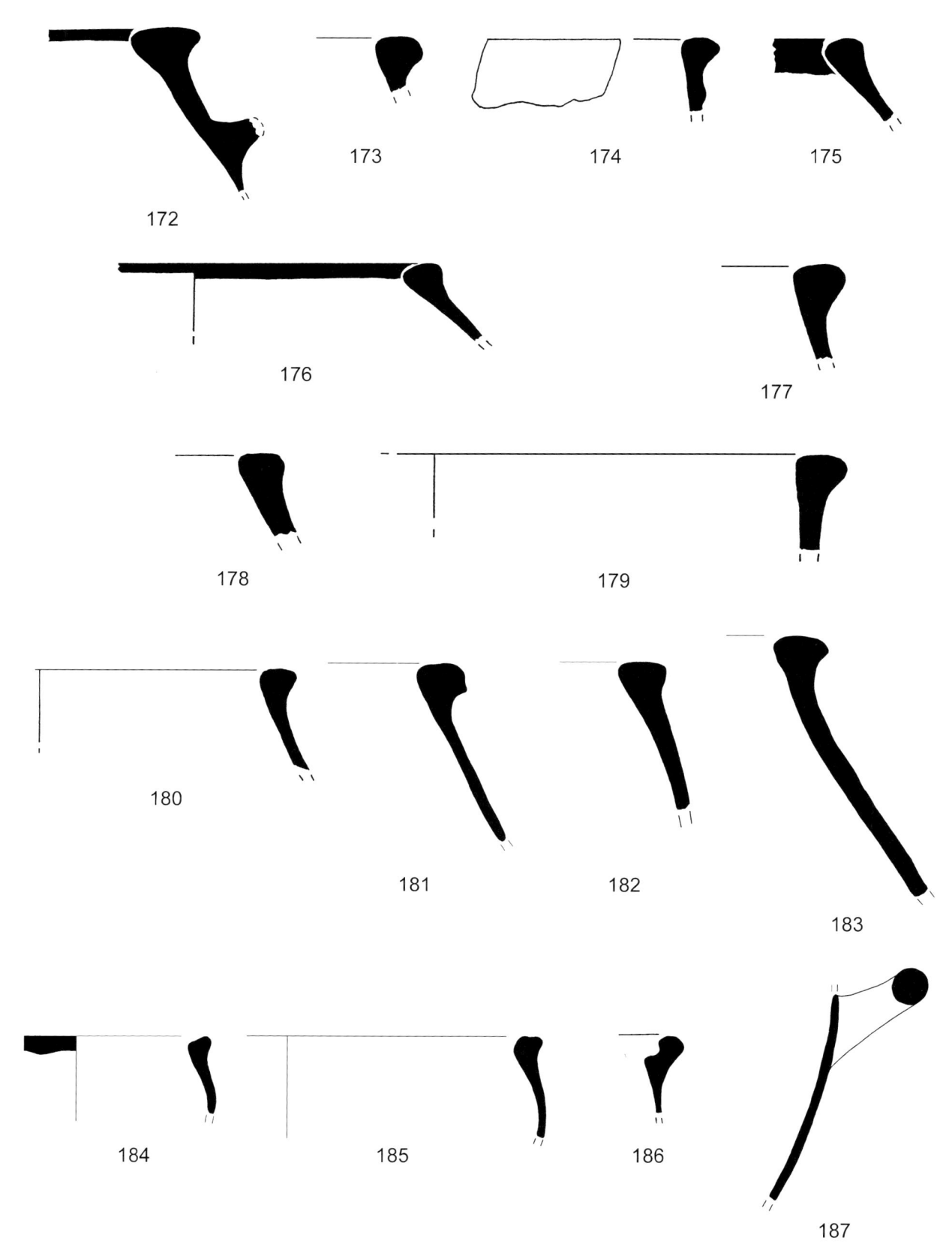

Figure 39. Plain pottery and pottery with slight decoration: pithoi (**172–174**) and pithoi or jars (**175–187**). Scale 1:3. Drawings D. Faulmann, L. Bonga, K. Chalikias, and P. Betancourt.

FIGURE 40

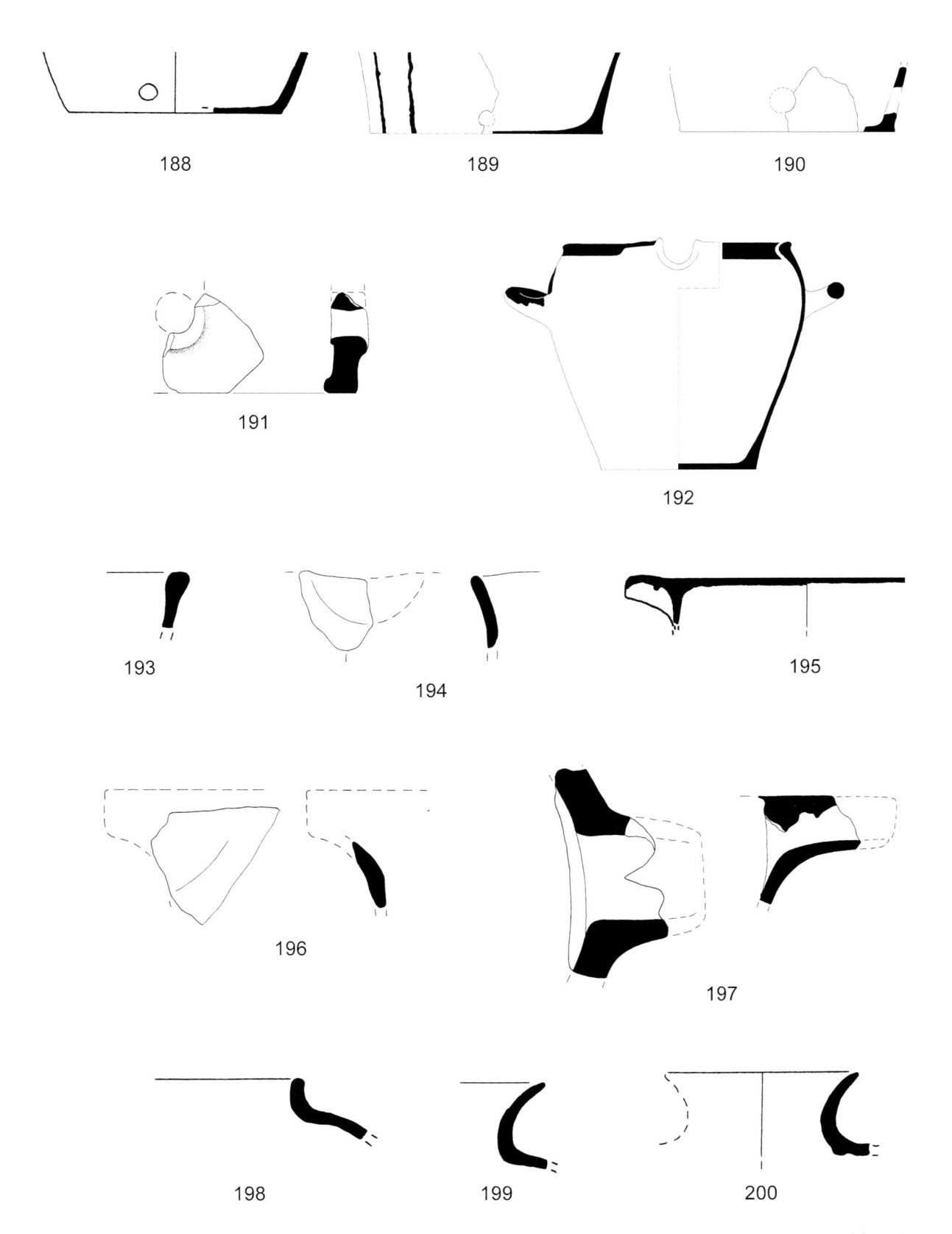

Figure 40. Plain pottery and pottery with slight decoration: pithoi or jars with drains (**188–191**), spouted jars (**192–197**), and collared jars (**198–200**). Scales 1:3 (**191, 193, 194, 196–200**) and 1:6 (**188–190, 192, 195**). Drawings D. Faulmann, L. Bonga, K. Chalikias, and P. Betancourt.

FIGURE 41

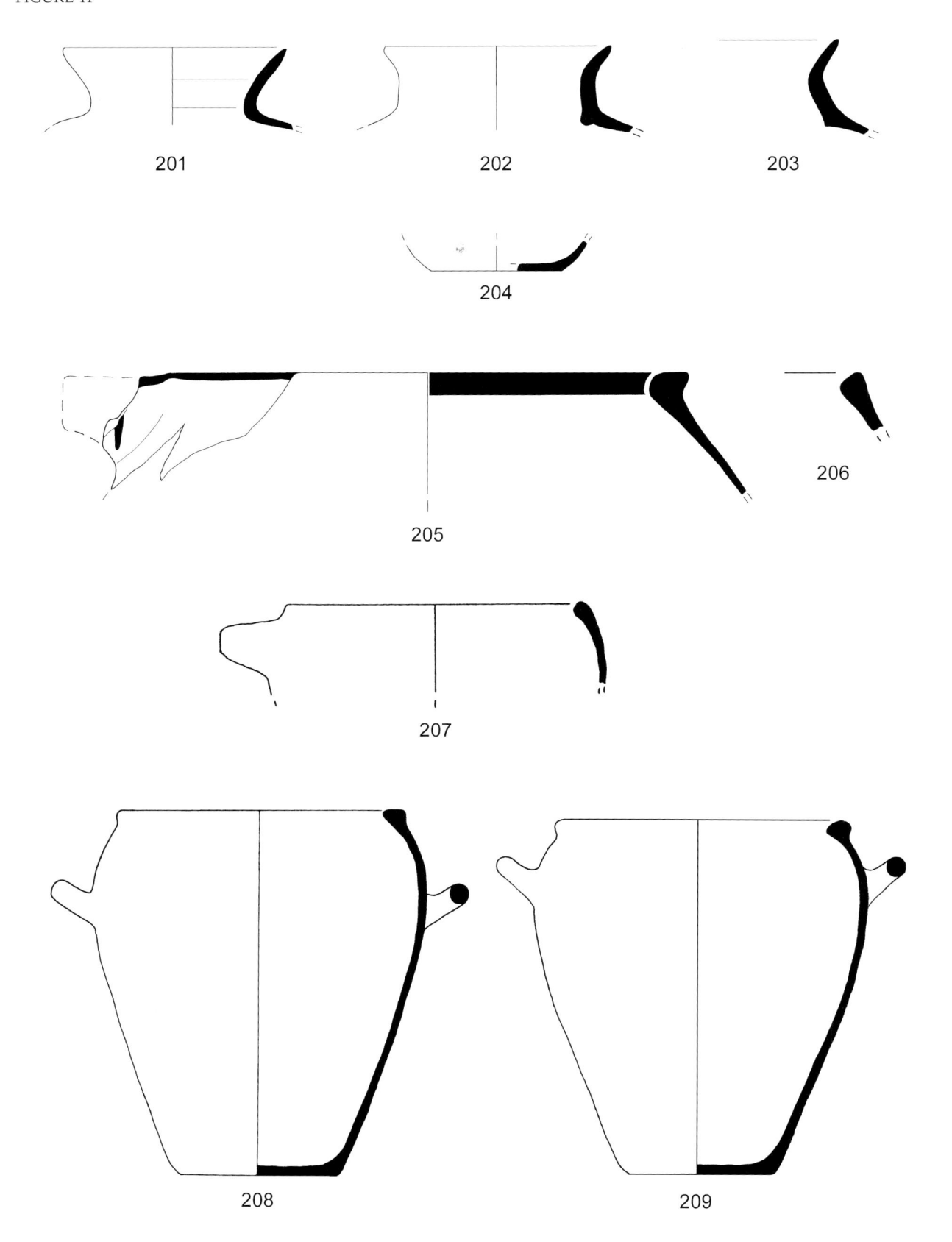

Figure 41. Plain pottery and pottery with slight decoration: collared jars (**201–203**), small vessel with knobs (**204**), bridge-spouted jars (**205–207**), and hole-mouthed jars (**208**, **209**). Scales 1:3 (**201–207**) and 1:6 (**208**, **209**). Drawings D. Faulmann, L. Bonga, K. Chalikias, and P. Betancourt.

FIGURE 42

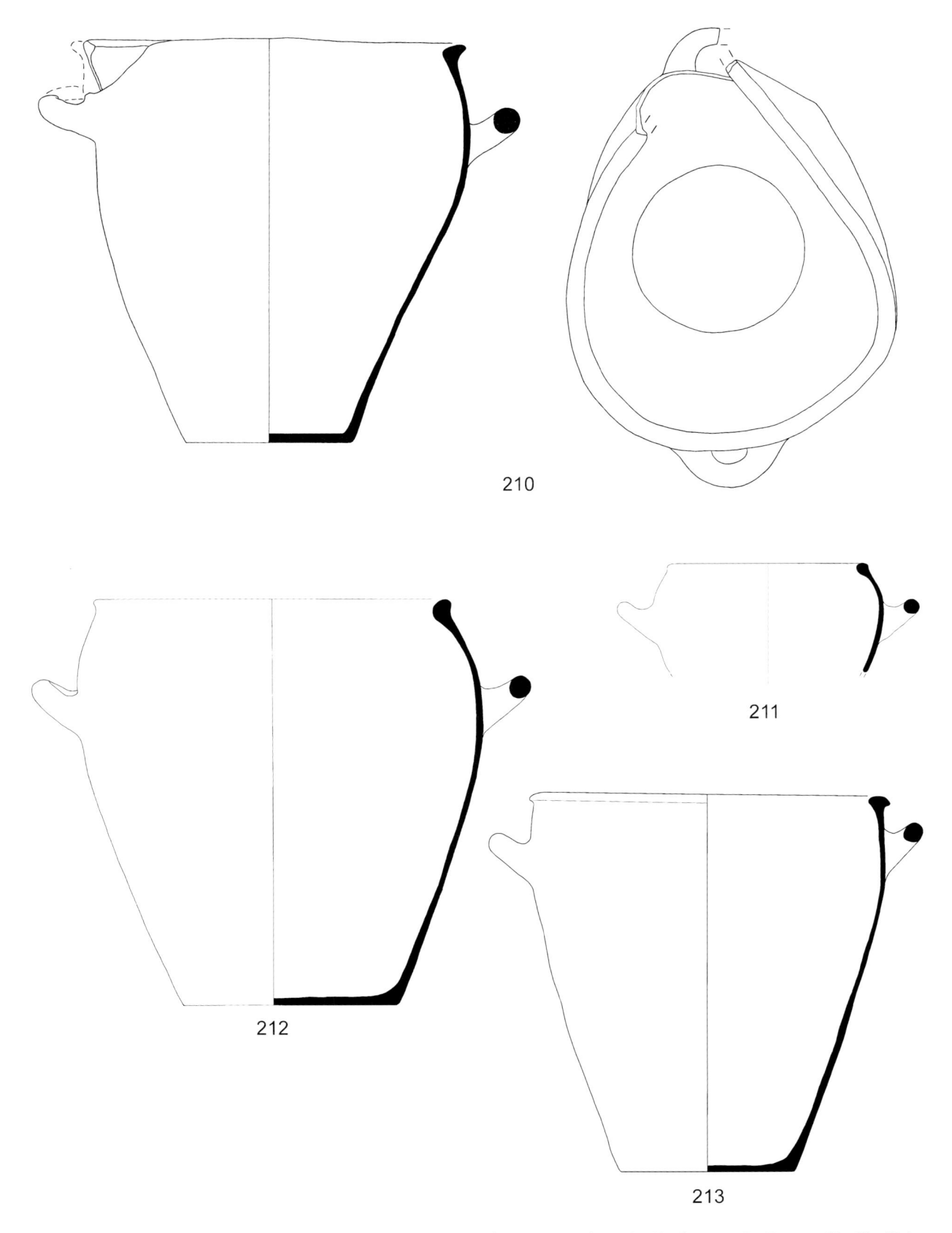

Figure 42. Plain pottery: hole-mouthed jars (**210**–**213**). Scale 1:6. Drawings D. Faulmann, L. Bonga, K. Chalikias, and P. Betancourt.

FIGURE 43

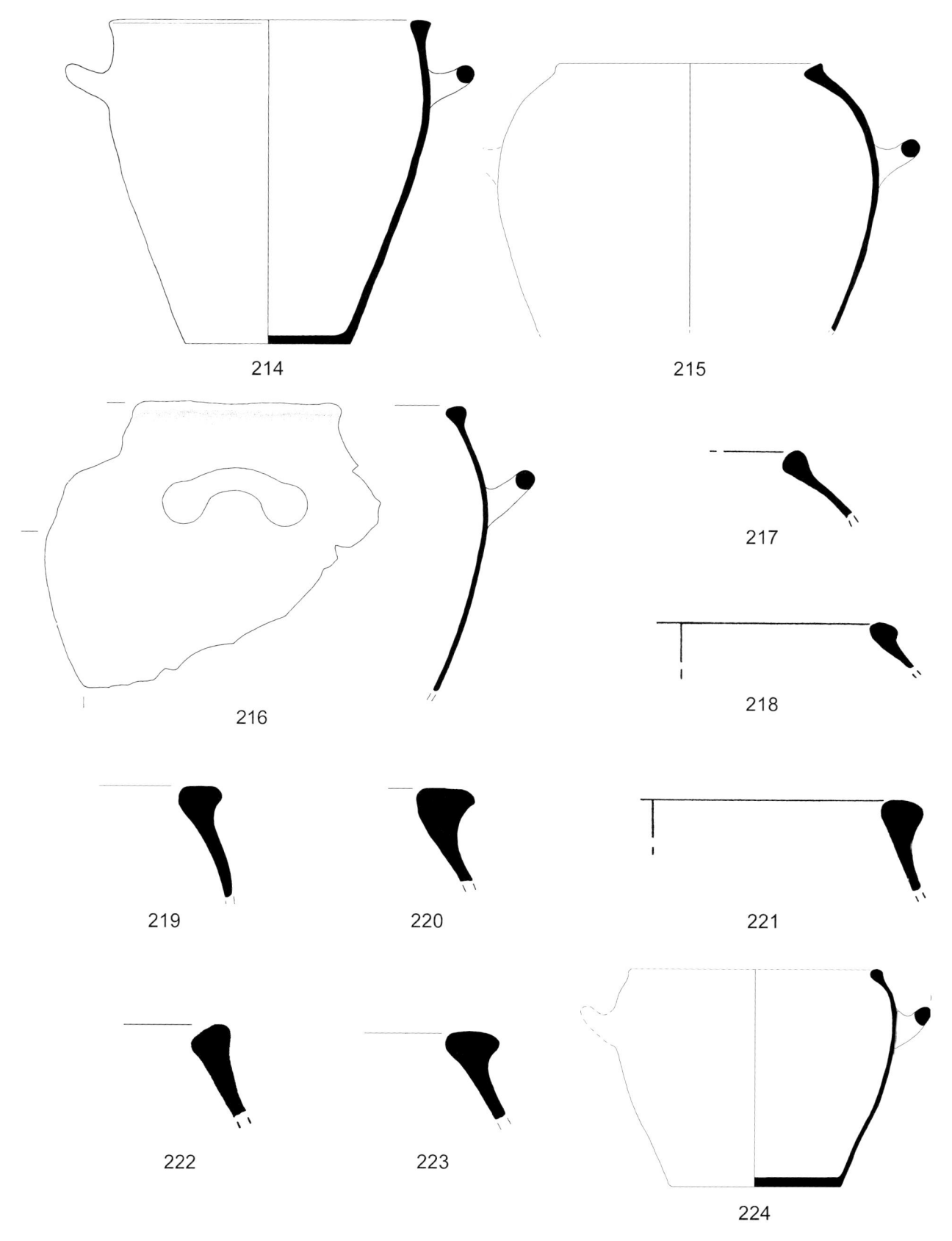

Figure 43. Plain pottery: hole-mouthed jars (**214–224**). Scales 1:6 (**214–216**, **224**) and 1:3 (**217–223**). Drawings D. Faulmann, L. Bonga, K. Chalikias, and P. Betancourt.

FIGURE 44

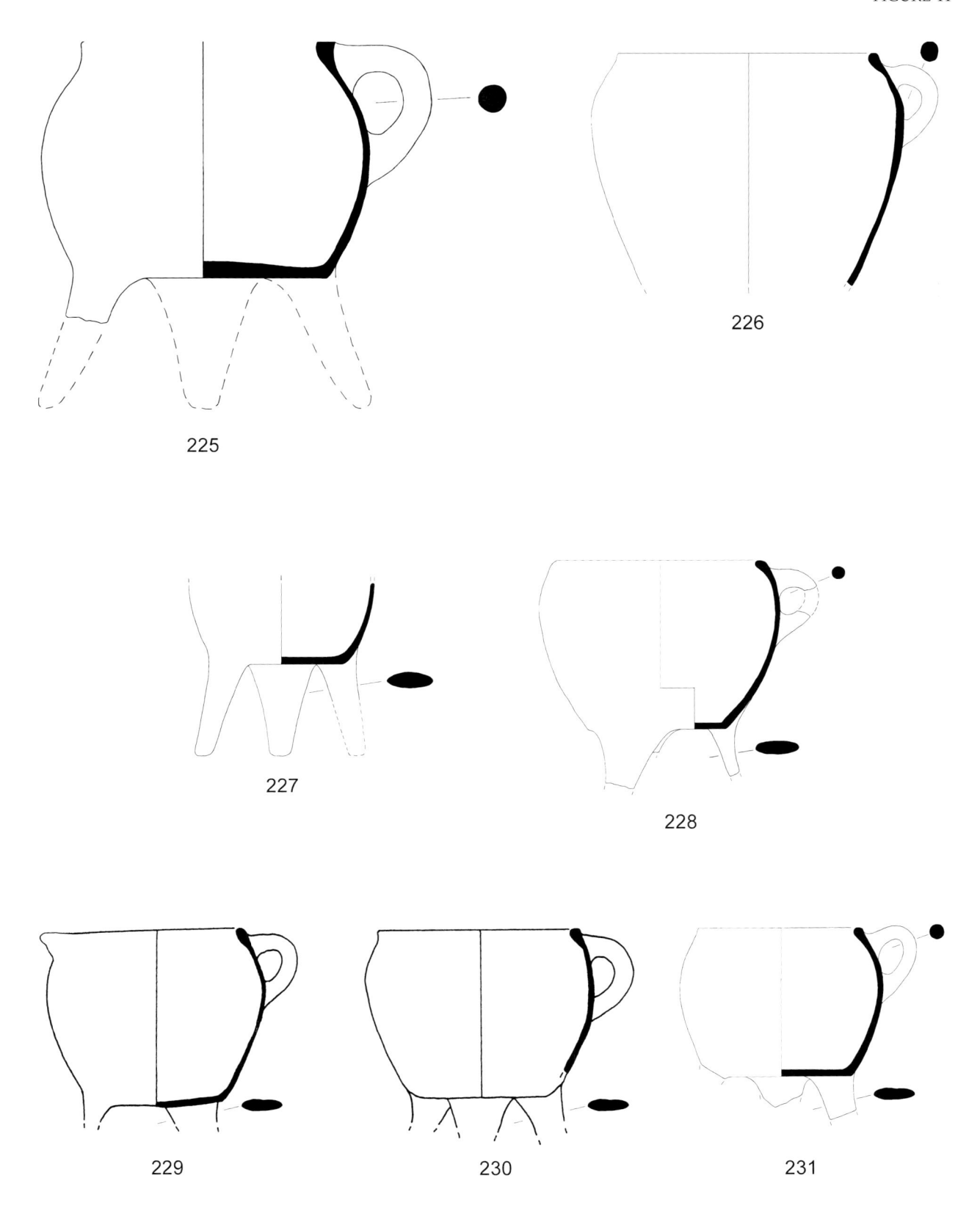

225

226

227

228

229   230   231

Figure 44. Plain pottery: tripod cooking pots (**225**–**231**). Scales 1:3 (**225**) and 1:6 (**226**–**231**). Drawings D. Faulmann, L. Bonga, K. Chalikias, and P. Betancourt.

FIGURE 45

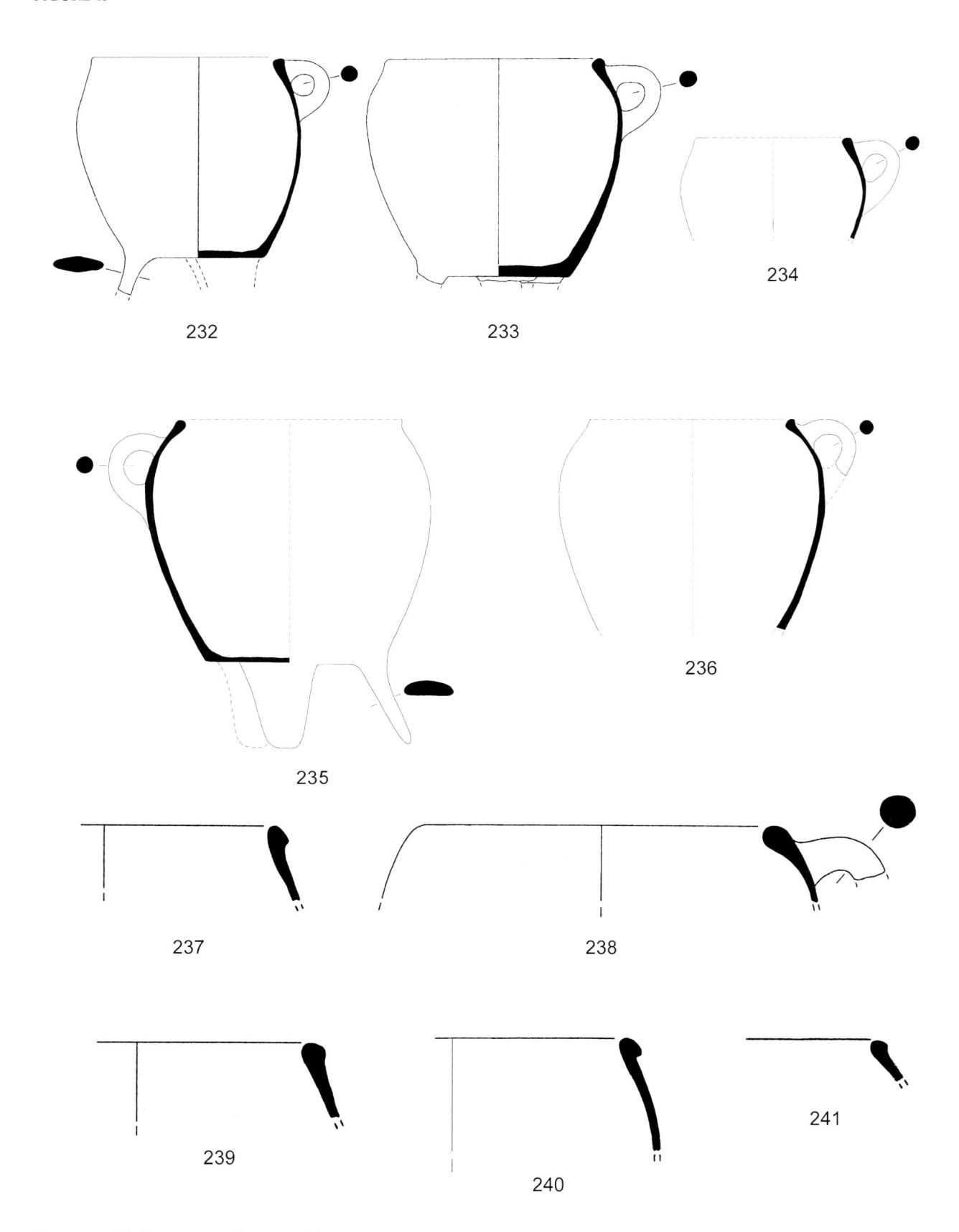

Figure 45. Plain pottery: tripod cooking pots (**232**–**241**). Scales 1:3 (**237**–**241**) and 1:6 (**232**–**236**). Drawings D. Faulmann, L. Bonga, K. Chalikias, and P. Betancourt.

FIGURE 46

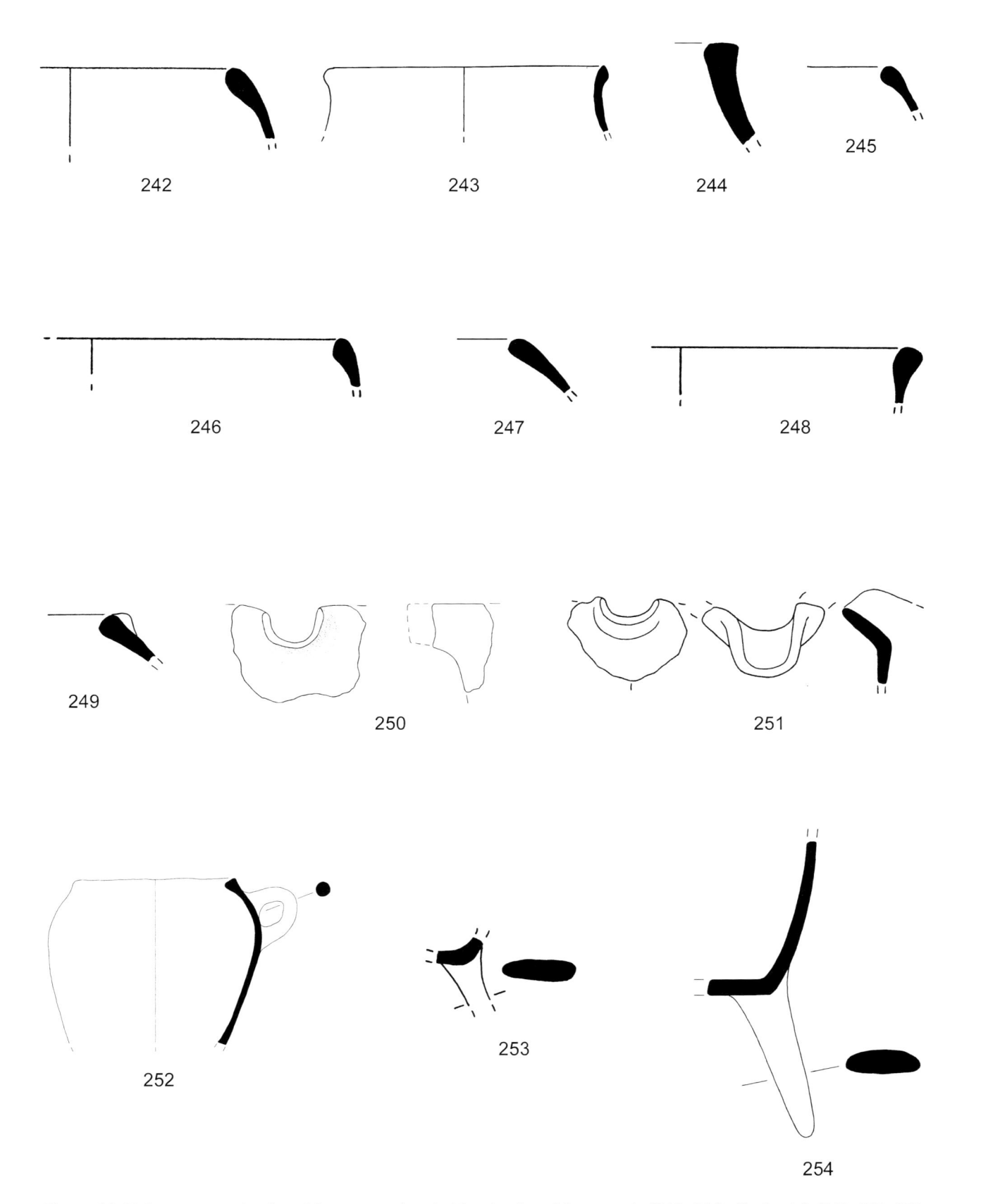

Figure 46. Plain pottery: tripod cooking pots and probable tripod cooking vessels (**242–254**). Scales 1:3 (**242–251**, **253**, **254**) and 1:6 (**252**). Drawings D. Faulmann, L. Bonga, K. Chalikias, and P. Betancourt.

FIGURE 47

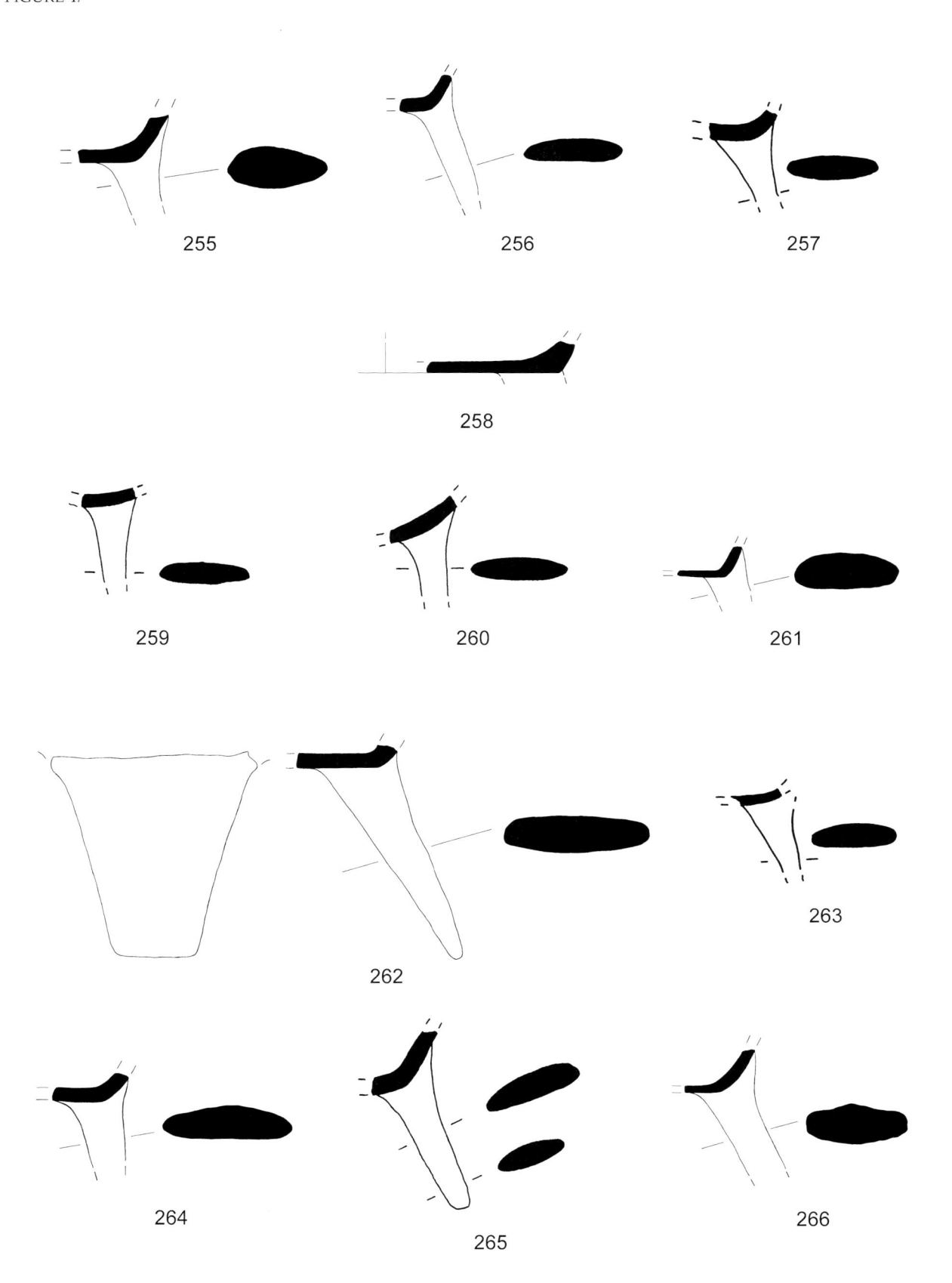

255 256 257

258

259 260 261

262 263

264 265 266

Figure 47. Plain pottery: tripod cooking pots (**255–266**). Scale 1:3. Drawings D. Faulmann, L. Bonga, K. Chalikias, and P. Betancourt.

FIGURE 48

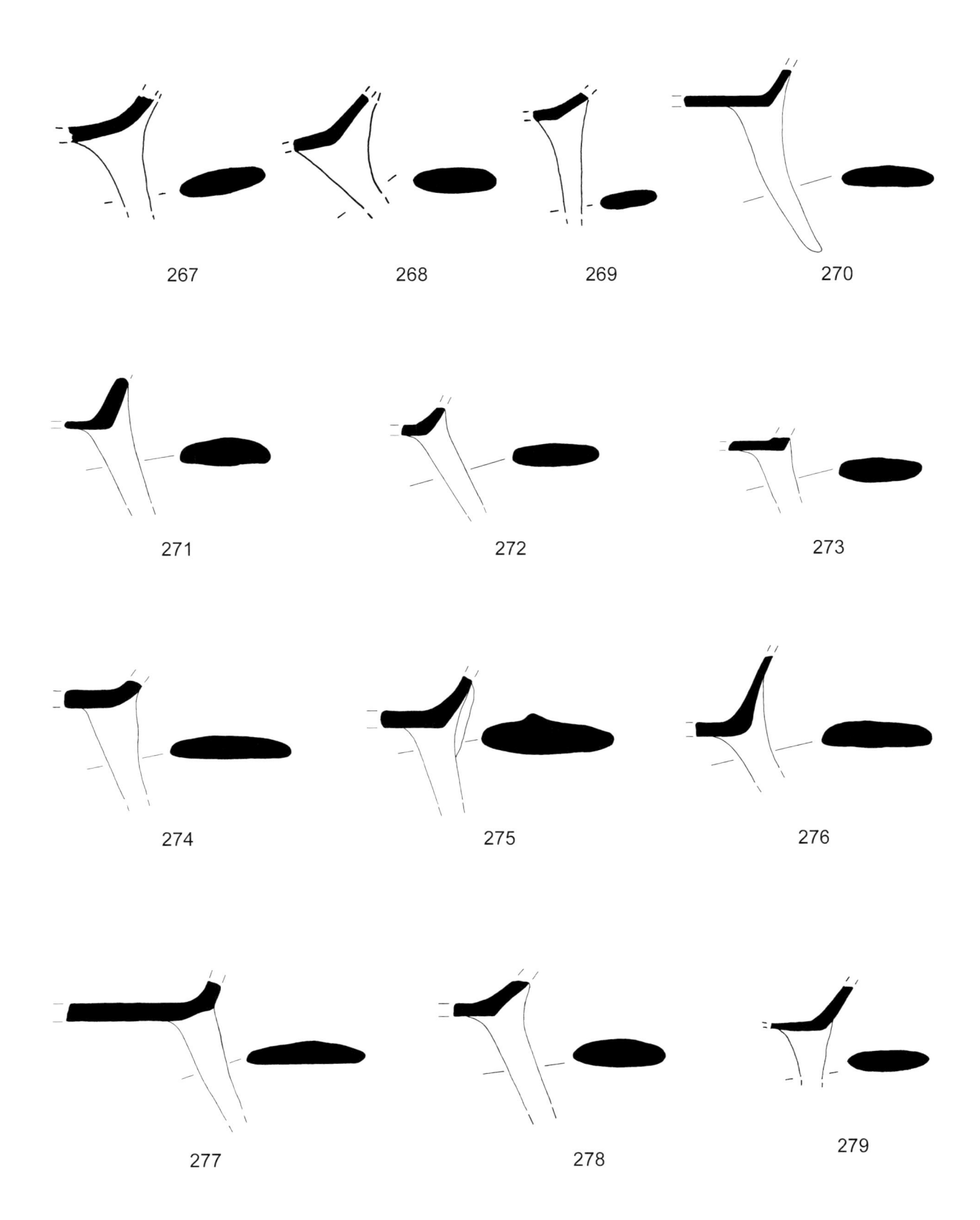

Figure 48. Plain pottery: tripod cooking pots (**267–279**). Scale 1:3. Drawings D. Faulmann, L. Bonga, K. Chalikias, and P. Betancourt.

FIGURE 49

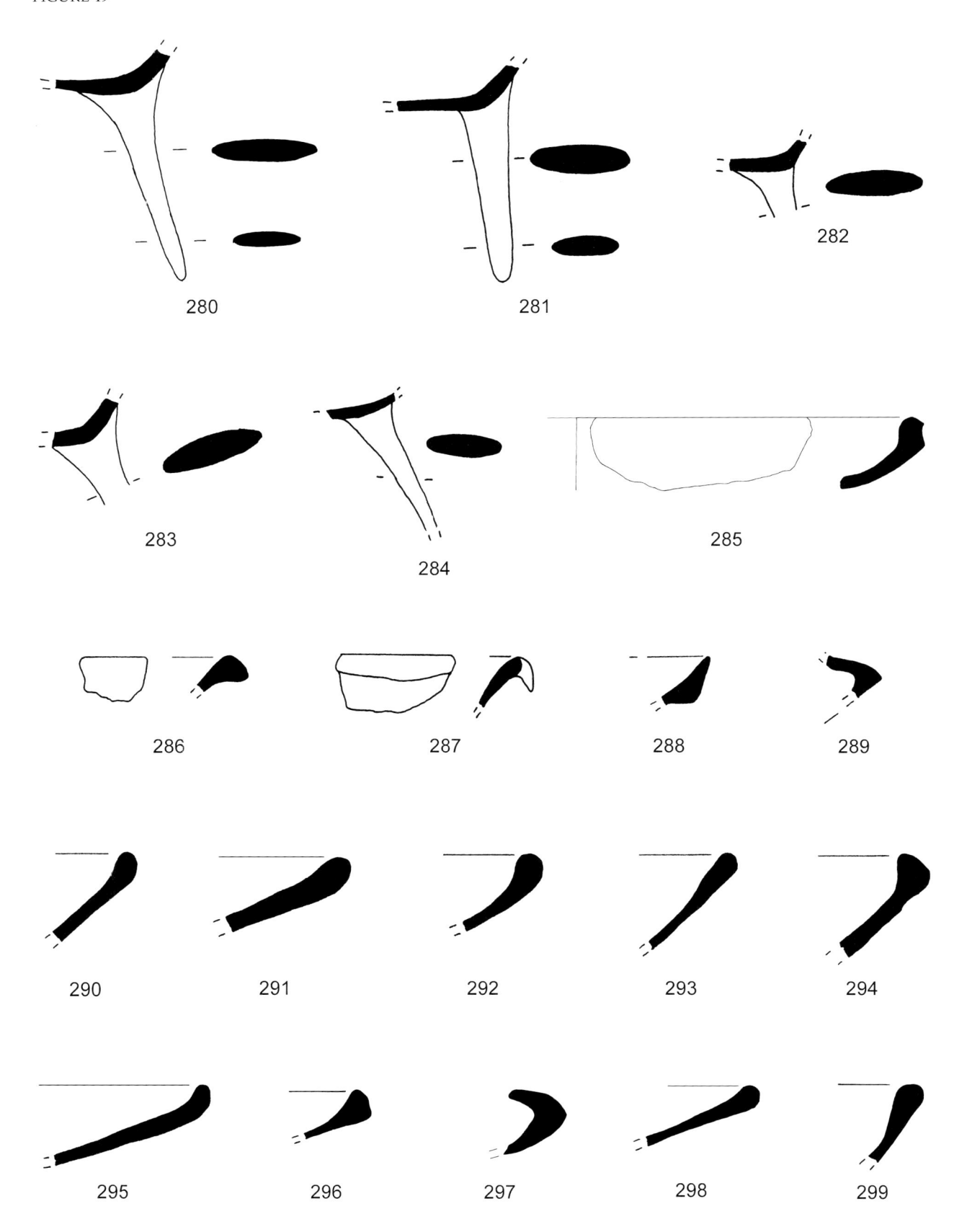

Figure 49. Plain pottery: tripod cooking pots (**280–284**) and cooking dishes (**285–299**). Scale 1:3. Drawings D. Faulmann, L. Bonga, K. Chalikias, and P. Betancourt.

FIGURE 50

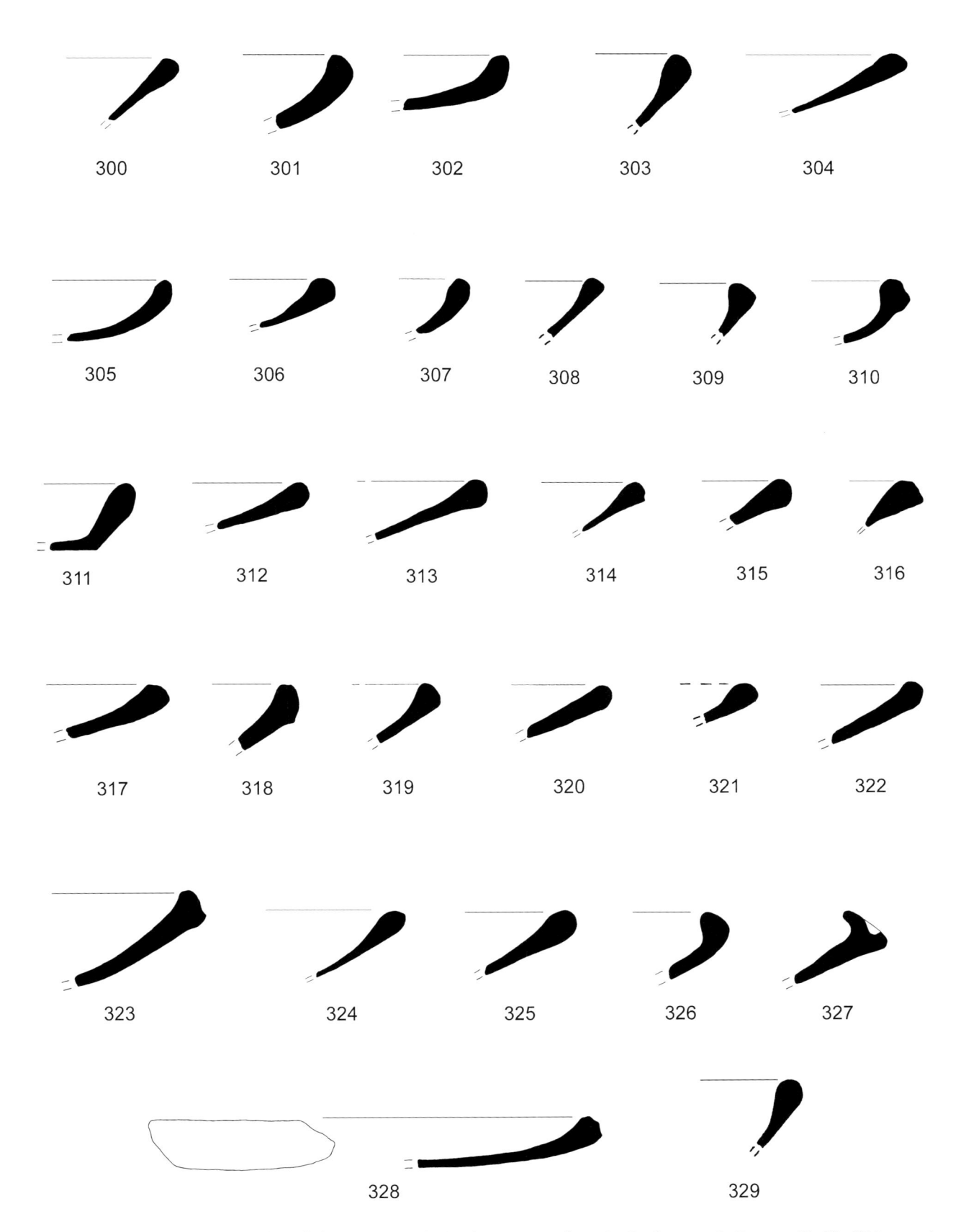

300 301 302 303 304

305 306 307 308 309 310

311 312 313 314 315 316

317 318 319 320 321 322

323 324 325 326 327

328 329

Figure 50. Plain pottery: cooking dishes (**300–329**). Scale 1:3. Drawings D. Faulmann, L. Bonga, K. Chalikias, and P. Betancourt.

FIGURE 51

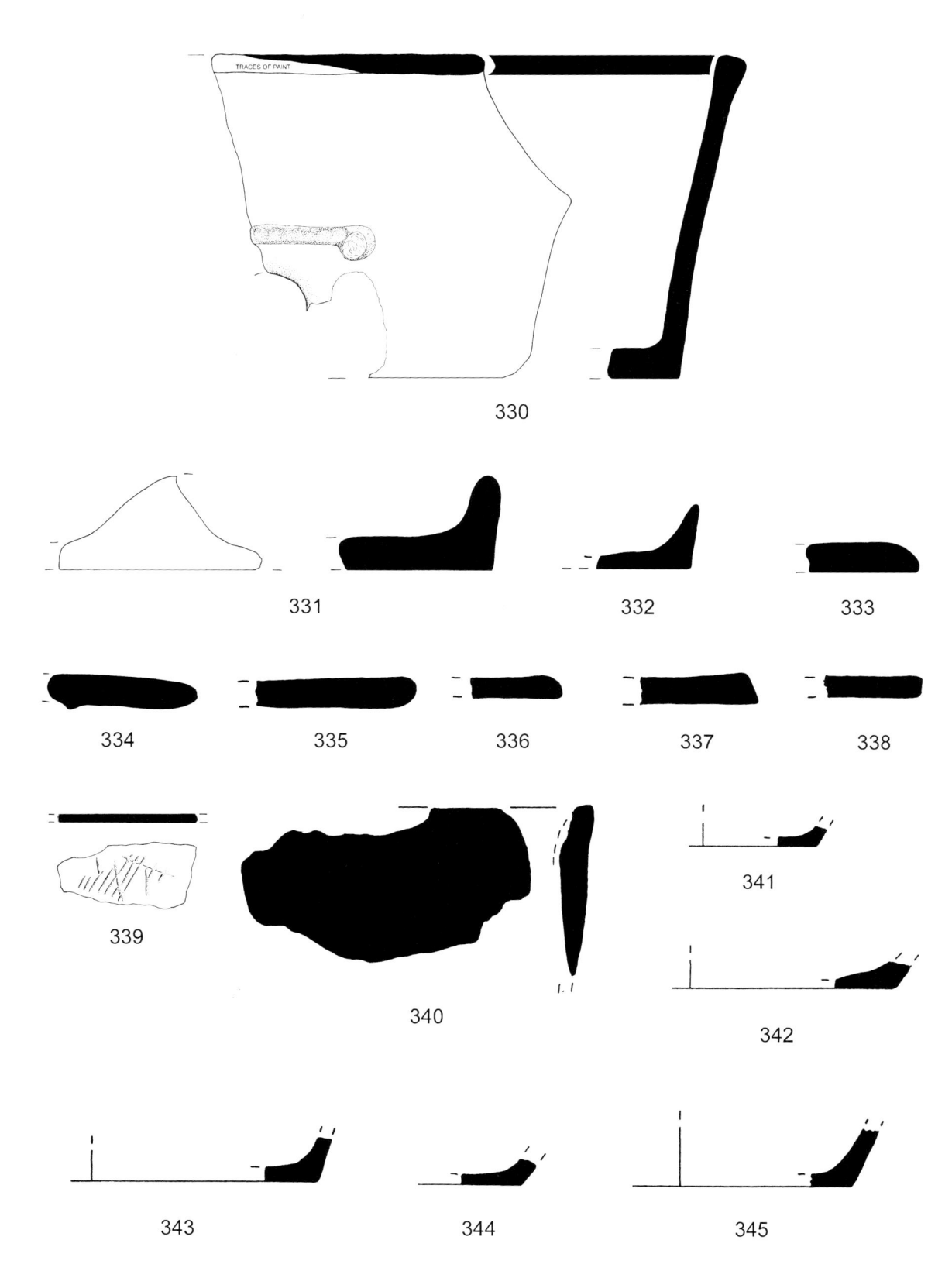

Figure 51. Plain pottery and pottery with slight decoration: large vats (**330–332**), flat lids (**333–338**), unknown shapes (**339**, **340**), and closed pottery vessels (**341–345**). Scales 1:6 (**330**) and 1:3 (**331–345**). Drawings D. Faulmann, L. Bonga, K. Chalikias, and P. Betancourt.

FIGURE 52

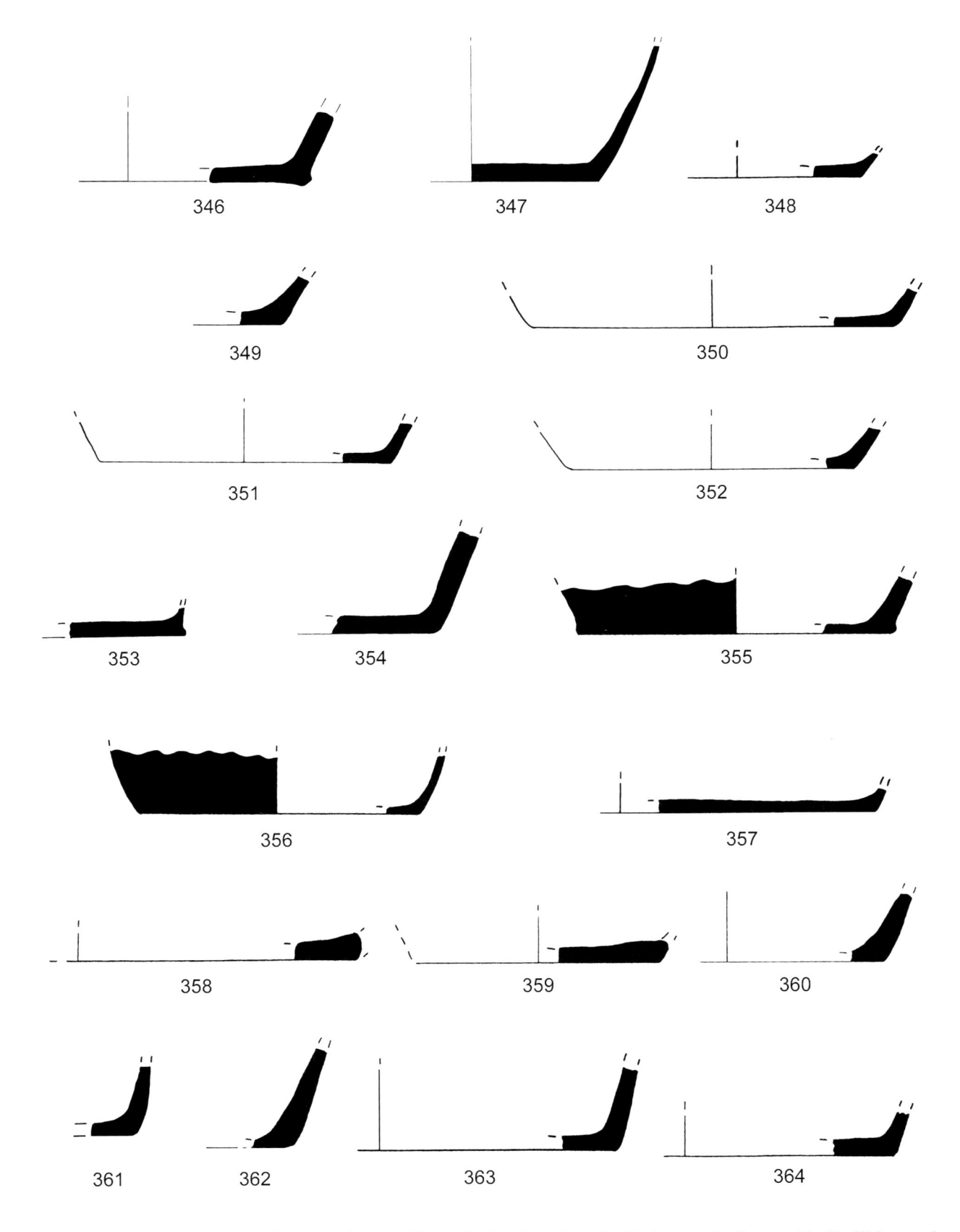

Figure 52. Plain pottery: closed vessels (**346–364**). Scale 1:3. Drawings D. Faulmann, L. Bonga, K. Chalikias, and P. Betancourt.

FIGURE 53

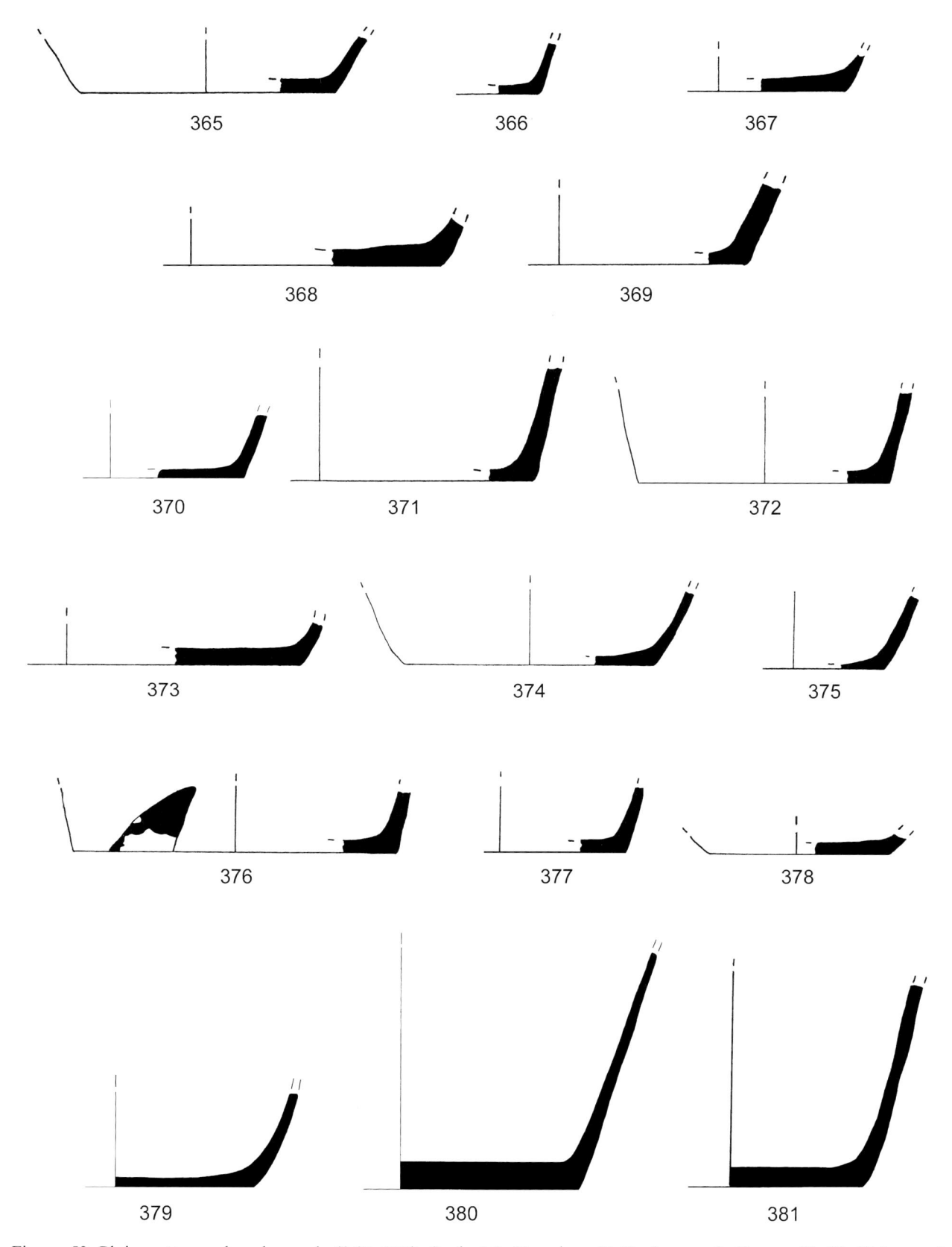

Figure 53. Plain pottery: closed vessels (**365–381**). Scale 1:3. Drawings D. Faulmann, L. Bonga, K. Chalikias, and P. Betancourt.

FIGURE 54

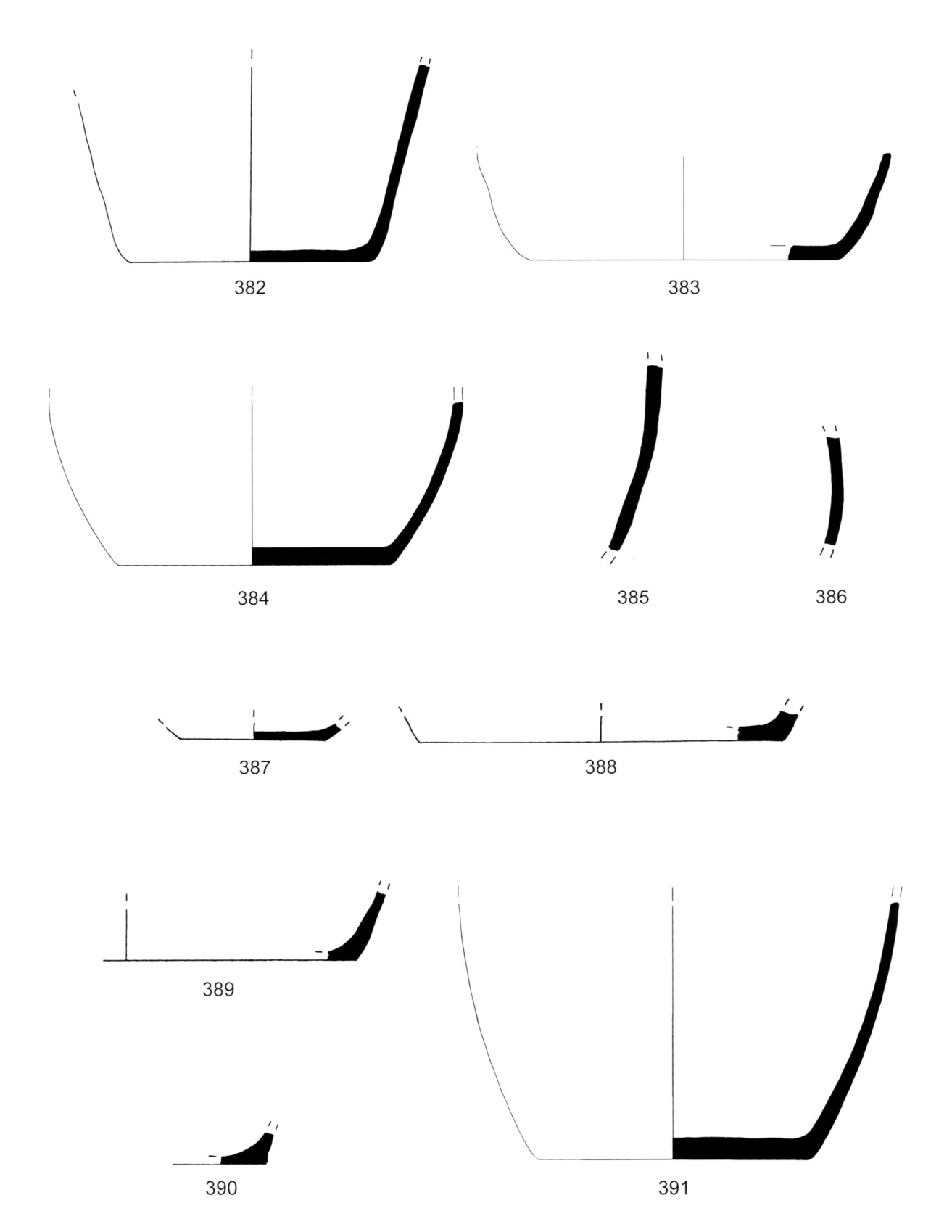

Figure 54. Plain pottery: closed vessels (**382–391**). Scale 1:3. Drawings D. Faulmann, L. Bonga, K. Chalikias, and P. Betancourt.

FIGURE 55

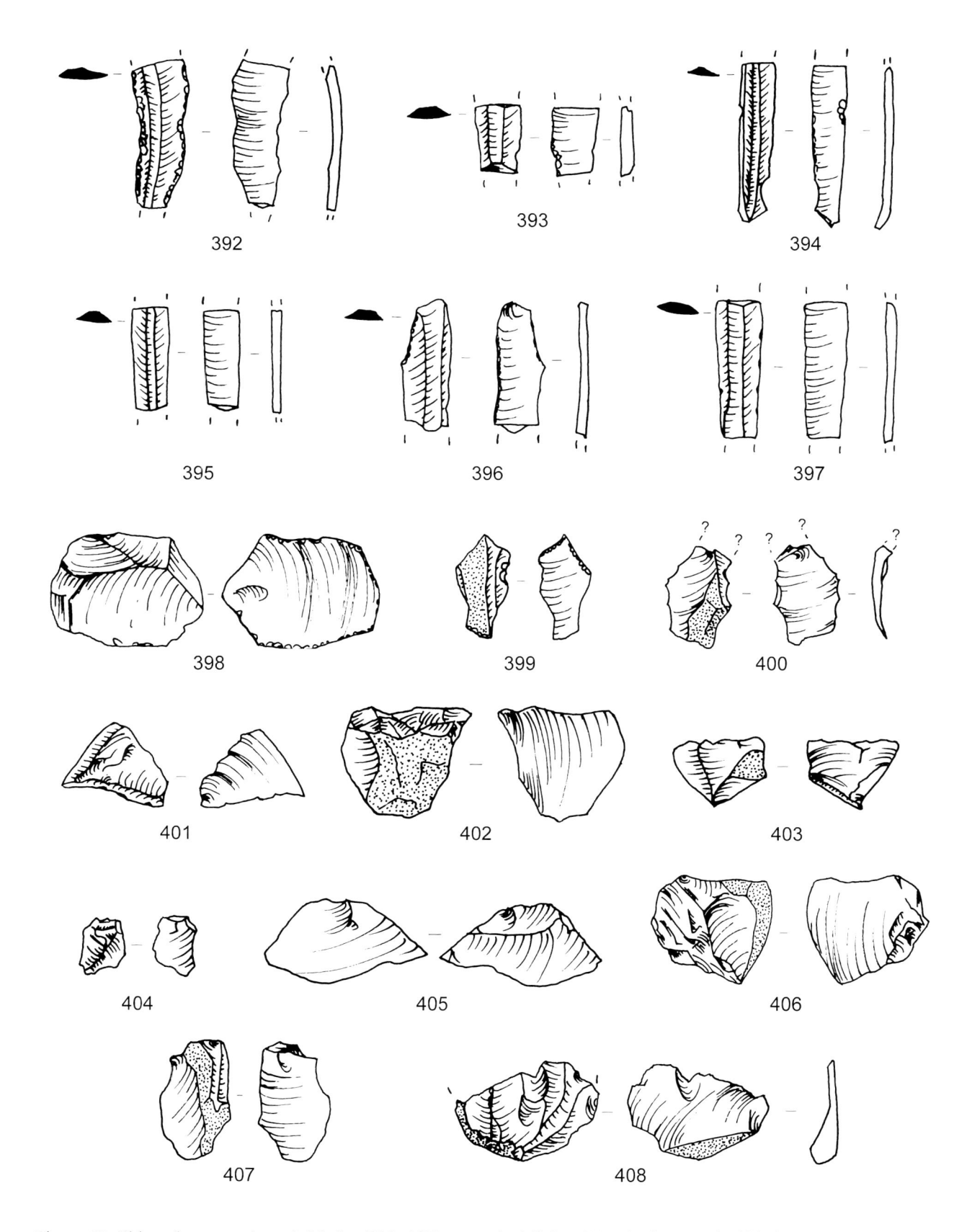

Figure 55. Chipped stone: prismatic blades (**392–397**), retouched flakes (**398**, **400**), retouched blade (**399**), and débitage (**401–408**). Scale 1:1. Drawings H. Dierckx.

FIGURE 56

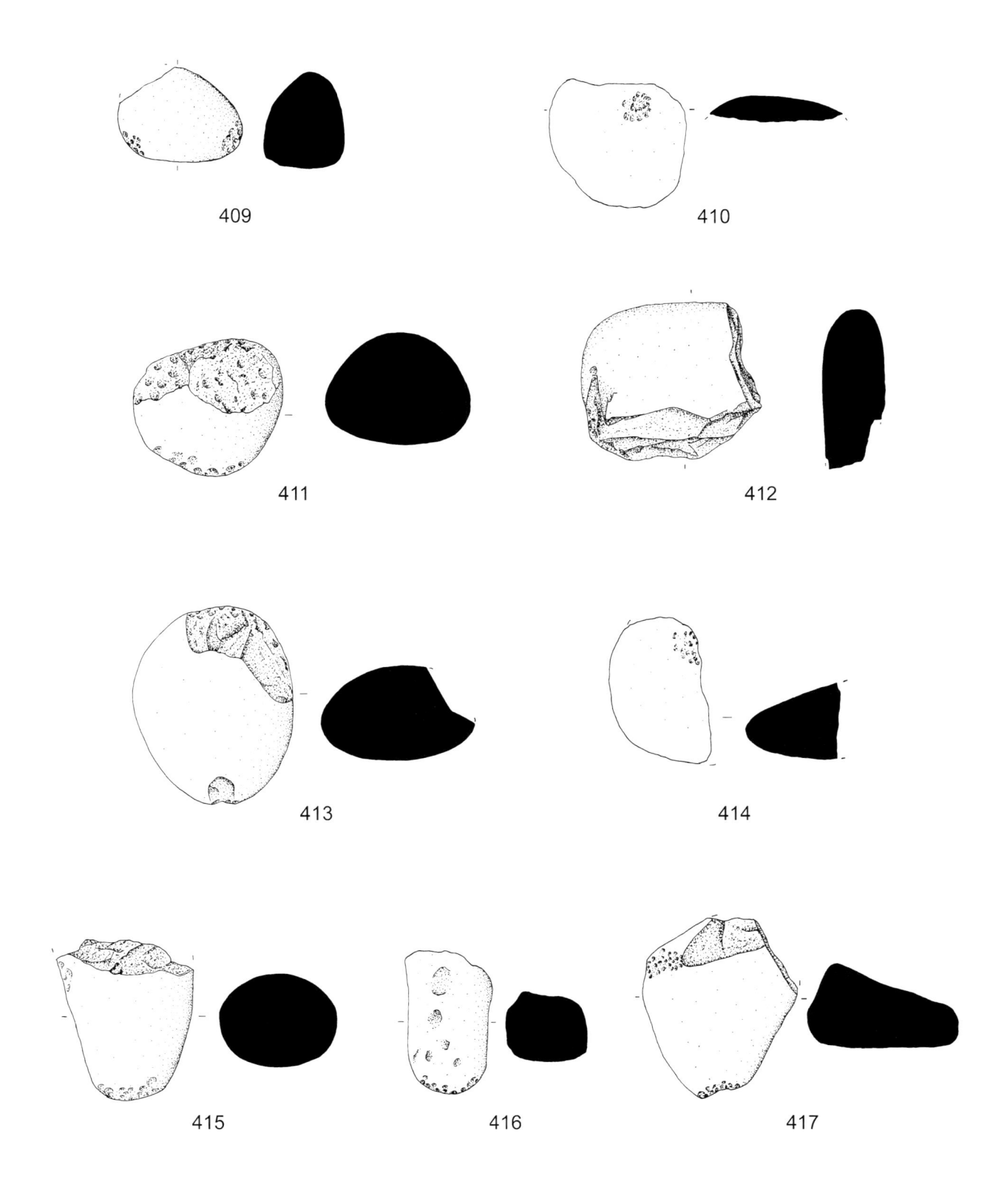

409

410

411

412

413

414

415

416

417

Figure 56. Pounders or hammer stones (Type 1: **409–417**). Scale 1:3. Drawings H. Dierckx.

FIGURE 57

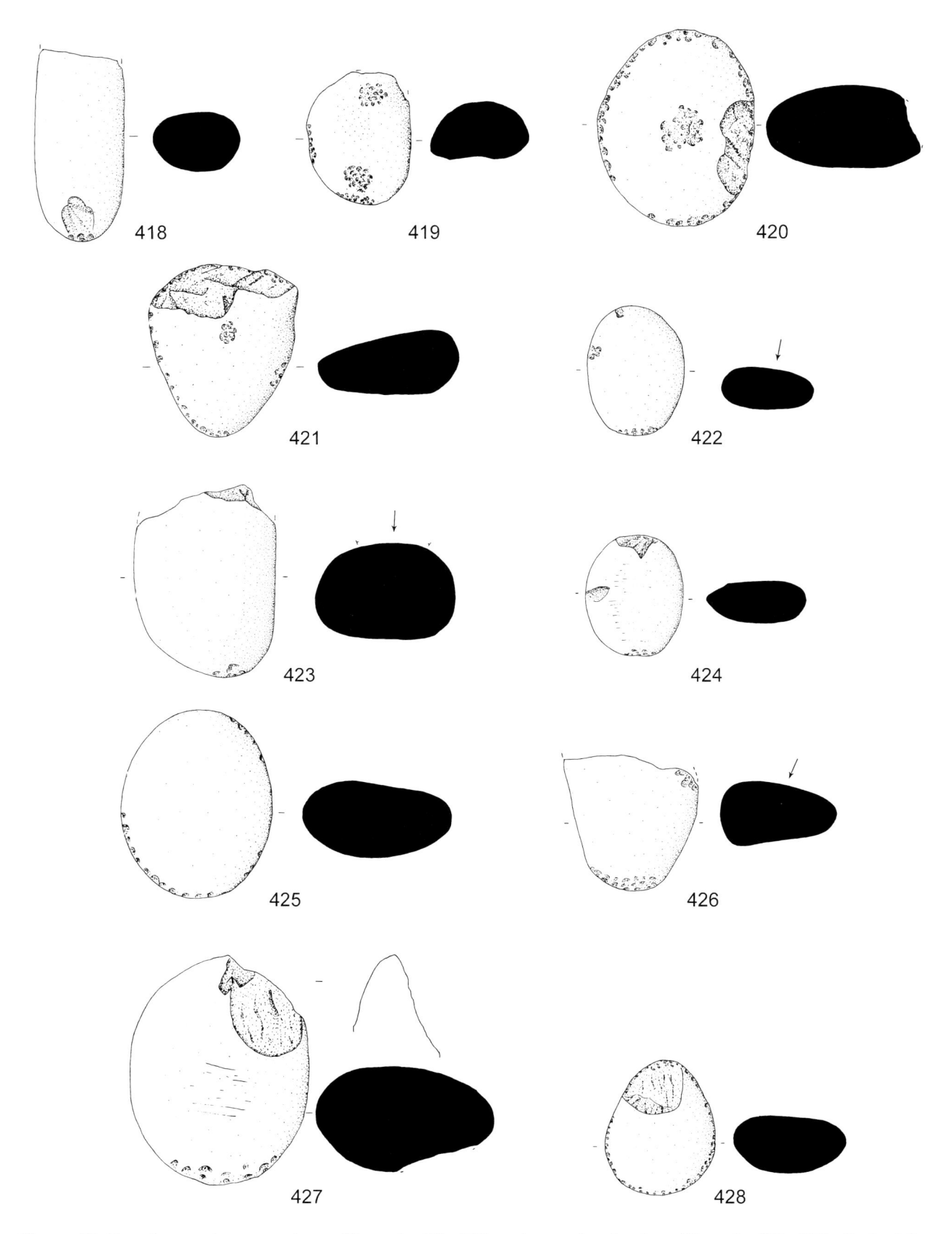

Figure 57. Pounders or hammer stones (Type 1: **418–421**) and pounder-abraders (Type 2: **422–428**). Scale 1:3. Drawings H. Dierckx.

FIGURE 58

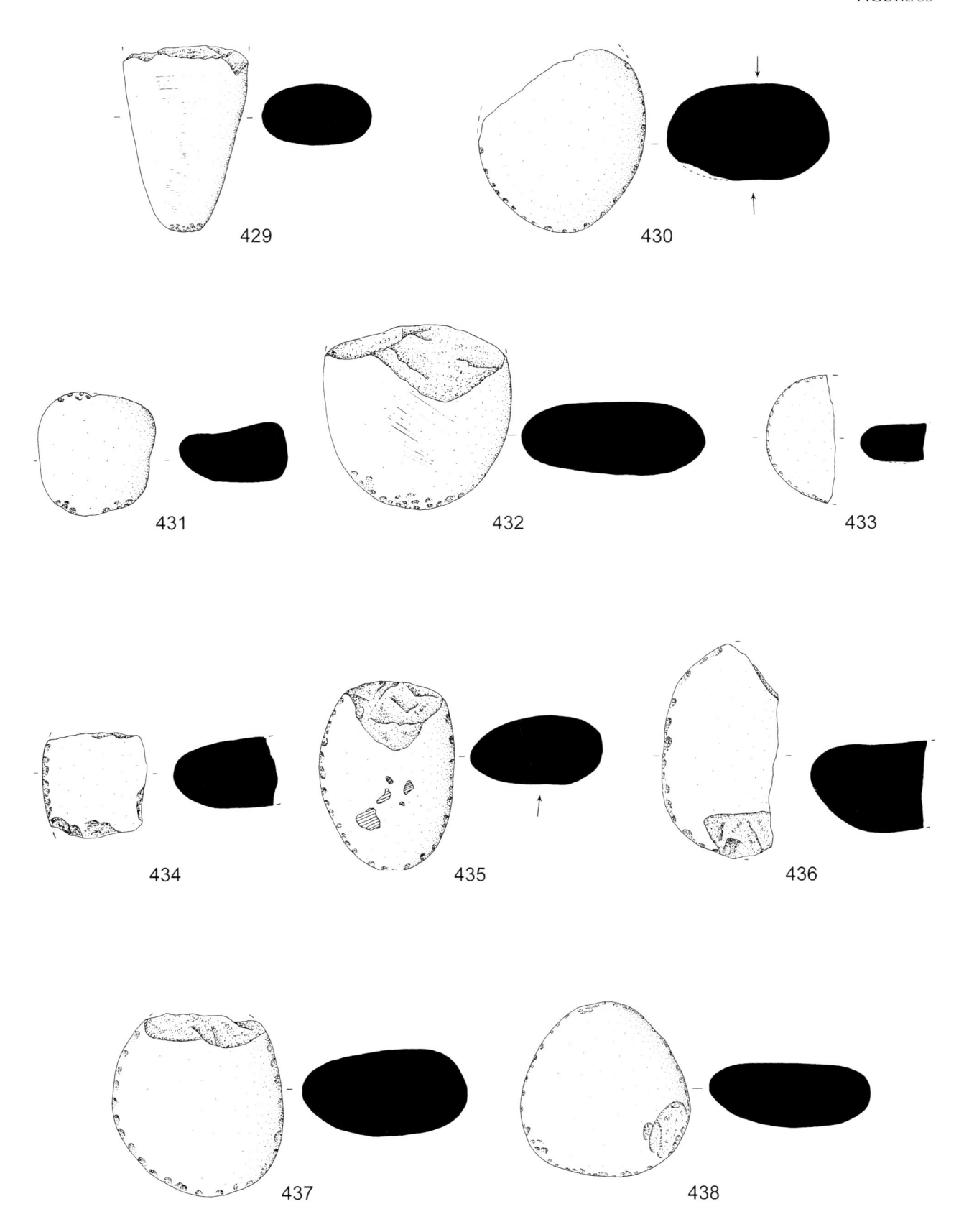

Figure 58. Pounder-abraders (Type 2: **429–438**). Scale 1:3. Drawings H. Dierckx.

FIGURE 59

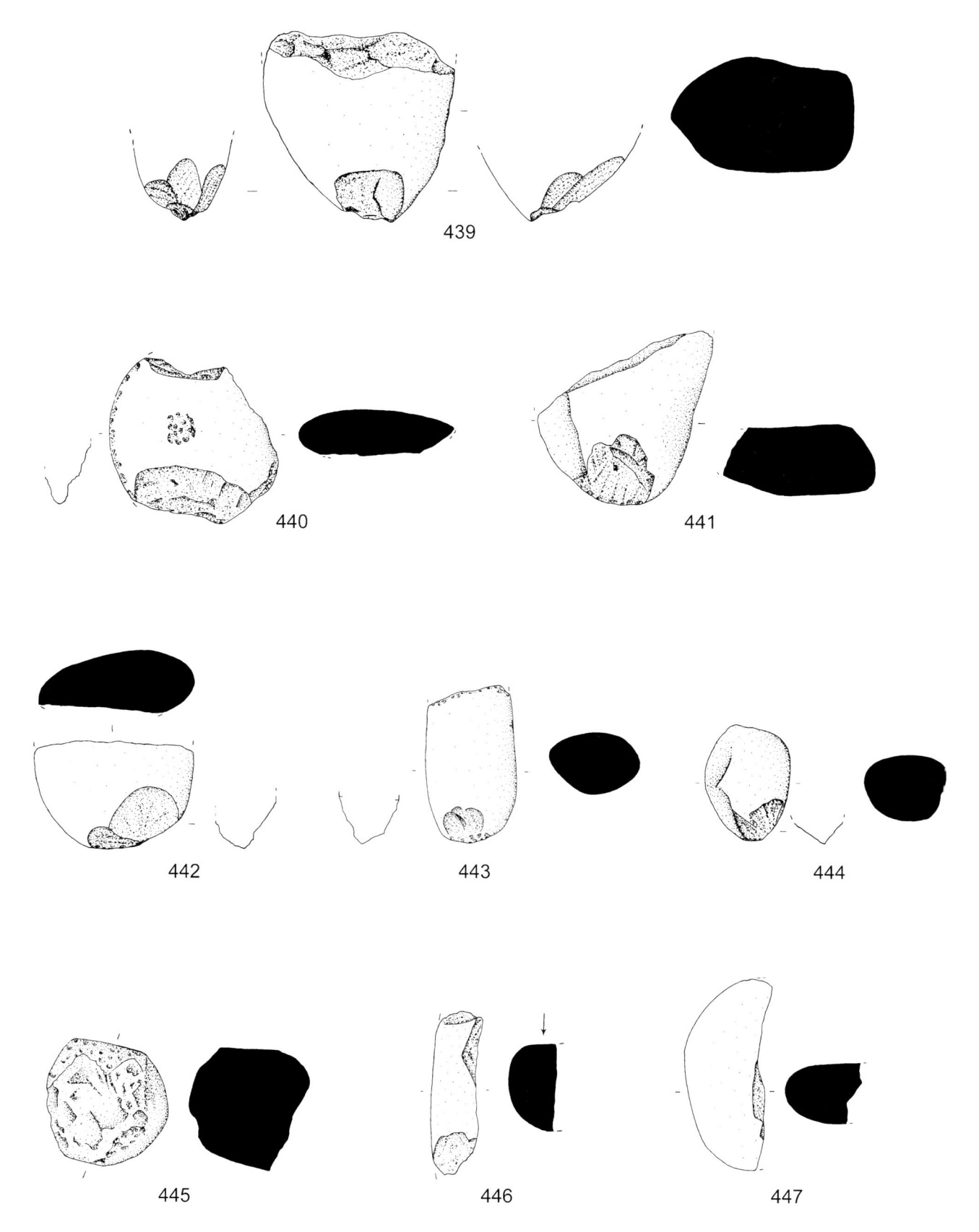

439

440 441

442 443 444

445 446 447

Figure 59. Choppers/hammers (Type 3: **439–444**), faceted tool (Type 4: **445**), and abraders/grinders (Type 5: **446**, **447**). Scale 1:3. Drawings H. Dierckx.

FIGURE 60

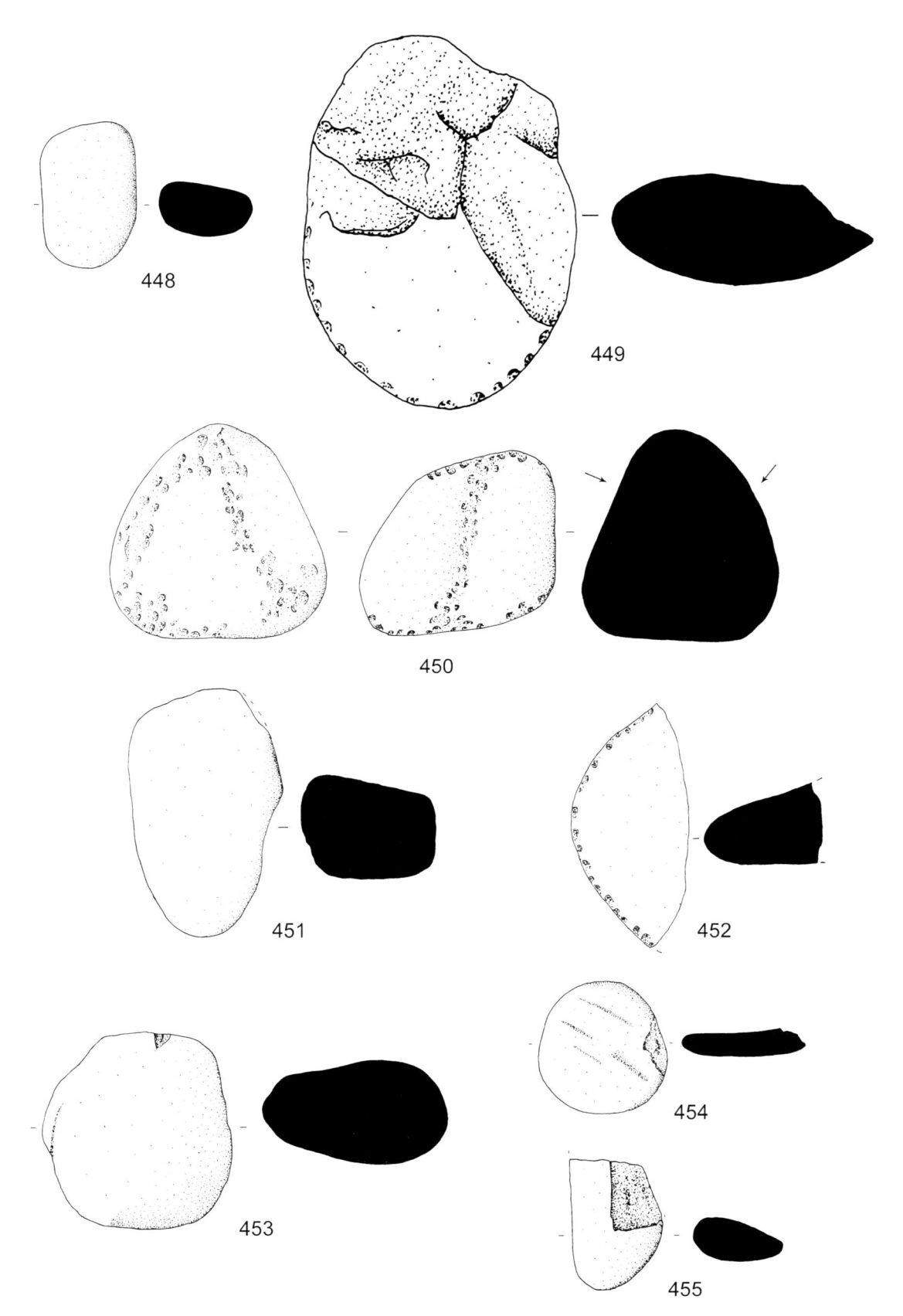

Figure 60. Abraders/grinders (Type 5: **448–453**) and whetstones (Type 6: **454**, **455**). Scale 1:3. Drawings H. Dierckx.

FIGURE 61

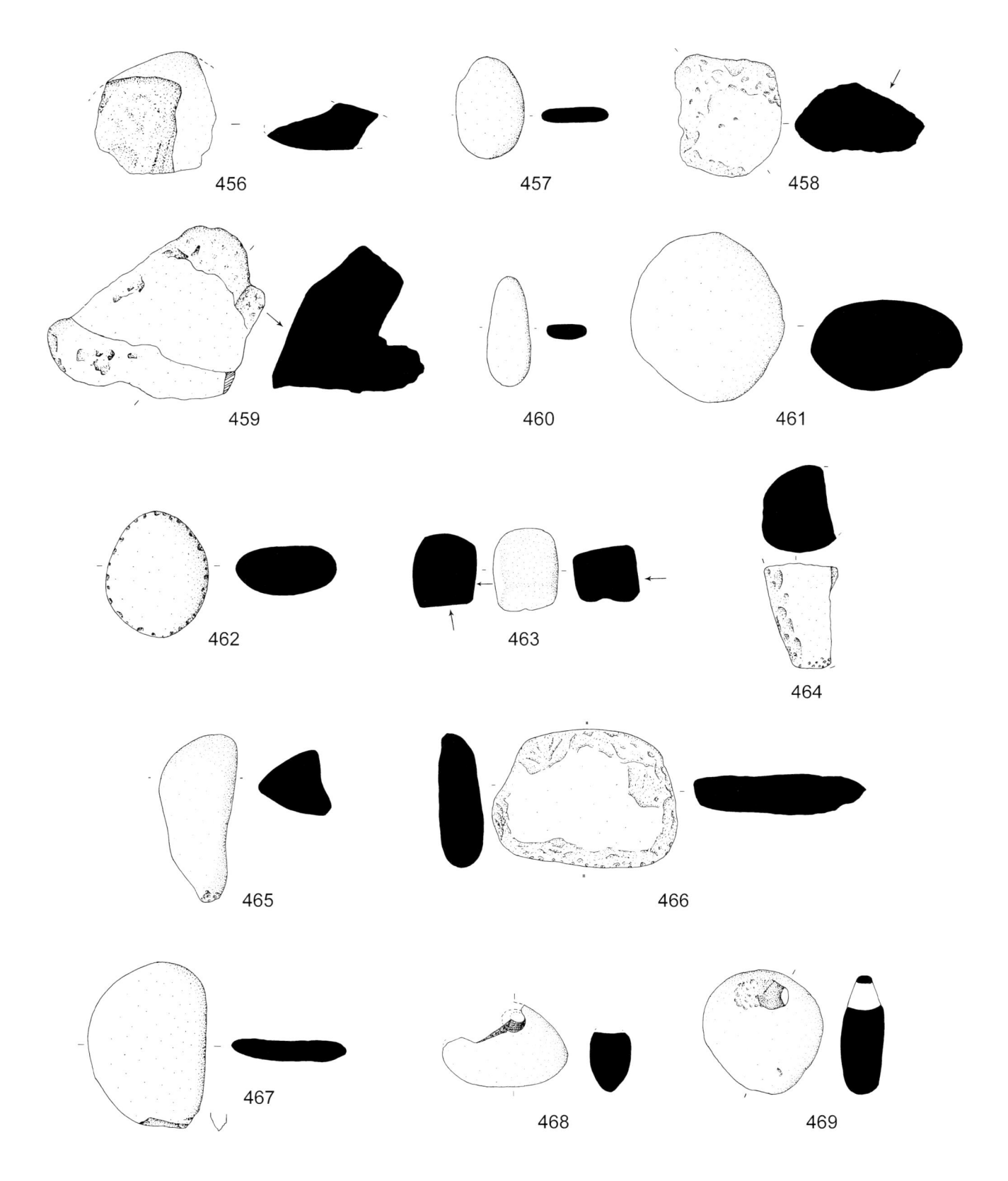

Figure 61. Whetstones (Type 6: **456**, **457**), pumice abrader/polishers (Type 7: **458**, **459**), polishers (Type 8: **460**–**463**), pestles (Type 9: **464**, **465**), scraper and piercer (Type 10: **466**, **467**), and weights (Type 11: **468**, **469**). Scale 1:3. Drawings H. Dierckx.

FIGURE 62

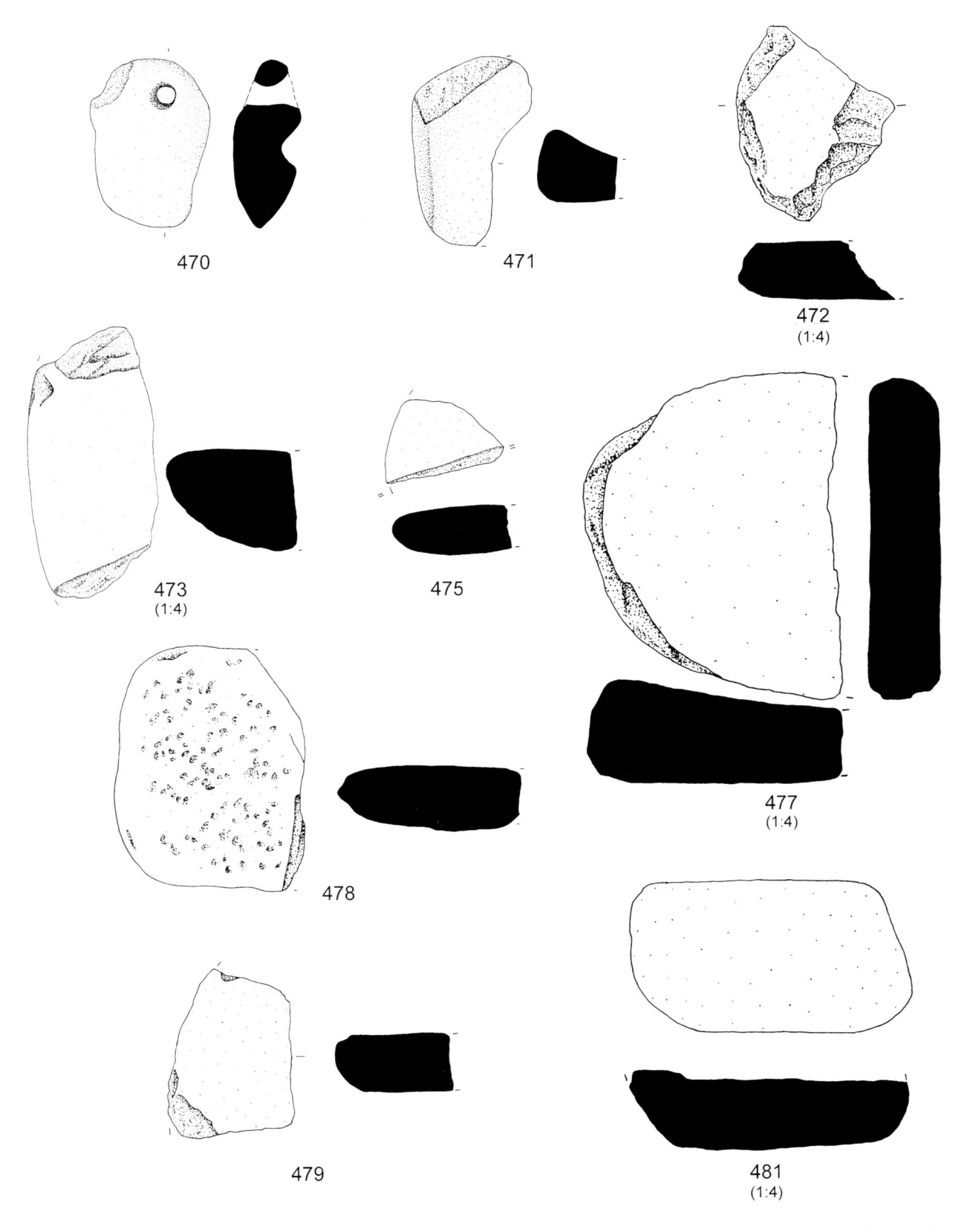

Figure 62. Weight (Type 11: **470**) and querns (Type 12: **471**–**473**, **475**, **477**–**479**, **481**). Scale 1:3 unless otherwise indicated. Drawings H. Dierckx.

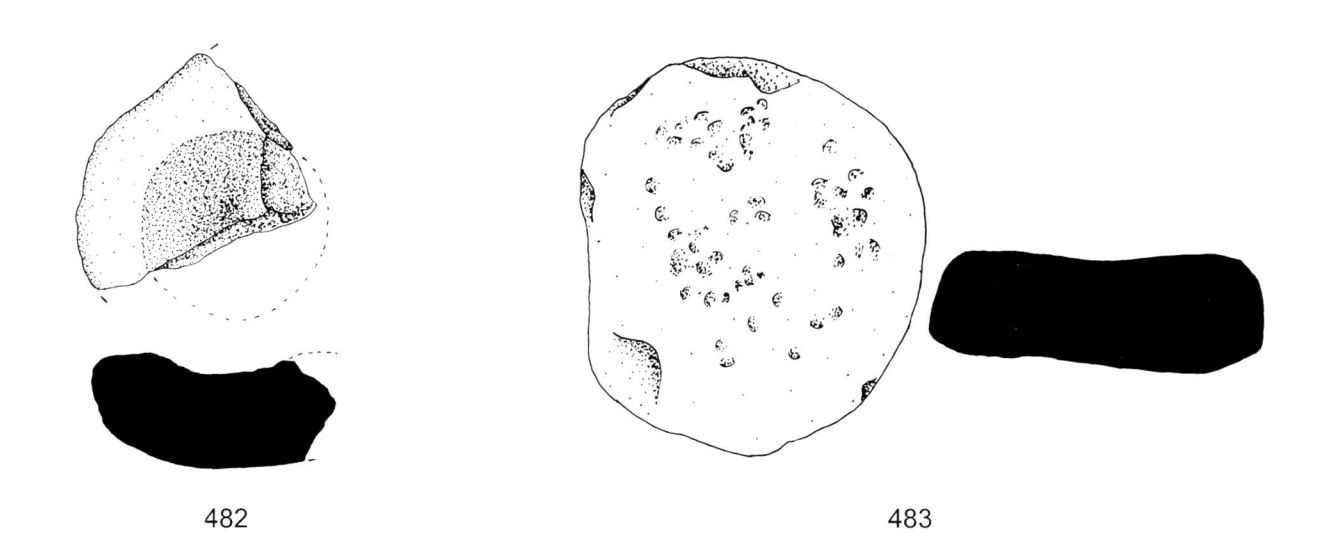

Figure 63. Mortar (Type 13: **482**) and working slab (Type 14: **483**). Scale 1:4. Drawings H. Dierckx.

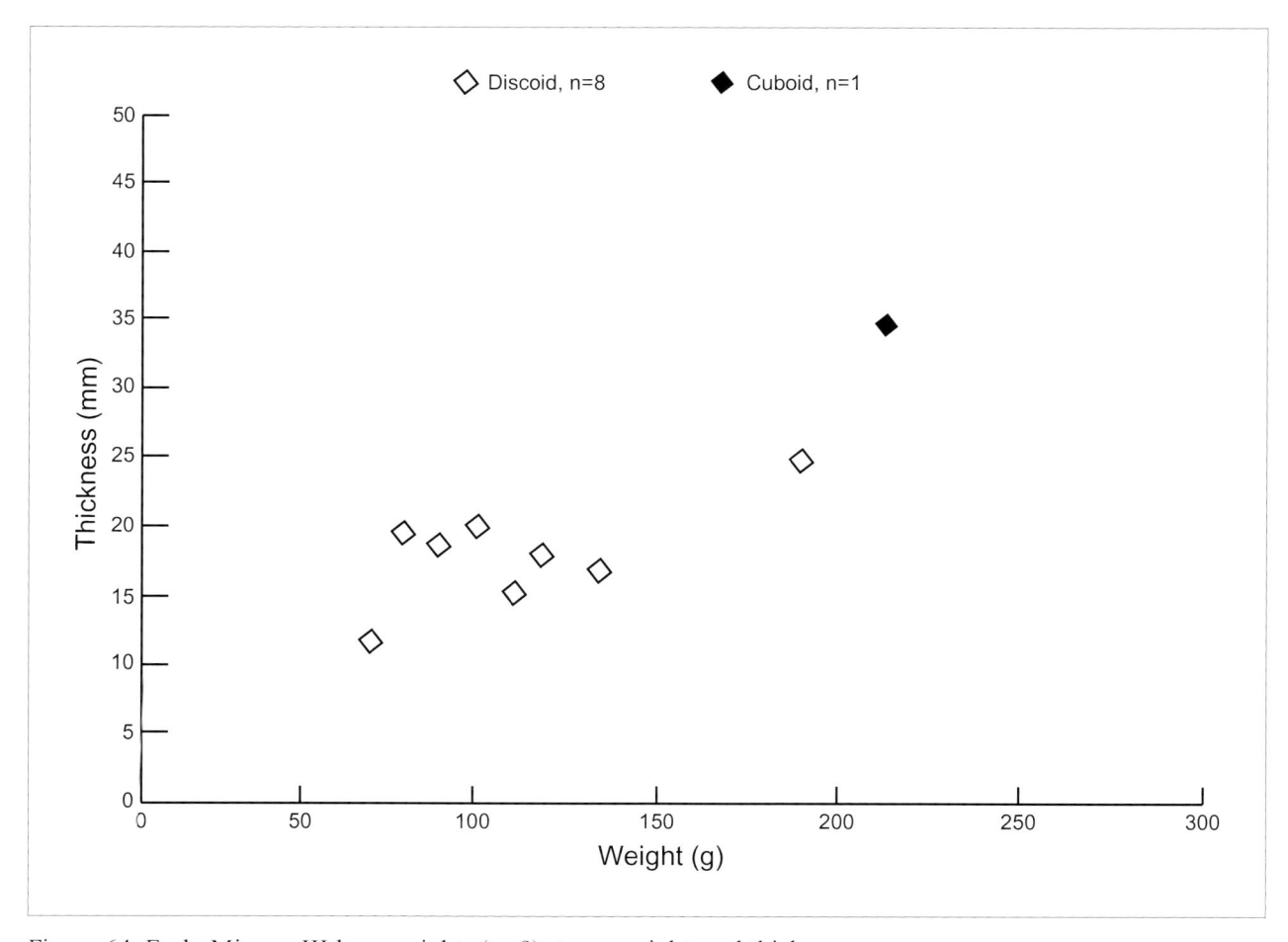

Figure 64. Early Minoan III loomweights (n=9): type, weight, and thickness.

FIGURE 65

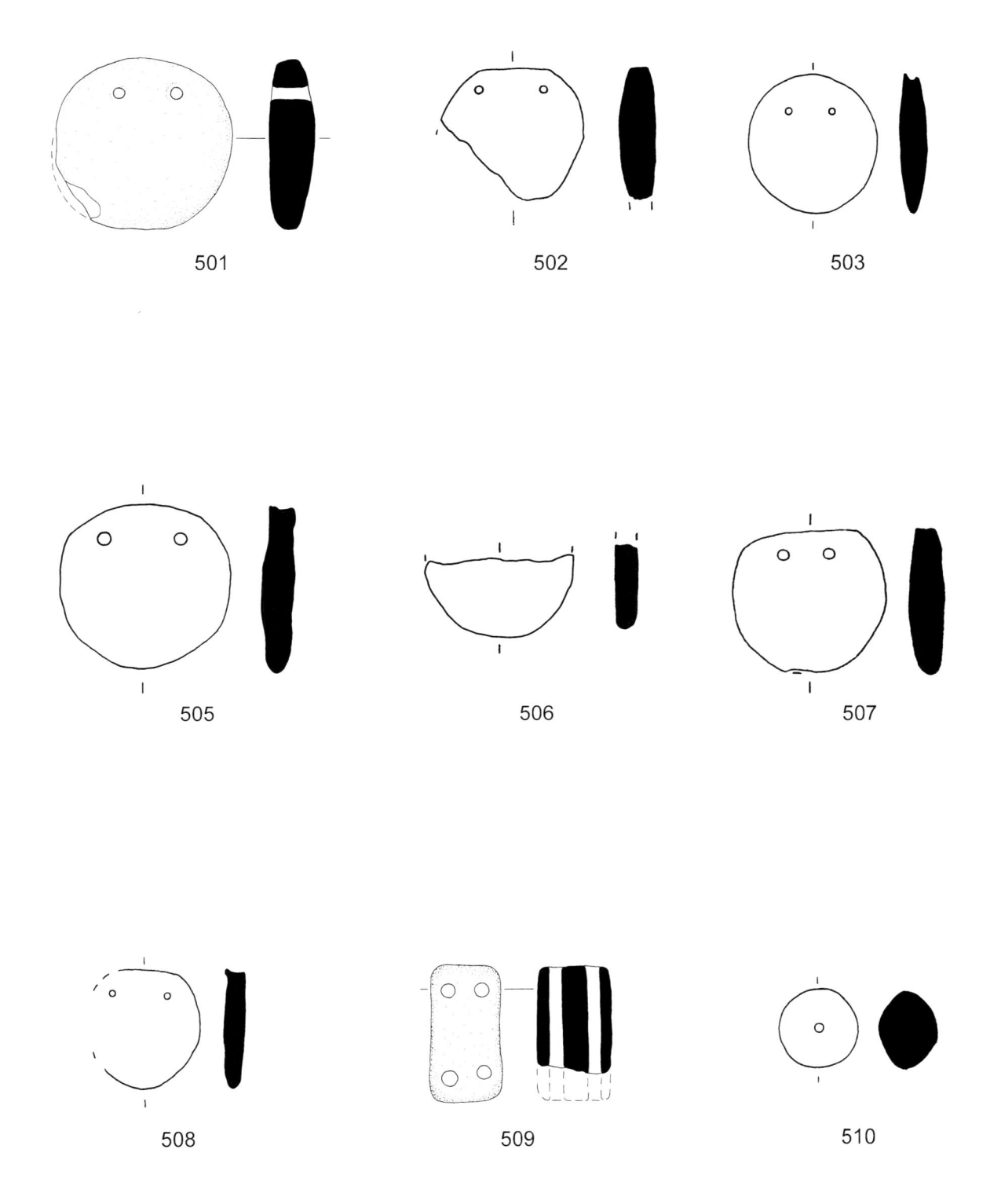

Figure 65. Textile tools from the rock shelter (**501**–**503**, **505**–**510**). Scale 1:3. Drawings L. Bonga.

FIGURE 66

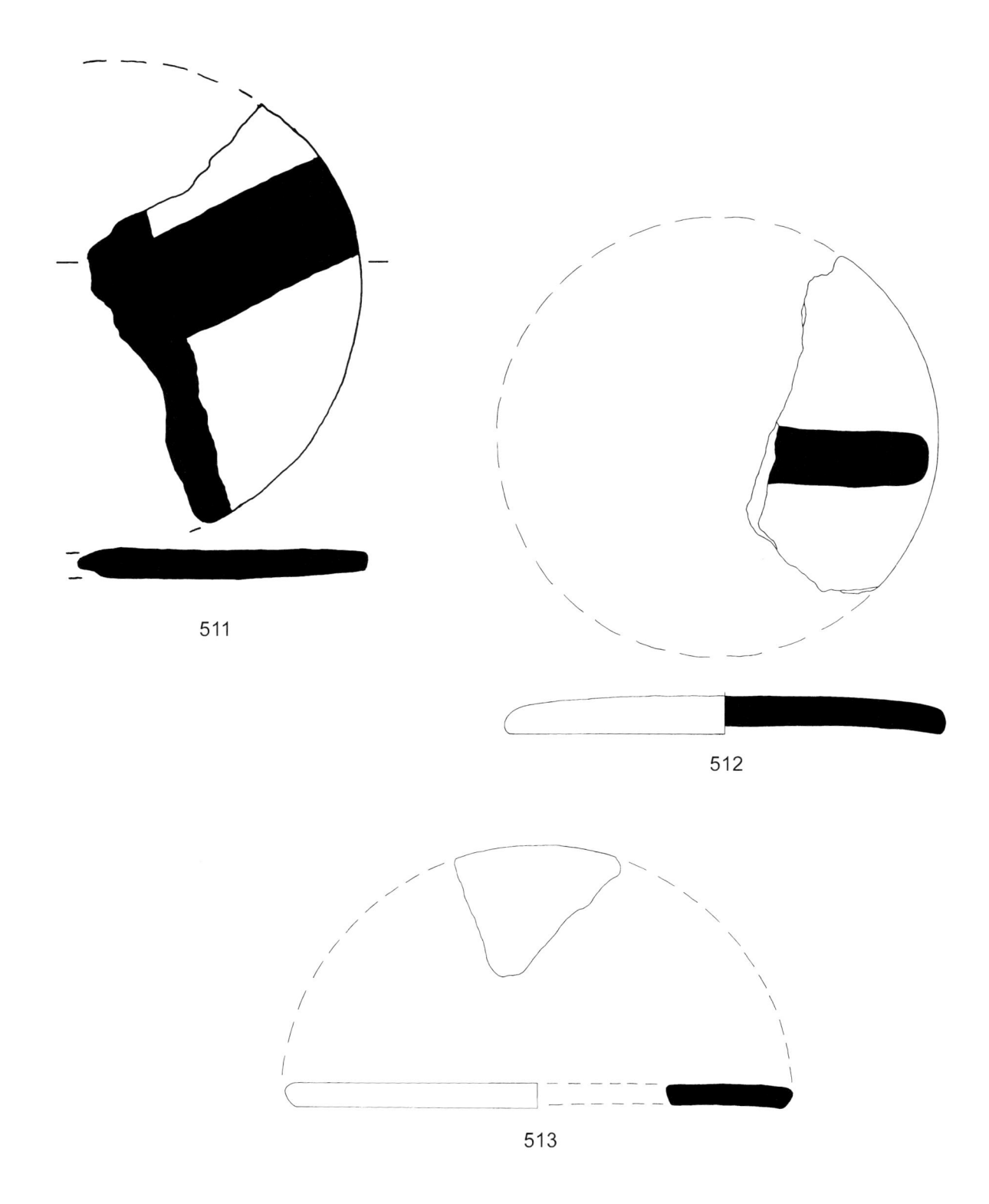

511

512

513

Figure 66. Clay potter's batts (**511**–**513**). Scale 1:3. Drawings L. Bonga.

FIGURE 67

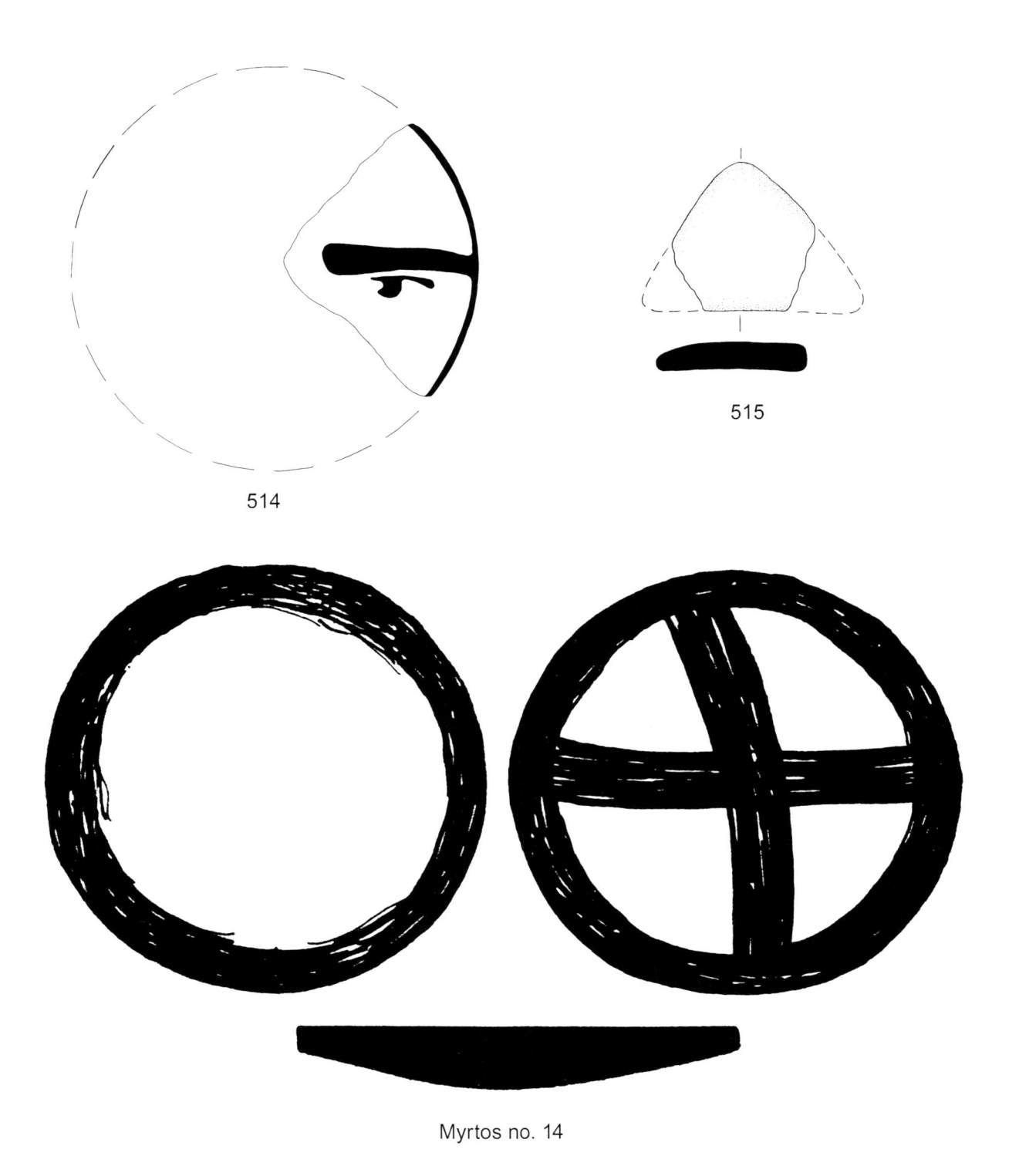

514

515

Myrtos no. 14

Figure 67. Clay potter's batt (**514**) and a clay triangle (**515**); scale 1:3. Potter's batt from Myrtos after Warren 1972, 245, fig. 98:14 (not to scale). Drawings L. Bonga.

FIGURES 68, 69, AND 70

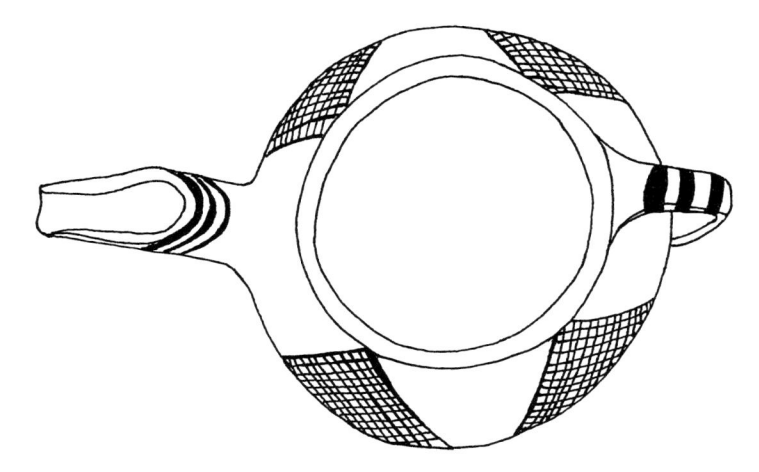

Figure 68. View, from above, of a teapot from Koumasa (HM 4107) showing the symmetry based on the vessel's structural parts. Drawing P. Betancourt.

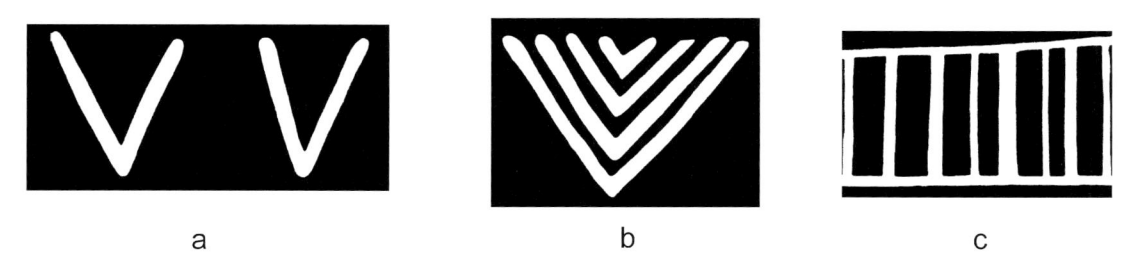

Figure 69. Motifs of White-on-Dark Ware from EM IIB Myrtos: (a) Warren 1972, 203, fig. 87:P 675; (b) Warren 1972, pl. 39D; (c) Warren 1972, pl. 62A. Drawings P. Betancourt.

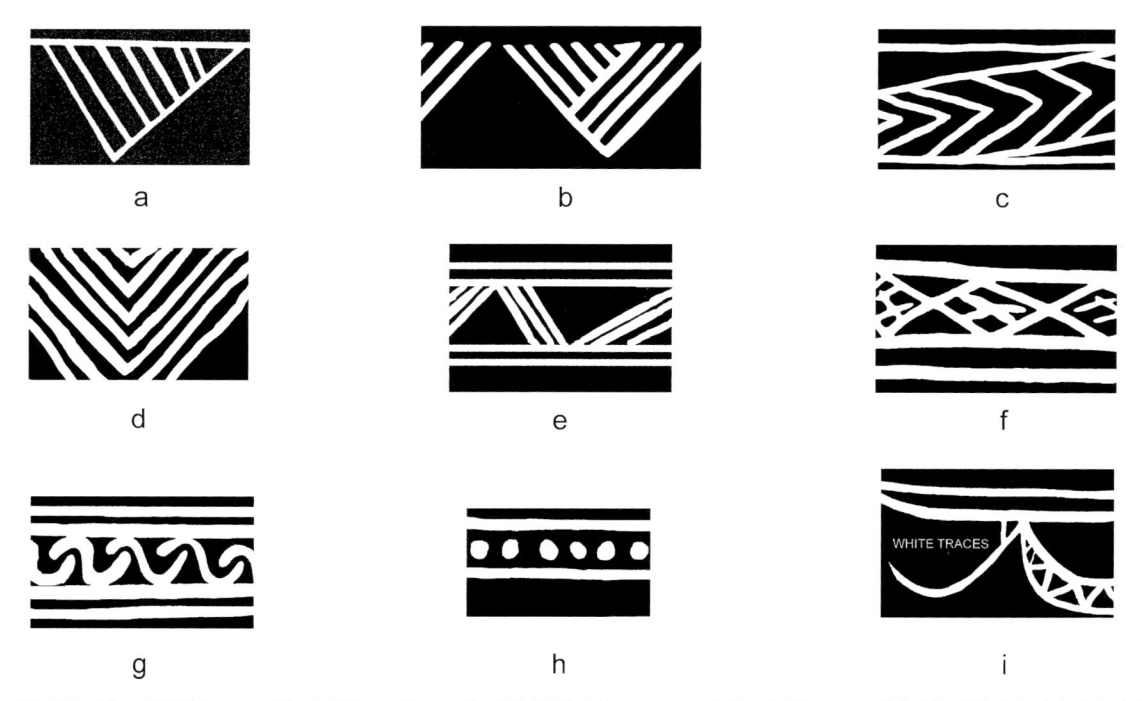

Figure 70. Motifs of White-on-Dark Ware from the EM III Alatzomouri Rock Shelter: (a) **23**; (b) **24**; (c) **46**; (d) **61**; (e) **62**; (f) **71**; (g) **56**; (h) **72**; (i) **96**. Drawings P. Betancourt.

FIGURE 71

Figure 71. Motifs of White-on-Dark Ware from MM IA. Vasiliki: (a) Seager 1906–1907, 122, fig. 5c. Gournia: (b) Hall 1904–1905, pl. 26:1; (c) Hall 1904–1905, pl. 27:6; (d) Hall 1904–1905, pl. 32:3. Vasiliki: (e) Seager 1906–1907, 120, fig. 3a; (f) Seager 1906–1907, 121, fig. 4d; (g) Maraghiannis and Karo 1907–1921, II, pl. 25:1. Gournia: (h) Hall 1904–1905, pl. 27:16; (i) Hall 1904–1905, pl. 27:18. Vasiliki: (j) Maraghiannis and Karo 1907–1921, II, pl. 25:10. Gournia: (k) Hall 1904–1905, pl. 27:7. Vasiliki: (l) Seager 1906–1907, 121, fig. 4a. Gournia: (m) Hall 1904–1905, pl. 29:12; (n) Betancourt and Silverman 1991, fig. 2:329. Vasiliki: (o) Seager 1908, pl. 12:34. Palaikastro: (p) Dawkins 1903–1904, 199, fig. 2f. Gournia: (q) Hall 1904–1905, pl. 33:4. Mochlos (r) Forsdyke 1925, fig. 96:A451-3. Gournia: (s) Hall 1904–1905, pl. 28:15. Vasiliki: (t) Seager 1906–1907, fig. 4e; (u) Maraghiannis and Karo 1907–1921, II, pl. 25:11. Gournia: (v) Hall 1904–1905, pl. 28:12; (w) Hall 1904–1905, pl. 28:28; (x) Hall 1904–1905, pl. 32:6. Drawings P. Betancourt.

Plates

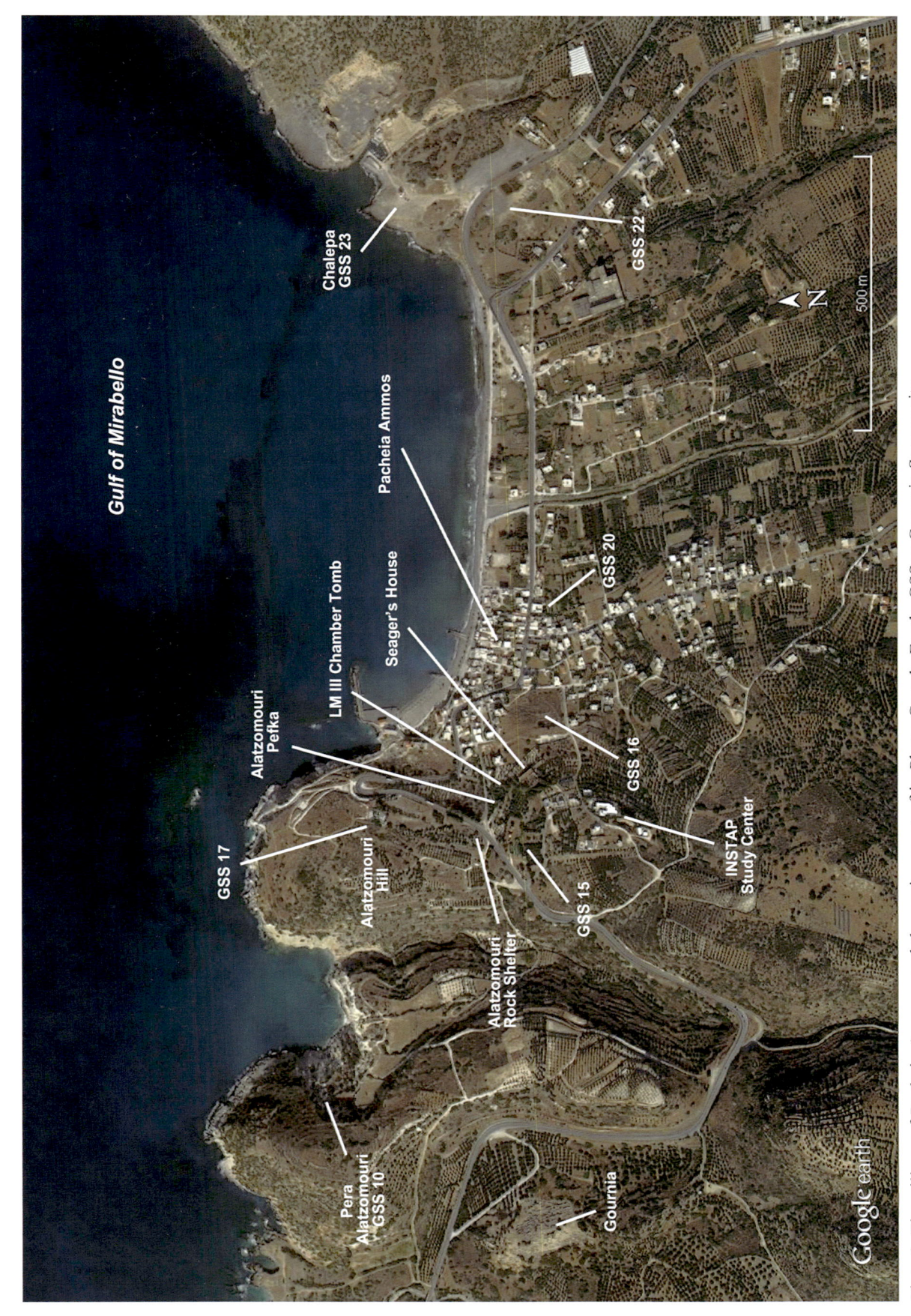

Plate 1. The village of Pacheia Ammos and the region west of it. Photo Google Earth. GSS = Gournia Survey site.

PLATE 2

Plate 2A. Area of excavation near Pacheia Ammos. Photo Google Earth.

Plate 2B. Alatzomouri Hill from the south. Photo T. Brogan.

Plate 3A. Alatzomouri Hill from the east. Photo T. Brogan.

Plate 3B. Pacheia Ammos from the west. Photo T. Brogan.

PLATE 4

Plate 4A. Pacheia Ammos from the south. Photographer unknown, published in 1904 D-DAI-ATH-Kreta 163, courtesy of the Photo Archive DAI Athens.

Plate 4B. Hand colored map of "Pachianamo" by Boschini, published in 1641, courtesy of American School of Classical Studies at Athens, Gennadion Library.

PLATE 5

Plate 5A. East side of Trenches A500–A800 during surface cleaning before excavation, from the north. Photo K. Chalikias.

Plate 5B. The hole-mouthed jar (left, **GRAVE 1A**) and jug (right, **GRAVE 1B**) recovered by the bulldozer. Photo Ch. Papanikolopoulos.

PLATE 6

Plate 6A. Stratum 1: cleaning plants growing in the trenches on the northwest side of Trench A700, from the north. Photo K. Chalikias.

Plate 6B. Stratum 1: after digging A600.1–2, from the south. Photo K. Chalikias.

PLATE 7

Plate 7A. Hole-mouthed jar (**208**) in A600.1. Photo K. Chalikias.

Plate 7B. Stratum 1: after digging A600/700.1. Photo K. Chalikias.

PLATE 8

Plate 8A. Stratum 1: after digging A700.1, from the north. Photo K. Chalikias.

Plate 8B. Stratum 2 upper: after digging A601.1 and A602.1, from the north. Photo K. Chalikias.

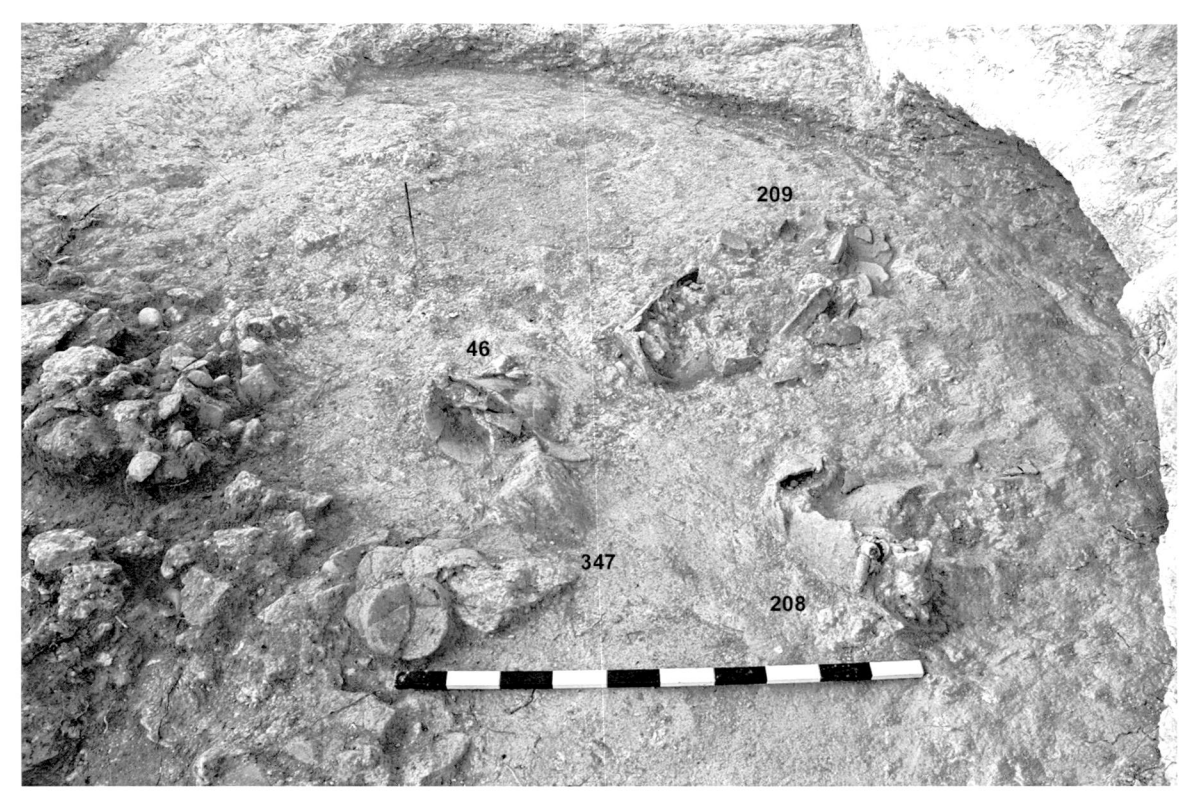

Plate 9A. Stratum 2 upper: deposit of broken vessels in A601.2–5 (**46**, **208**, **209**, **347**), from the north. Photo K. Chalikias.

Plate 9B. Stratum 2 upper: after digging A601.7 and A604.1, from the north. Photo K. Chalikias.

PLATE 10

Plate 10A. Stratum 2 upper: deposit of broken vessels in A604.2–5 (**8**, **96**, **208**, **209**, **229**, **231**, **347**), from the east. Photo K. Chalikias.

Plate 10B. Stratum 2 upper: EM IIB jug from the south coast of Crete (**8**) in situ. Photo K. Chalikias.

Plate 11A. Stratum 2 upper: a tripod cooking pot (**231**) and a teapot (**96**) in situ. Photo K. Chalikias.

Plate 11B. Stratum 2 upper: a tripod cooking pot (**229**) in situ. Photo K. Chalikias.

Plate 11C. Stratum 2 lower interior: after digging A604.11–13, from the west. Photo K. Chalikias.

PLATE 12

Plate 12A. Stratum 2 lower interior: after digging A604.13, from the west. Photo K. Chalikias.

Plate 12B. Stratum 2 lower interior: spouted conical jar (**46**) in situ. Photo K. Chalikias.

PLATE 13

Plate 13A. Stratum 2 lower interior: closed vessel (**347**) in situ. Photo K. Chalikias.

Plate 13B. Stratum 2 lower interior: tripod cooking pot (**229**) in situ. Photo K. Chalikias.

PLATE 14

Plate 14A. Stratum 2 lower interior: tripod cooking pot (**230**), teapot (**96**), and tripod cooking pot (**231**) in situ left to right. Photo K. Chalikias.

Plate 14B. Stratum 2 lower interior: after digging A604.17, from the east. Photo K. Chalikias.

PLATE 15

Plate 15A. Stratum 2 lower interior: hole-mouthed jar (**208**). Photo K. Chalikias.

Plate 15B. Stratum 2 lower interior: hole-mouthed jars (**209**, **214**). Photo K. Chalikias.

PLATE 16

Plate 16A. Stratum 2 lower interior: after excavating A604.19–20, from the west. Photo K. Chalikias.

Plate 16B. Stratum 2 lower interior: after excavating A604.19–20, from the north. Photo K. Chalikias.

PLATE 17

Plate 17A. Stratum 2 lower interior: after excavating A604.27–29, from the west. Photo K. Chalikias.

Plate 17B. Stratum 2 lower interior: after excavating A604.27–29, from the north. Photo K. Chalikias.

PLATE 18

Plate 18A. Stratum 2 lower interior: tripod cooking pot (**230**) at left. Photo K. Chalikias.

Plate 18B. Stratum 2 lower interior: vat (**330**). Photo K. Chalikias.

Plate 19A. Stratum 2 lower interior: hole-mouthed jar (**210**). Photo K. Chalikias.

Plate 19B. Stratum 2 lower interior: jug (**87**), conical bowl (**23**), spouted conical jar (**47**), rounded cup (**125**), and loomweight (**509**), from the west. Photo K. Chalikias.

Plate 20A. Stratum 2 lower interior: conical bowl with spout (**24**). Photo K. Chalikias.

Plate 20B. Stratum 2 lower exterior: after digging A604.7, from the north. Photo K. Chalikias.

PLATE 21

Plate 21A. Stratum 2 lower exterior: after digging A604.6–7, from the south. Photo K. Chalikias.

Plate 21B. Stratum 2 lower exterior: after digging A604.11–12, from the north. Photo K. Chalikias.

PLATE 22

Plate 22A. Stratum 2 lower exterior (**388**, **316**, **171**), from the south. Photo K. Chalikias.

Plate 22B. Stratum 2 lower exterior: loomweight (**505**), from the northwest. Photo K. Chalikias.

Plate 23A. Stratum 2 lower exterior: loomweight (**506**), from the northwest. Photo K. Chalikias.

Plate 23B. Stratum 2 lower exterior: after digging A604.9, from the north. Photo K. Chalikias.

PLATE 24

Plate 24A. Stratum 2 lower exterior: after digging A604.11, from the west. Photo K. Chalikias.

Plate 24B. Stratum 2 lower exterior: after digging A604.19, from the west. Photo K. Chalikias.

Plate 25A. Stratum 2 lower exterior: after digging A604.19, from the north. Photo K. Chalikias.

Plate 25B. Stratum 2 lower exterior: spindle whorl (**510**), from the east. Photo K. Chalikias.

PLATE 26

Plate 26A. Stratum 2 lower exterior: after digging A604.29, from the north. Photo K. Chalikias.

Plate 26B. Stratum 2 lower exterior: after digging A604.29, from the south. Photo K. Chalikias.

PLATE 27

Plate 27A. Stratum 3a fill: after digging A604.46–47, from the north. Photo K. Chalikias.

Plate 27B. Stratum 3a fill: after digging A604.46–47, from the south. Photo K. Chalikias.

PLATE 28

Plate 28A. Stratum 3a fill showing a sherd of a bowl (**26**), from the northeast. Photo K. Chalikias.

Plate 28B. Stratum 3a fill showing a basin (**114**), from the west. Photo K. Chalikias.

PLATE 29

Plate 29A. Stratum 3a fill with the sherd of a basin (**97**), from the east. Photo K. Chalikias.

Plate 29B. Stratum 3a fill: detail of southeast corner of the rock shelter, which continues under the modern road, from the northwest. Photo K. Chalikias.

PLATE 30

Plate 30A. Stratum 3b fill: after digging A604.56, from the north. Photo K. Chalikias.

Plate 30B. Stratum 3b fill: after digging A604.55–58, from the north. Photo K. Chalikias.

PLATE 31

Plate 31A. Stratum 3b fill: after digging A604.55–58, from the south. Photo K. Chalikias.

Plate 31B. Detail of Stratum 3b fill, from the west. Photo K. Chalikias.

PLATE 32

Plate 32A. Stratum 3c fill: after digging A604.59–64, from the north. Photo K. Chalikias.

Plate 32B. Stratum 3c fill: after digging A604.59–64, from the south. Photo K. Chalikias.

PLATE 33

Plate 33A. Stratum 3c fill: after digging A604.59–64, from the west. Photo K. Chalikias.

Plate 33B. Stratum 3c fill: pottery concentration in the middle of the fill after digging A604.59–64, from the west. Photo K. Chalikias.

Plate 34A. Stratum 3c fill: pottery in A604.61 in the southeast corner of the chamber. Photo K. Chalikias.

Plate 34B. Stratum 3d fill: after digging A604.65–66, from the north. Photo K. Chalikias.

PLATE 35

Plate 35A. Stratum 3d fill: after digging A604.65–66, from the south. Photo K. Chalikias.

Plate 35B. Stratum 3d fill: after digging A604.65–66, from the west. Photo K. Chalikias.

PLATE 36

Plate 36A. Stratum 3d fill: southeast corner of the chamber after digging A604.65–66, from the west. Photo K. Chalikias.

Plate 36B. Stratum 3d fill: after digging A604.67, from the north. Photo K. Chalikias.

Plate 37A. Stratum 3d fill: after digging A604.67, from the south. Photo K. Chalikias.

Plate 37B. Stratum 3d fill: after digging A604.67, from the west. Photo K. Chalikias.

PLATE 38

Plate 38A. Stratum 3d fill: after digging A604.65–66 (**151**), from the north. Photo K. Chalikias.

38B. Stratum 3d fill: after digging A604.65–66 (**137**, **380**), from the north. Photo K. Chalikias.

PLATE 39

Plate 39. South Coast Class of pottery: jug (**8**). East Cretan White-on-Dark Ware: conical bowls (**23–25**) and spouted conical jars (**46–50**). Scale 1:4. Photo Ch. Papanikolopoulos.

PLATE 40

Plate 40. East Cretan White-on-Dark Ware: spouted conical jar (**52**), spouted two-handled cup (**71**), bridge-spouted jar (**80**), and teapot (**96**). Plain pottery and pottery with slight decoration: basin (**97**) and rounded cup (**124**). Scale 1:4. Photo Ch. Papanikolopoulos.

PLATE 41

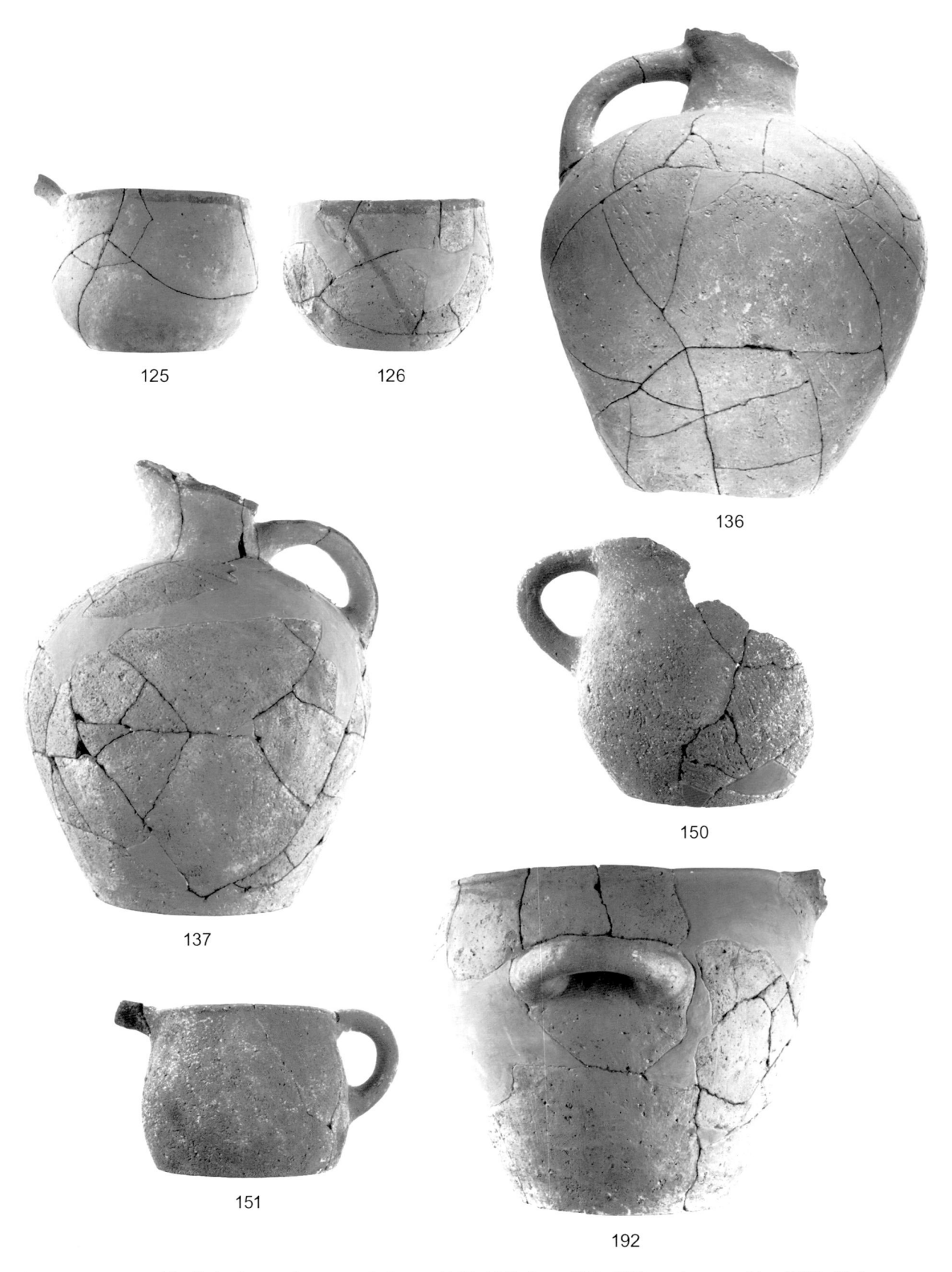

Plate 41. Pottery with slight decoration: rounded cups (**125**, **126**), jugs (**136**, **137**), and spouted jar (**192**). Plain pottery: wide-mouthed jugs (**150**, **151**). Scale 1:4. Photo Ch. Papanikolopoulos.

PLATE 42

208

209

Plate 42. Plain pottery: hole-mouthed jars (**208**, **209**). Scale 1:4. Photo Ch. Papanikolopoulos.

PLATE 43

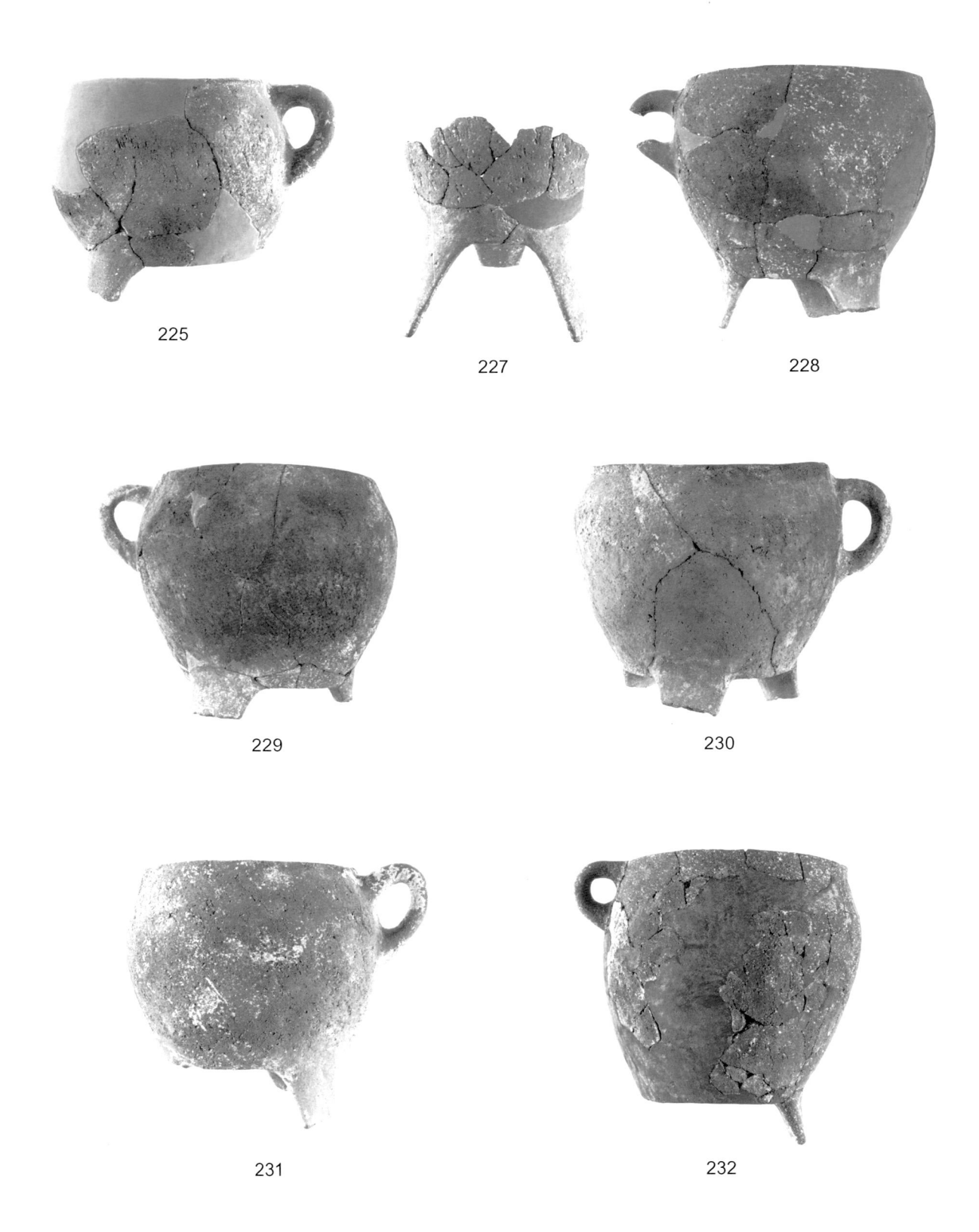

225

227

228

229

230

231

232

Plate 43. Plain pottery: tripod cooking pots (**225**, **227**–**232**). Scale 1:4 except for **227**–**228**, which are 1:6. Photo Ch. Papanikolopoulos.

PLATE 44

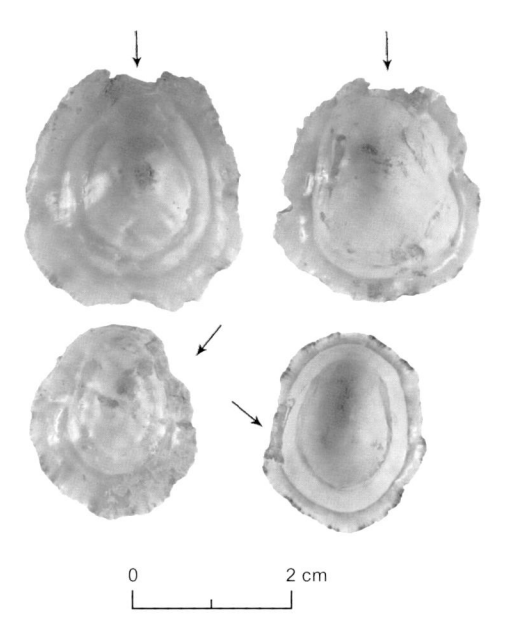

Plate 44A. "Nicking" and breaking of the limpets' lip (A604.1). Photo D. Mylona.

Plate 44B. A limpet with broken tip (A604.65). Photo D. Mylona.

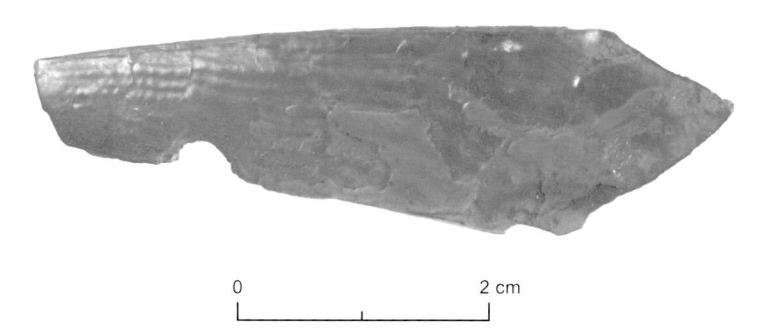

Plate 44C. Perforated fragment of a *Pinna nobilis* shell (A604.67). Photo D. Mylona.

Plate 44D. Olive pit (left) from locus A604.22 and grape pip (right) from locus A604.53. Photo E. Margaritis.

PLATE 45

Plate 45A. Fabric Group 1, the jar fabric, sample PAR 09/29 (**207**), x25. Photo E. Nodarou.

Plate 45B. Fabric Group 1, the jar fabric, sample PAR 09/08 (**73**), x50. Photo E. Nodarou.

PLATE 46

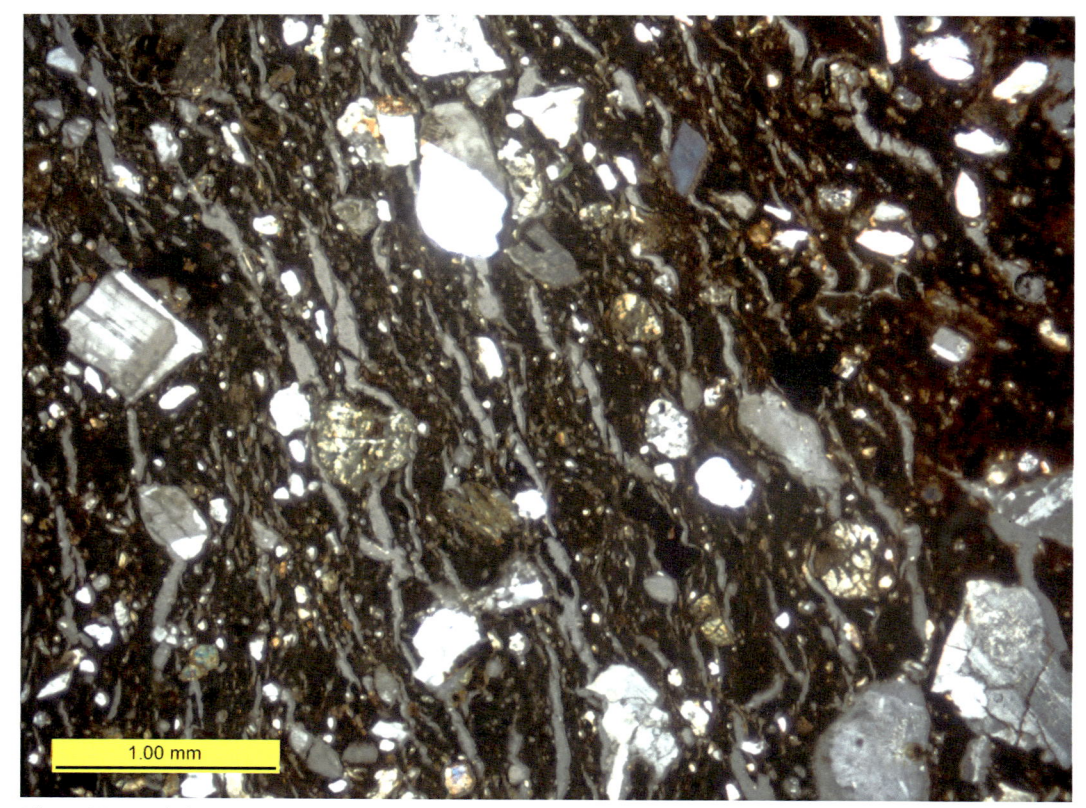

Plate 46A. Fabric Group 1, the jar fabric, sample PAR 09/14 (**183**), x25. Note elongated and oriented voids indicative of tempering with organic matter. Photo E. Nodarou.

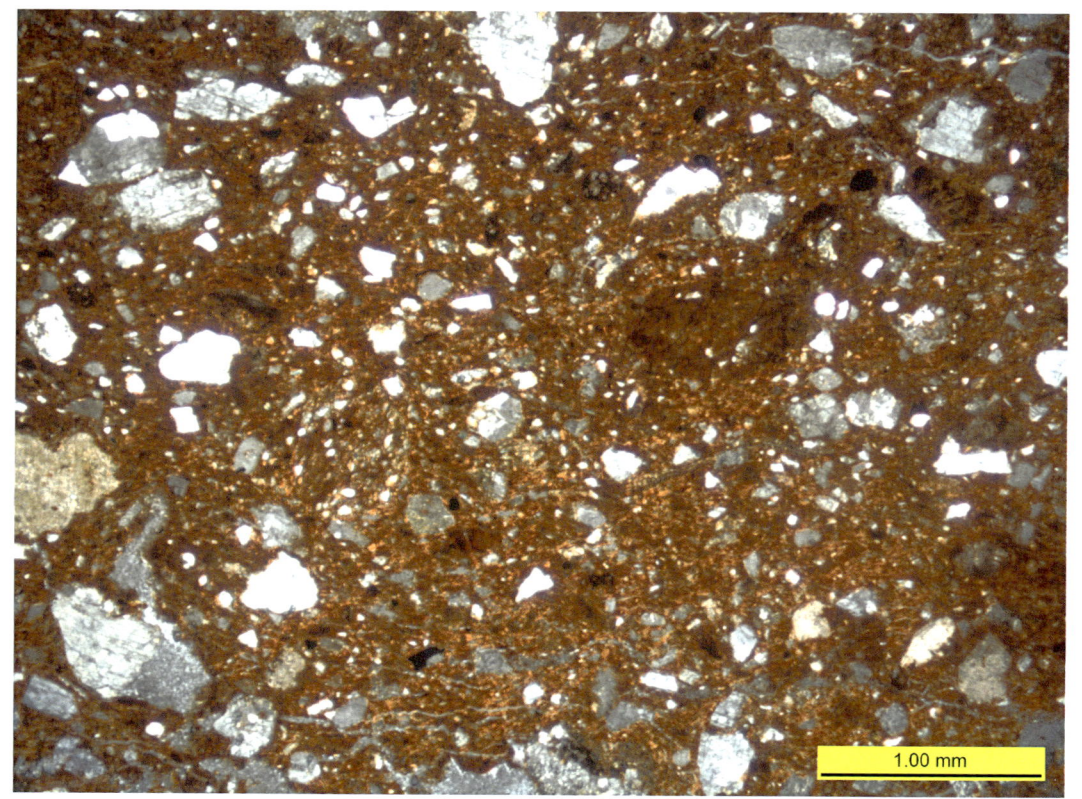

Plate 46B. Fabric Group 2, Subgroup A: the cooking fabric, sample PAR 09/28 (**246**), low fired, x25. Photo E. Nodarou.

PLATE 47

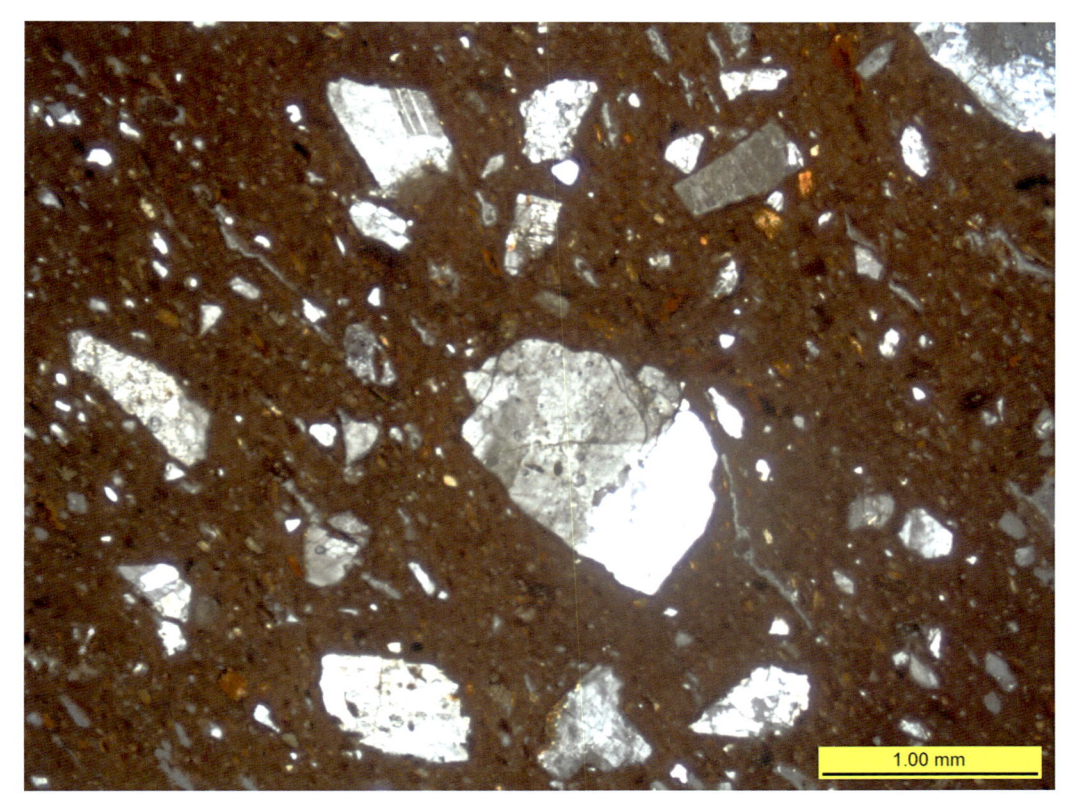

Plate 47A. Fabric Group 2, Subgroup B: the cooking fabric, sample PAR 09/26 (**262**), high fired with black mica (biotite), x25. Photo E. Nodarou.

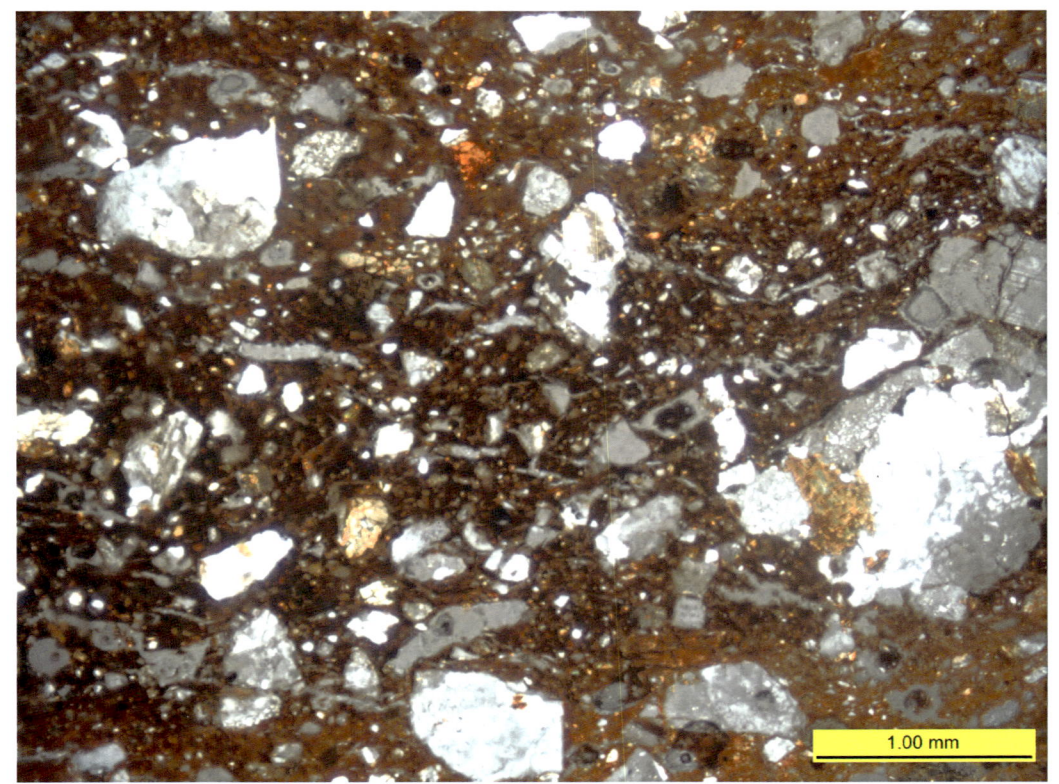

Plate 47B. Fabric Group 2, Subgroup C: the cooking fabric, sample PAR 09/25 (**276**), with weathered granodiorite, x25. Photo E. Nodarou.

PLATE 48

Plate 48A. Sample PAR 09/06 (**72**), x25. Photo E. Nodarou.

Plate 48B. Sample PAR 09/01 (**6**), x25. Photo E. Nodarou.

PLATE 49

Plate 49A. Sample PAR 09/02 (**11**), x50. Photo E. Nodarou.

Plate 49B. Clay sample CS 13/26 from experimental briquette, x25. Note similarity with the jar fabric. Photo E. Nodarou.